Health and Well-Being in Animals

Health and Well-Being in Animals

Edited by Dennis Cameron

SYRAWOOD
PUBLISHING HOUSE
New York

Published by Syrawood Publishing House,
750 Third Avenue, 9th Floor,
New York, NY 10017, USA
www.syrawoodpublishinghouse.com

Health and Well-Being in Animals
Edited by Dennis Cameron

International Standard Book Number: 978-1-64740-259-4 (Hardback)

Cataloging-in-Publication Data

Health and well-being in animals / edited by Dennis Cameron.
p. cm.
Includes bibliographical references and index.
ISBN 978-1-64740-259-4
1. Animals. 2. Animal health. 3. Animal welfare. I. Cameron, Dennis.
QL45.2 .H43 2022
590--dc23

TABLE OF CONTENTS

PREFACE

The world is advancing at a fast pace like never before. Therefore, the need is to keep up with the latest developments. This book was an idea that came to fruition when the specialists in the area realized the need to coordinate together and document essential themes in the subject. That's when I was requested to be the editor. Editing this book has been an honour as it brings together diverse authors researching on different streams of the field. The book collates essential materials contributed by veterans in the area which can be utilized by students and researchers alike.

The branch of science which seeks to ensure that the animals are healthy and well cared for is known as veterinary science. It involves various practices such as preventive care, psychological analysis and surgical procedures. Some of the areas of study within this field are nutrition, anatomy, animal behavior, animal genetics, cell biology, physiology, animal husbandry and epidemiology. The well-being of animals falls under animal welfare. It makes use of diverse measures such as longetivity, immunosuppression, physiology and reproduction to assess the welfare of animals. Animal welfare deals with numerous issues such as animal testing, abandoned pets and cruelty to animals. This book includes some of the vital pieces of work being conducted across the world, on various topics related to the health and well-being in animals. It strives to provide a fair idea about this discipline and to help develop a better understanding of the latest advances within this field. Students, researchers, experts and all associated with this field will benefit alike from this book.

Each chapter is a sole-standing publication that reflects each author´s interpretation. Thus, the book displays a multi-facetted picture of our current understanding of application, resources and aspects of the field. I would like to thank the contributors of this book and my family for their endless support.

Editor

1

Effects of Elevated Crude Glycerin Concentrations on Feedlot Performance and Carcass Characteristics in Finishing Steers

P. Chanjula*, T. Raungprim[1], S. Yimmongkol[2], S. Poonko[1], S. Majarune[1], and W. Maitreejet[1]
Department of Animal Science, Faculty of Natural Resources,
Prince of Songkla University, Songkhla 90112, Thailand

ABSTRACT: Twenty crossbred steers (400±40.1 kg of initial body weight) were used to assess the effects of a dietary supplementation with crude glycerin (CG) as a substitute for corn grain on performance, carcass traits, and meat quality. Four isocaloric and isonitrogenous diets were offered to the experimental animals (5 steers per treatment) for 121 days using randomized complete block design. The steers individually received dietary treatments containing 0%, 7%, 14%, and 21% of CG (88.91% pure) on a dry matter (DM) basis. The diets were offered *ad libitum* as total mixed rations twice daily. Weight gain and carcass traits were determined. At the end of the experimental period, the harvest data and carcass characteristics of the steers were recorded, and meat quality was determined. No significant effect of CG inclusion was observed in any of the growth performance and carcass characteristics traits studied. Also, there were no apparent effects of diets ($p>0.05$) on meat quality (pH, water holding capacity, drip losses, and cooking losses). The study concluded that CG could be used as a substitute for corn grain up to the level of approximately 21% of DM in the diets of finishing steers. (**Key Words:** Crude Glycerin, Growth Performance, Carcass Characteristic, Cattle)

INTRODUCTION

Feed is the biggest cost in livestock production, and the use of alternative feeds, such as by-products of biodiesel, may be a viable alternative both in economic and nutritional terms to increase profitability. For example, crude glycerin (CG) is a by-product of biodiesel production resulting from the formation of methyl esters of fatty acids from triglycerides (Chanjula et al., 2015). CG is available because of the expansion of the biodiesel industry, and might be an optimal product for animal feeding (Donkin et al., 2009). Approximately, 7.9 kg of CG is generated per 100 L of biodiesel produced. Therefore, the increase of biodiesel production has led to an increase of glycerin stocks with a subsequent price reduction making glycerin a potential high energy feed source for ruminants (Avila-Stagno et al., 2013). The energy values of corn, wheat, and glycerin are similar. Thus, glycerin may be an attractive feedstuff to replace corn or wheat and to enhance physical properties of pelleted diets. Several researchers have estimated the energy value of glycerol in beef, and concluded that it is similar to corn grain (Mach et al., 2009). Similarly, Lammers et al. (2008) found that CG containing of 86.95% pure glycerol had a metabolizable energy (ME) content of 3,207 kcal/kg, which was 94% the ME content of corn (NRC, 2000). Therefore, glycerin could be used as an energetic ingredient in animal diets instead of cereals (which are usually more expensive than glycerin).

In ruminants, different quantities of glycerin are converted to volatile fatty acids. Particularly, propionate and butyrate, at the expense of acetate, are directly absorbed from the digestive system and act as precursors for gluconeogenesis in the liver and can provide energy for cellular metabolism (Krehbiel, 2008). CG is an appealing

* Corresponding Author: P. Chanjula.
E-mail: pin.c@psu.ac.th
[1] Buffalo and Beef Production Research Center, Suwanvajokkasikit Animal Production Research and Development Institute, Kasetsart University, Kamphaeng Saen Campus 73140, Thailand.
[2] Department of Animal Science, Faculty of Agriculture, Kasetsart University, Kamphaeng Saen Campus 73140, Thailand.

by-product in feedlot diets because it is hypothesized that CG is primarily converted to propionate in the rumen and thus, is acting as a precursor for glucose synthesis. Researchers reported that 35% to 69% of the CG administered was used to produce propionate (Krehbiel, 2008). If CG increased propionate concentration, then an increased gain-to-feed (GF) would be expected. In addition, feeding glycerin may also improve feed digestibility and growth performance of cattle in a dose-dependent manner (Miller et al., 2001). However, GF responses have been variable. As dietary CG increased, average daily gain (ADG) and GF either did not change (Mach et al., 2009) or decreased (Pyatt et al., 2007) and responded quadratically (Parsons et al., 2009). Previous studies had assessed the effects of the inclusion of CG (above 86% of glycerol) in diets on intake, performance, carcass, and meat quality traits of beef cattle and reported acceptable inclusion of 10%, 12%, and 8% respectively in diet dry matter (DM) (Pyatt et al., 2007; Mach et al., 2009; Parsons et al., 2009). Additionally, inclusions of 10% to 15% in diet DM have been used without adverse effects on milk production or milk composition (Donkin et al., 2009). However, there is little information available on feeding rates and production responses in finishing steers fed moderate to high amounts of glycerin regarding the effects of this by-product on their performance, carcass characteristics, and meat quality. This study hypothesized that CG was less fermented than corn starch, and this should be translated in animal performance and would produce similar growth performance and carcass characteristics as that of steers fed a diet without CG. Therefore, the objectives were to determine feedlot performance, carcass characteristics, and meat quality of finishing steers fed diets with 0%, 7%, 14%, or 21% CG formulated to an equal energy and crude protein (CP) content.

MATERIALS AND METHODS

All procedures involving animals in the metabolism and finishing studies were approved by the Ethical Principles for the Use of Animals for Scientific Purposes of the National Research Council of Thailand (NRCT).

Animals, housing and experimental diets

Twenty crossbred steers (approximately 25% Thai native breed with the remainder represented by approximately 25% Brahman, and 50% Charolais breeds) with an average initial body weight (BW) of 400±40.1 kg and 24±4 months of age were used in a randomized complete block design. They were distributed in 5 blocks to evaluate their intake and digestibility of nutrients, performance, carcass characteristics, and meat quality under feedlot conditions. Steers were blocked into 5 groups based on initial BW and allotted randomly to 1 of 4 treatments (n = 5 steers per treatment) and were adapted to the experimental diets for 14 d before beginning of data collection. The steers were treated for internal and external parasites with Ivomex-F intramuscularly, vaccinated against Foot and Mouth Disease, and were kept in individual pens of approximately 12-m^2 (3.0×4.0 m) with concrete floor and free access to feed and water at all time. The experiment was conducted for 135 d (14 d for diet adaptation and 4 periods of 30 d for data collection). Four 4 corn-based dietary treatments consisted of 0%, 7%, 14%, and 21% of CG (DM basis) and were formulated to be isonitrogenous (DM basis) to meet or exceed the NRC (2000) requirements of fattening steers. The ingredients and determined chemical composition of the components of each diet are presented in Table 1. The CG was produced in a palm-diesel

Table 1. Ingredients and chemical composition of diets containing increasing amounts of crude glycerin (% DM basis)

Item	Dietary crude glycerin (% of dietary DM)			
	0	7	14	21
Ingredients composition (%)				
Crude glycerin[1]	0.0	7.0	14.0	21.0
Ground corn	40.0	33.0	26.0	19.0
Cassava chip	26.6	22.6	18.5	14.4
Palm kernel cake	12.4	16.4	20.5	24.6
Leucaena leave meal	10.0	10.0	10.0	10.0
Napier hay	1.0	1.0	1.0	1.0
Molasses	5.0	5.0	5.0	5.0
Salt	0.2	0.2	0.2	0.2
Urea	2.0	2.0	2.0	2.0
Mineral and vitamin mix[2]	1.0	1.0	1.0	1.0
Dicalcium phosphate	1.0	1.0	1.0	1.0
Sodium bicarbonate ($NaHCO_3$(	0.5	0.5	0.5	0.5
Analyzed nutrient content[3] (% DM)				
DM	87.91	86.25	84.47	83.39
Ash	4.59	4.71	4.86	5.20
OM	95.41	95.29	95.14	94.48
CP	13.95	13.98	13.95	13.92
EE	2.64	2.49	2.34	2.22
NDF	44.07	42.33	38.24	39.08
ADF	19.44	19.97	20.00	19.07
ME[4] (Mcal/kg DM)	2.59	2.58	2.58	2.57

DM, dry matter; OM, organic matter; CP, crude protein; EE, ether extract; NDF, neutral detergent fiber; ADF, acid detergent fiber; ME, metabolizable energy, TDN, total digestible nutrient.

[1] Contained 88.91% glycerol, 5.60% water, 2.24% sodium, and 0.52% methanol (Colorless, odorless, viscous liquid obtained from Biodiesel Producers, New Biodiesel, Surat Thani Province, Thailand).

[2] Minerals and vitamins (each kg contains): Vitamin A, 10,000,000 IU; Vitamin E, 70,000 IU; Vitamin D, 1,600,000 IU; Fe, 50 g; Zn, 40 g; Mn, 40 g; Co, 0.1 g; Cu, 10 g; Se, 0.1 g; I, 0.5 g.

[3] Based on analysis of composite feed sample.

[4] ME = TDN×0.04409×0.82 (NRC, 2000). Calculated with an estimated ME for glycerol of 3.47 Mcal/kg of DM (Mach et al., 2009).

facility (New Biodiesel, Surat Thani Province, Thailand) and contained 88.91% of glycerin, 5.60% of water, 2.24% of sodium, and 0.52% of methanol (Table 2). Glycerin fed in the current study was used as an energetic ingredient. Therefore, to obtain 4 isoenergetic concentrates, the increase in glycerin content was counter balanced mainly by a decrease in cereal grain content. Palm-derived glycerin from single batch was added to the total mixed rations (TMR) as liquid.

After 14 d adaptation, the animals received diets in the form of a TMR twice daily in two equal portions at 0800 and 1600 h for 121 days. The amount of TMR offered and refused was recorded for each steer once daily, and the offered amount was adjusted to ensure approximately 10% of refusals after feeding. Feed refusals were weighed daily, analyzed for DM, recorded, and discarded to calculate dry matter intake (DMI), accurately. Individual feed ingredients were analyzed weekly for DM in order to adjust the diet composition for ingredient moisture content. Composite feed samples were collected weekly and dried in a forced-air oven at 60°C for 48 h to analyze DM. Dried samples were ground to pass a 1-mm screen (Cyclotech Mill, Tecator) and then analyzed for DM, CP, ether extract, and ash (AOAC, 1995). The tests for the evaluation of neutral detergent fiber and acid detergent fiber were determined according to Van Soest et al. (1991).

Table 2. Chemical composition of crude glycerin used in this experiment [1, 2]

Items	Content	Analytical method
Total glycerin (%)	88.91	ASTM D 6584-00E01, titration assay (AOAC,1995)
Moisture (%)	5.60	AOAC[3] method 984.20
DM (%)	94.40	Determined by difference
Methanol[4] (%)	0.52	Gas chromatography
Ash[4] (%)	3.51	AOAC method 942.05
Sodium[4] (%)	2.24	AOAC methods 956.01, 9.15.01
Sulfur[4] (%)	0.10	AOAC method 956.01
Free fatty acid[4] (%)	0.09	AOAC method Ca 5a-40
Crude protein[4] (%)	0.01	AOAC method 990.03
Gross energy (kcal/kg)	3,961	Adiabatic bomb calorimeter

[1] Crude glycerin was obtained from New Biodiesel Co., Ltd., Surat Thani Province.

[2] Analysis by Central Laboratories (Songkhla, SK), Co., Ltd., Songkhla 90110, Thailand.

[3] AOAC (1995).

[4] Expressed as a percentage of crude glycerin DM.

Animal performance, slaughter, and sample collection

All steers were weighed before the morning feeding at the beginning of the experiment and thereafter continued monthly at the same time of day and before transportation to the abattoir of the Institute of Animal Science (final BW). Daily DMI by each steer was estimated by summing the monthly intake and dividing by the number of days of the month. The ADG was calculated as the difference between the final and initial BW (16 h of fasting, only water was provided) divided by the number of days of feeding (121 d). Gain efficiency was calculated as ADG divided by DMI.

At the end of the experiment, the day before slaughter, the steers were transported to the experimental slaughterhouse after 121 d of study with a final BW of 514±46.9 kg and fasted overnight (only water was provided). All animals were slaughtered at a commercial beef plant. The animals were stunned by a captive bolt pistol, suspended by their hind legs and exsanguinated. The steers were slaughtered according to industrial practices in Thailand at a commercial slaughterhouse. After the slaughter, carcass dressing followed a standardized protocol without electrical stimulation as recommends by the National Livestock and Meat Board (Ziegler, 1977). Immediately following slaughter, hot carcass weight (HCW) was recorded, and hot dressing percentage (HDP) was defined as the ratio of HCW to slaughter BW×100. All carcasses were refrigerated at 4°C for approximately 24 h, and then the cold carcass weight (CCW) was recorded. Chilled dressing percentage (CDP) was calculated by CCW to slaughter BW×100.

After the postmortem chill period, carcass pH (pH_{24}), 12th rib fat thickness (RFT), and 12th rib longissimus muscle (LM) area were measured on the left side of each carcass after a cross-section cut was made between the 12th and 13th. Marbling score was measured in the LM between the 12th and 13th ribs by using the Thailand scoring system (1 = no marbling and 5 = highest marbling). Meat color was determined by using a Subjective Color Score divided into 7 levels (1 = pale pink, 2 = soft pink, 3 = pink, 4 = light red, 5 = red, 6 = medium dark red, and 7 = dark red) (Smith et al., 2001). Meat colors preferred by consumers are ranges from soft pink to red color (from 2 to 4).

The muscular *longissimus dorsi* (LD) area was made on the left cut surface (of the chilled carcass) between rib 12th and 13th. The LD (the section between the last lumbar and the first sacral vertebrae) were collected. These cuts of meat, two per animal, were labeled and frozen immediately after collection for later measurement of color coordinates, water holding capacity (WHC), drip losses, cooking losses, and shear force characteristics.

Meat quality

Muscle surface color was measured objectively using a HunterLab Miniscan Plus Spectrocolorimeter on the same cut surface of the LD. Instrumental color measurements were recorded for L* (measures darkness to lightness when lower L* indicates a dark color), a* (measures redness when higher a* value indicates a redder color), and b* (measures yellowness when higher b* value indicates a yellower color) at 3 locations of exposed lean to obtain a representative reading. To determine Warner–Bratzler shear force (WBSF), samples were defrosted at room temperature until their internal temperature reached 2°C to 5°C. After weighing, samples were trimmed, and thin sections from the lateral and extremities were removed. Four LD samples, parallel to the muscle fibers and having 1 cm of thickness and 5 cm of length, were obtained to measure the shear force in a texture analyzer (TA-XT*Plus*-Texture Analyzer with a Warner-Bratzler Blade probe, Texture Expert Exponent-Stable Micro Systems software, Ltd in Godalming, Surrey, UK. SMS). For each sample, 6 shear force results were obtained. For the drip loss, the samples (2 cm thick) were packaged in clear trays of crystal polystyrene covered with a permeable film and stored at 4°C. Drip loss was expressed as a percentage of the initial sample weight. The thawing loss and cooking loss were calculated as described by Vergara et al. (2003). The meat samples (two steaks, 2 cm thick) were placed in polyethylene bags and were heated at 75°C for 20 min in a water bath up to an internal temperature of 72°C. The cooking loss was expressed as a percentage of the initial sample weight. Thawing loss was calculated as the difference between the weight of the steaks before and after thawing. All meat quality measurements were made in triplicate.

Laboratory analyses

Samples of feed and LD muscle were subjected to proximate analysis following the standard methods of AOAC (1995). Dry matter was determined by oven drying in a forced air oven at 105°C for 24 h. The N content of feed and LD muscle was determined using a Kjeltec Auto Analyzer (Tecator, Hoganas, Sweden). Ether extract (EE) was determined in petroleum ether using a Soxtec Auto Analyzer (Tecator, Sweden). The ash content was determined by ashing the samples in a muffle furnace at 550°C for 5 h. Neutral detergent fiber (NDF), acid detergent fiber (ADF), and lignin concentrations were determined by methods of Van Soest et al. (1991). NDF was analyzed without α-amylase, and the values of NDF and ADF were expressed inclusive of residual ash. Lignin was obtained by treatment of ADF residue with 72% of sulfuric acid (Van Soest et al., 1991).

Statistical analysis

All data were analyzed using SAS (Cary, NC, USA) software. The MIXED procedure was used to analyze the fixed effects of treatment and block on performance and carcass characteristics, with animal serving as the experimental unit. Orthogonal contrasts were used to determine linear and quadratic effects, as well as the effect of the 0% of CG diet vs. the average of all diets containing CG. Treatment means were statistically compared using Duncan's Multiple Range Test to identify differences between means. Contrasts were considered significant when the p-value was ≤0.05, with tendencies declared at p-values between 0.05 and 0.10.

RESULTS AND DISCUSSION

Chemical composition of feeds

The ingredients and chemical compositions of the experimental diets are presented in Table 1. The four experimental diets contained similar concentrations of ash, organic matter, CP, and EE. CG-based diets had a slightly lower NDF as proportion of CG in diets increased due to feeding of less of corn grain and cassava chip, ranging from 38.24% to 44.07%, respectively. Palm kernel cake content increased as proportion of CG in diets decreased. The similarity in ADF for the CG 7%, 14%, and 21% diets was unexpected. Additionally, the reason for the increase in ADF from 7% to 14% CG is not known. The differences among TMR in fibrous components can be related to differences in the ingredients used in diet formulation (Table 1). The CG used in the present feeding trial was produced from crude palm oil contained 88.91% of glycerin, 2.24% of sodium, 0.52% of methanol, and less other compounds. Similar values for CG have been previously reported by Mach et al. (2009), Gunn et al. (2010a), Avila-Stagno et al. (2013), Chanjula et al. (2015). The potential problematic compound in CG is methanol. Methanol toxicity in humans and animals is characterized by central nervous system depression, weakness, headache, vomiting, metabolic acidosis and optic disc oedema with the clinical consequences being blindness and/or Parkinsonian-like motor disease (Chanjula et al., 2015). However, methanol concentration can vary widely according to the manufacturing processes and should be monitored. To the authors' knowledge, no previous specification for the use of CG (with methanol) in animal feed has been published. Early studies of assessing the effects of feeding pure or crude glycerol to pigs (Schieck et al., 2010), and lambs (Avila-Stagno et al., 2013) provided initial evidence that glycerol can be used as a source of dietary energy for livestock. Based on the experimental data, no steers demonstrated clinical symptoms of methanol toxicity in the present study even though the diet with 21% of CG would contain 0.109%, assuming that all the methanol in the CG remained in the feed. Moreover, the high-risk to health associated to methanol consumption due to inclusion of CG in diets of ruminant animals is not expected since methanol is naturally produced in the ruminal environment as a result of pectin digestion (Chanjula et al., 2015).

Intake and feedlot performance

The effects of CG on growth performance of finishing steers are presented in Table 3. Overall initial BW, final BW, and BW gain were not significantly affected (p>0.05) by CG content for all dietary treatments as compared between the experimental diets (7% to 21% of CG) with the control diet. No significant differences attributable to dietary treatment were observed in terms of total DMI (% BW and g/kg $BW^{0.75}$); although the average DMI was numerically lower in glycerin-fed groups. This could imply that DMI was potentially regulated by the energy density of the diet. Literature data has established a correlation between dietary energy concentration and DMI. When cattle are consuming energy-dense high-concentrate diets, DMI is thought to be controlled by the animal's energetic demands and metabolic factors (NRC, 2000). Furthermore, conversion of glycerol to propionate by ruminal microbes (Roger et al., 1992) may lead to an observed decrease in DMI. Infusion of propionate into the portal vein or into the rumen (Oba and Allen, 2003) has been shown to reduce intake in sheep and cattle. Results of this study are in agreement with the study by Mach et al. (2009) who fed diets containing different contents of glycerin (up to 12% of DM) to Holstein bulls for 91 d. Gunn et al. (2010a) and Chanjula et al. (2015) reported that there were no changes in DMI and digestibility when increasing concentrations of CG (0% to 20% of DM) to replace dry rolled corn or corn grain in lamb and goat diets. Also, the addition of glycerin at levels up to 30% of DM fed to heavier lambs (Gomes et al., 2011) had no effect on their growth performance. Likewise, Avila-Stagno et al. (2013) used CG (up to 21% of DM) for finishing lambs without adverse effect on nutrient intakes and digestibility. In contrast, a decreased DM intake was reported when a diet containing 10% of glycerin as a corn replacement was fed to feedlot steers (Pyatt et al., 2007). Also, increasing glycerin to 4%, 8%, 12%, and 16% of DM reduced DM intake in finishing heifers (Parsons et al., 2009). DM intakes particularly decreased when glycerin was fed to finishing lambs in high amounts up to 45% (Gunn et al., 2010b). According to Trabue et al. (2007), increasing lactic acid concentrations could depress CG fermentation in the rumen, thus altering (decreasing) DMI. However, it remains unclear whether this was due to the dietary treatment. Indeed, substituting corn with high levels of glycerin was reported to adversely affect ruminal fermentation through reducing fiber digestion, acetate production, and bacterial populations (Roger et al., 1992). Roger et al. (1992) demonstrated that introducing glycerol to the ruminal environment reduced cellulolytic activity of ruminal bacteria. Paggi et al. (2004) also reported that digestibility of other substrates in the diet might be inhibited with the inclusion of glycerol in an *in vitro* environment. However, there is more recent digestibility data which support results from the current study. Krehbiel (2008) reported that microorganisms adapted rapidly to glycerol feeding because elevated disappearance rates of glycerol were noted with increased days of glycerol feeding. Additionally, Hess et al. (2008) reported that CG could be added at 15% of DM to ruminant diets without negatively affecting to DM or fiber digestibility. These data, coupled with data from the current study, suggest that the ruminal environment, and concurrent decrease in DMI, might not be affected until CG concentrations exceeded to 21% of dietary DM. Further research, however, is needed to test this hypothesis and pinpoint the exact causes of decreased feedlot performance associated with elevated amounts (>21%) of CG in the diet.

Table 3. Effects of dietary crude glycerin on performance and DMI of finishing steers

Item	Dietary crude glycerin (%)				SEM	Contrasts, p-value[1]			
	0	7	14	21		L	Q	C	0 vs glycerin[2]
No. of steer	5	5	5	5	-	-	-	-	-
Days on feed	121	121	121	121	-	-	-	-	-
BW (kg)									
Initial BW (kg)	404.0	403.2	404.4	388.9	7.41	0.61	0.70	0.83	0.81
Final BW (kg)	521.0	515.0	516.0	499.0	10.60	0.53	0.81	0.80	0.68
Weight gain (kg)	117.0	111.8	111.6	110.2	14.49	0.75	0.89	0.92	0.73
DMI									
kg/d	7.79	7.50	7.55	7.25	0.39	0.47	0.99	0.75	0.52
% BW	1.69	1.62	1.65	1.63	0.08	0.67	0.80	0.73	0.57
g/kg of $BW^{0.75}$	78.28	75.42	76.25	74.80	3.75	0.58	0.85	0.73	0.53
ADG (kg/d)	0.968	0.932	0.920	0.920	0.12	0.77	0.88	0.98	0.75
ADG (g/kg $W^{0.75}$)	9.81	9.32	9.26	9.41	1.19	0.81	0.79	0.96	0.73
Feed efficiency	0.124	0.122	0.121	0.125	0.01	0.96	0.83	0.92	0.95

DMI, dry matter intake; SEM, standard error of the mean (n = 5); BW, body weight; ADG, average daily gain.

[1] Treatment and contrast p-values; p-value for L, linear effect; Q, quadratic effect; C, cubic effect.

[2] Compares the effects of 0% glycerin with the combined glycerin treatment.

Likewise, partial dietary replacement of corn grain with CG did not significantly affect the average daily gain (ADG) (0.935±0.02 kg/d) and feed efficiency of steers (0.123±0.00 kg/kg) in this study. Therefore, there were no differences in average daily gain in the overall fattening period as indicated by the treatment effect. Similarly, feeding CG up to 10% of DM did not affect ADG and feed efficiency in cattle (Pyatt et al., 2007; Mach et al., 2009; Parsons et al., 2009). Pyatt et al. (2007) reported an 11.4% and 21.9% of improvement in ADG and efficiency when glycerin replaced 10% of the dry-rolled corn in the diet but ADG and efficiency improved by only 2.5% and 16.4% when glycerin replaced 10% of the dry-rolled corn in diets also containing 30% of distillers grains. Similarly to the results of this study, glycerin as an energy ingredient effectively replaced dry-rolled corn in the diet up to 20% of DM for finishing lambs had no negative impact on cumulative ADG and feed efficiency (Gunn et al., 2010a). Also, the addition of glycerin at levels up to 30% of DM fed to heavier lambs had no effect on their growth performance (Gomes et al., 2011). Conversely, Parsons et al. (2009), and Gunn et al. (2010b) demonstrated that feeding CG to finishing ruminants (finishing heifers and lambs) above 15% of DM decreased feed efficiency through decreased ADG. Data from the current study demonstrated that feeding CG up to 21% of dietary DM might have a positive impact on steer performance.

Carcass characteristics and meat quality

Carcass characteristics are presented in Table 4. No significant effect of dietary CG was observed on fasted live weight, HCW, CCW, dressing percentage, and weight loss. The lack of effects of CG inclusion on HCW and dressing percentage are in accordance with previous reports in lambs and goats (Gunn et al., 2010a; Avila-Stagno et al., 2013; Chanjula et al., 2015) and beef cattle (Mach et al., 2009; Françozo et al., 2013) that no effects on carcass traits found when replacing corn and barley grain with CG in concentrations of up to 21% and 18% of DM, respectively. CG seems to provide a similar amount of metabolizable energy as barley when CG is converted into volatile fatty acids (VFA) in the rumen (Mach et al., 2009). VFA provide energy to the animal. This fact permits normal growth and normal carcass values which were found in this study. The dressing percentage was similar in all steers and within the previously published range of 61.5% to 62.1% (Egea et al., 2014) in feedlot with a high energy density diet. Thus, the inclusion of CG in the studied levels had no effect on the dressing percentage in beef cattle finished in feedlot. However, Parsons et al. (2009) found that HCW increased by 8.1, 5.1, and 3.2 kg when glycerin was fed at 2%, 4%, and 8%, respectively, but HCW decreased by 1.2 and 9.1 kg when glycerin was fed at 12% and 16%, respectively. An explanation for this might be the lower intake, water intake, and digestibility of the diet and nutrients of feed when glycerin is used up to 10% (Chanjula et al., 2015). Likewise,

Table 4. Effects of dietary crude glycerin on carcass characteristics of finishing steers

Item	Dietary crude glycerin (%)				SEM	Contrasts, p-value[1]			
	0	7	14	21		L	Q	C	0 vs glycerin[2]
Fasted live weight (kg)	474.6	487.8	484.2	471.2	11.76	0.89	0.59	0.94	0.81
HCW (kg)	289.2	297.8	293.8	285.6	9.17	0.82	0.57	0.89	0.85
CCW (kg)	283.2	291.5	287.9	279.7	8.93	0.82	0.57	0.91	0.85
Dressing percentage (%)	60.8	60.9	60.8	61.1	1.02	0.89	0.89	0.84	0.93
Cold dressing percentage (%)	59.63	59.70	59.58	59.88	1.01	0.87	0.90	0.88	0.93
Weight loss (kg)	5.94	6.26	5.84	5.86	0.31	0.74	0.73	0.55	0.92
Weight loss (%)	1.24	1.27	1.20	1.25	0.05	0.77	0.85	0.30	0.95
LM area[3] (cm^2)	71.60	75.22	72.30	74.50	2.75	0.77	0.87	0.56	0.64
WBSF (kg)	5.82	6.49	6.36	6.75	0.35	0.12	0.71	0.43	0.11
KPH fat[4] (kg)	22.72	24.48	24.42	25.68	1.84	0.43	0.91	0.77	0.45
KPH fat (%)	4.81	5.00	5.00	5.43	0.41	0.32	0.77	0.73	0.48
Back fat thickness (cm)	1.80	1.92	1.76	1.62	0.09	0.14	0.22	0.52	0.78
Marbling score[5]	2.0	2.0	2.0	1.8	0.09	0.19	0.33	0.66	0.57
Meat color[6]	3.86	3.32	3.76	3.75	0.13	0.67	0.61	0.79	0.31

SEM, standard error of the mean (n = 5); HCW, hot carcass weight; CCW, cold carcass weight; LM, *longissimus dorsi*; WBSF, Warner-Bratzler shear force;

[1] Treatment and contrast p-values; p-value for L, linear effect; Q, quadratic effect; C, cubic effect.

[2] Compares the effects of 0% glycerin with the combined glycerin treatment.

[3] LM, longissimus muscle area, cm^2.

[4] KPH (kidney, pelvic, and heart fat) as a percentage of carcass weight.

[5] Marbling score from 1 to 5; 1 = no marbling and 5 = highest marbling (Sethakul and Opatpatanakit, 2005).

[6] Meat color score from 1 to 7; 1 = pale pink, 2 = soft pink, 3 = pink, 4 = light red, 5 = red, 6 = medium dark red, and 7 = dark red (Smith et al., 2001).

the area and WBSF of LM, kidney, pelvic, and heart (KPH) fat, and fat thickness were not affected by treatments. Similarly, Bartoň et al. (2013) found that a long-term dietary treatment with CG as a replacement of barley meal up to the level of 10% of DM had no significantly effect on any of the bull carcass and meat quality traits studied. On the other hand, Parsons et al. (2009) observed a linear reduction in the LM area and subcutaneous fat when increasing the amounts of glycerin fed (up to 16%). Nevertheless, the obtained WBSF results (<4.0 kg) ensure a tenderness that should result in high consumer acceptance (Miller et al., 2001).

No differences (p>0.05) were reported in marbling and color scores when corn was replaced by CG in the diets of steers finished in feedlot. Marbling score was classified as "light" or "small" (1.95 points). Although medium marbled meat is well accepted within the home market, beef should feature more accentuated marbling to be acceptable in foreign markets. Parsons et al. (2009) observed that the inclusion of glycerin (16%) in the diets for heifers had led to a linear decrease in marbling scores. Because glycerin reduced subcutaneous fat, it is conceivable that glycerin may alter fat deposition, which might explain the observed reductions in marbling scores. Glucose was previously shown to be quantitatively the primary lipid precursor in intramuscular adipose tissue whereas the relative contribution of acetate to lipogenesis was greatest in subcutaneous adipose tissue (Parsons et al., 2009). Previous research suggests that increasing the glucogenic substrates (e.g., glycerin) fed to cattle results in increased marbling scores (Mach et al., 2009). Unlike in our study, it has been previously reported that glycerin increased ruminal propionic and butyric acid concentrations at the expense of acetic acid concentration (Chanjula et al., 2015). Therefore, lower concentrations of acetic acid as a lipogenic precursor could have been the reason why glycerin supplemented diets reduced in both subcutaneous fat and marbling scores in feedlot heifers fed increasing quantities of CG (Parsons et al., 2009) and decreased LD ether extract values in finishing lambs (Gunn et al., 2010b). However, glycerin showed no positive effects on marbling when various concentrations were fed to feedlot heifers. The meat color was similar for all treatments. According to Mancini and Hunt (2005), meat color is an important commercial characteristic that influences consumer behavior. Meat color was considered good (3.67 points), ranging between "red" and "slightly dark red". Adequate nutrition and low age may have affected meat color (Mancini and Hunt, 2005).

Table 5 shows the effect of CG inclusion in the animal diet on meat quality parameters. CG inclusion had no effect (p>0.05) on the pH of meat 45 min and 24 h after slaughter or on the colorimetric parameters of LD among treatments. These results agreed with Mach et al. (2009) who supplemented CG in Holstein bull diets. Similarly, Françozo et al. (2013) found no differences between control and CG (5% and 12%) groups in Nellore bull meat. Pearce et al. (2011) reported that the lightness (L*) was influenced by the amount of water on the meat surface and was a consequence of water retention capacity which in turn affected the pH. Therefore, LM water loss was not affected by diet when CG was supplemented in the diets. Lightness, redness, and yellowness on LM were normal for bulls finished in feedlot (Bartoň et al., 2013). According to Mancini and Hunt (2005), changes in muscle color for L* and b* can be attributed to diet and can affect the marbling score and muscle glycogen levels in the pre-slaughter.

Table 5. Physico-chemical characteristics of beef of steers fed different levels of crude glycerin

Item	Dietary crude glycerin (%)				SEM	Contrasts, p-value[1]			
	0	7	14	21		L	Q	C	0 vs glycerin
45 min pH[3]	6.54	6.41	6.46	6.50	0.07	0.78	0.20	0.57	0.27
24 h pH[4]	6.10	6.09	6.11	6.11	0.02	0.86	0.83	0.75	0.94
Color of LM[5]									
L*	38.64	38.81	36.04	39.92	1.79	0.89	0.31	0.25	0.85
a*	17.32	18.03	18.57	17.92	0.79	0.56	0.45	0.80	0.42
b*	8.67	10.66	8.71	10.24	1.01	0.53	0.81	0.10	0.30
WHC	73.10	72.81	71.32	72.21	2.56	0.71	0.83	0.55	0.61
Drip loss (%)	1.68	1.70	1.66	1.70	0.12	0.97	0.93	0.80	0.96
Thawing loss (%)	11.57	10.64	11.62	12.19	1.46	0.63	0.57	0.69	0.95
Cooking loss (%)	21.70	24.53	21.22	20.71	2.26	0.50	0.42	0.33	0.84

SEM, standard error of the mean (n = 5); LM, *longissimus dorsi*; WHC, water holding capacity

[1] Treatment and contrast p-values; p-value for L, linear effect; Q, quadratic effect; C, cubic effect.

[2] Compares the effects of 0% glycerin with the combined glycerin treatment.

[3] pH measurements taken at 45 min after slaughter.

[4] pH measurements taken at 24 h after slaughter.

[5] L* values are a measure of lightness (higher value indicates a lighter color); a* values are a measure of redness (higher value indicates a redder color); b* values are a measure of yellowness (higher value indicates a more yellow color), by CIE, complete international commission on illumination (Hunter color flex).

However, the possible increase in energy levels, resulting from supplementation with glycerin and interference in the final characteristics of the meat, was not enough to give result in color changes. There was no dietary effect ($p>0.05$) of CG inclusion on WHC, drip loss, thawing loss, or cooking loss. In contrast, studies in non-ruminants found different results for these parameters. Also, Mourot et al. (1994) found a reduction in water losses and cooking losses in pigs fed 5% of glycerol because the glycerol increased cell osmotic pressure, increasing the intracellular water content, which would increase the WHC. These differences between species may be explained with the fact that glycerol is absorbed without being transformed the pig stomach, while in ruminants, 80% of glycerol is transformed in the rumen into volatile fatty acids (Mach et al., 2009), so suggesting a low absorption of the unchanged glycerol molecule. Consequently, water holding parameters in ruminant meat may not be altered by glycerol feeding, as is demonstrated in the current study and previously in beef (Mach et al., 2009; Françozo et al., 2013).

CONCLUSION

CG was a good an alternative energy source to substitute for corn grain in the diets. Results from current study demonstrated that a diet with up to 21% of CG could be fed to finishing steers with no effect on growth performance, carcass characteristics, and meat quality traits studies. Thus, in the case of competitive pricing, CG may be effectively used as an alternative energy source to substitute for cereals in the diets of finishing steers. However, the effects of glycerin metabolism on LM fatty acid composition needs further research.

ACKNOWLEDGMENTS

The authors would like to express the most sincere gratitude and appreciation to the Faculty of Natural Resources, Prince of Songkla University for financial support of this research (Project no. NAT550821S) and to the Buffalo and Beef Production Research Center, Suwanvajokkasikit Animal Production Research and Development Institute at Kasetsart University for allowing use of the research facilities. This work was also supported by the Tropical Feed Resources Research and Development Center (TROFREC), Department of Animal Science, Faculty of Agriculture, Khon Kaen University.

REFERENCES

AOAC. 1995. Official Methods of Analysis. 16th ed. Assoc. Official Analytical Chemist, Arlington, VA, USA.

Avila-Stagno, J., A. V. Chaves, M. L. He, O. M. Harstad, K. A. Beauchemin, S. M. McGinn, and T. A. McAllister. 2013. Effects of increasing concentrations of glycerol in concentrate diets on nutrient digestibility, methane emissions, growth, fatty acid profiles, and carcass traits of lambs. J. Anim. Sci. 91:829-837.

Bartoň, L., D. Bureš, P. Homolka, F. Jančík, M. Marounek, and D. Řehák. 2013. Effects of long-term feeding of crude glycerine on performance, carcass traits, meat quality, and blood and rumen metabolites of finishing bulls. Livest. Sci. 155:53-59.

Chanjula, P., P. Pakdeechanuan, and S. Wattanasit. 2015. Effects of feeding crude glycerin on feedlot performance and carcass characteristics in finishing goats. Small Rumin. Res. 123:95-102.

Donkin, S. S., S. L. Koser, H. M. White, P. H. Doane, and M. J. Cecava. 2009. Feeding value of glycerol as a replacement for corn grain in rations fed to lactating dairy cows. J. Dairy Sci. 92:5111-5119.

Egea, M., M. B. Linares, M. D. Garrido, C. Villodre, J. Madrid, J. Orengo, S. Martínez, and F. Hernández. 2014. Crude glycerine inclusion in Limousin bull diets: Animal performance, carcass characteristics and meat quality. Meat Sci. 98:673-678.

Françozo, M. C., I. N. Prado, U. Cecato, M. V. Valero, F. Zawadzki, O. L. Ribeiro, R. M. Prado, and J. V. Visentainer. 2013. Growth performance, carcass characteristics and meat quality of finishing bulls fed crude glycerine-supplemented diets. Braz. Arch. Biol. Technol. 56:327-336.

Gomes, M. A. B., G. V. DeMoraes, M. Mataveli, F. D. F. DeMacedo, C. Carneiro, and R. M. Rossi. 2011. Performance and carcass characteristics of lambs fed on diets supplemented with glycerin from biodiesel production. Rev. Bras. Zootec. 40:2211-2219.

Gunn, P. J., M. K. Neary, R. P. Lemenager, and S. L. Lake. 2010a. Effects of crude glycerin on performance and carcass characteristics of finishing wether lambs. J. Anim. Sci. 88:1771-1776.

Gunn, P. J., A. F. Schultz, M. L. Van Emon, M. K. Neary, R. P. Lemenager, C. P. Rusk, and S. L. Lake. 2010b. Effects of elevated crude glycerin concentrations on feedlot performance, carcass characteristics, and serum metabolite and hormone concentrations in finishing ewe and wether lambs. Prof. Anim. Sci. 26:298-306.

Hess, B. W., S. L. Lake, and S. A. Gunter. 2008. Using glycerin as a supplement for forage-fed ruminants. J. Anim. Sci. 86(E-Suppl. 2):392 (Abstr.)

Krehbiel, C. R. 2008. Ruminal and physiological metabolism of glycerin. J. Anim. Sci. 86(E-Suppl. 2):392 (Abstr.).

Lammers, P. J., B. J. Kerr, T. E. Weber, W. A. Dozier, III, M. T. Kidd, K. Bregendahl, and M. S. Honeyman. 2008. Digestible and metabolizable energy of crude glycerol for growing pigs. J. Anim. Sci. 86:602-608.

Mach, N., A. Bach, and M. Devant. 2009. Effects of crude glycerin supplementation on performance and meat quality of Holstein bulls fed high-concentrate diets. J. Anim. Sci. 87:632-638.

Mancini, R. A. and M. C. Hunt. 2005. Current research in meat color. Meat Sci. 71:100-121.

Miller, M., M. Carr, C. Ramsey, K. Crocket, and L. Hoover. 2001. Consumer thresholds for establishing the value of beef tenderness. J. Anim. Sci. 79:3062-3068.

Mourot, J., A. Aumaitre, A. Mounier, P. Peiniau, and A. C. François. 1994. Nutritional and physiological effects of dietary glycerol in the growing pig. Consequences on fatty tissues and post mortem muscular parameters. Livest. Prod. Sci. 38:237-244.

NRC. 2000. Nutrient Requirements of Beef Cattle, 7th ed. Natl. Acad. Press, Washington, DC, USA.

Oba, M. and M. S. Allen. 2003. Intraruminal infusion of propionate alters feeding behavior and decreases energy intake of lactating dairy cows. J. Nutr. 133:1094-1099.

Paggi, R. A., J. P. Fay, and C. Faverin. 2004. *In vitro* ruminal digestibility of oat hay and cellulolytic activity in the presence of increasing concentrations of short-chain acids and glycerol. J. Agric. Sci. 142:89-96.

Parsons, G. L., M. K. Shelor, and J. S. Drouillard. 2009. Performance and carcass traits of finishing heifers fed crude glycerin. J. Anim. Sci. 87:653-657.

Pearce, K. L., K. Rosenvold, H. J. Andersen, and D. L. Hopkins. 2011. Water distribution and mobility in meat during the conversion of muscle to meat and ageing and the impacts on fresh meat quality attributes - A review. Meat Sci. 89:111-124.

Pyatt, A., P. H. Doane, and M. J. Cecava. 2007. Effect of crude glycerin in finishing cattle diets. J. Anim. Sci. 85(Suppl. 1):530 (Abstr.).

Roger, V., G. Fonty, C. Andre, and P. Gouet. 1992. Effects of glycerol on the growth, adhesion, and cellulolytic activity of ruminal cellulolytic bacteria and anaerobic fungi. Curr. Microbiol. 25:197-201.

Schieck, S. J., B. J. Kerr, S. K. Baidoo, G. C. Shurson, and L. J. Johnston. 2010. Use of crude glycerol, a biodiesel coproduct, in diets for lactating sows. J. Anim. Sci. 88:2648-2656.

Smith, G. C., D. B. Griffin, and H. K. Johnson. 2001. Meat Evaluation Handbook. American Meat Science Association, Champaign, IL, USA.

Trabue, S., K. Scoggin, S. Tjandrakusuma, M. A. Rasmussen, and P. J. Reilly. 2007. Ruminal fermentation of propylene glycol and glycerol. J. Agric. Food Chem. 55:7043-7051.

Van Soest, P. J., J. B. Robertson, and B. A. Lewis. 1991. Methods for dietary fiber, neutral detergent fiber, and nonstarch polysaccharides in relation to animal nutrition. J. Dairy Sci. 74:3583-3597.

Vergara, H., L. Gallego, A. García, and T. Landete-Castillejos. 2003. Conservation of *Cervus elaphus* meat in modified atmospheres. Meat Sci. 65:779-783.

Ziegler, J. H. 1977. The Meat We Eat. The Interstate Printers, Danville, IL, USA.

2

Effects of Rice Bran, Flax Seed, and Sunflower Seed on Growth Performance, Carcass Characteristics, Fatty Acid Composition, Free Amino Acid and Peptide Contents, and Sensory Evaluations of Native Korean Cattle (Hanwoo)

Chang Bon Choi*, Hana Kwon[1], Sung Il Kim[2], Un Mok Yang[3], Ju Hwan Lee[3], and Eun Kyu Park[3]
School of Biotechnology, Yeungnam University, Gyeongsan 38541, Korea

ABSTRACT: This study was conducted to evaluate the effect of dietary supplementation with rice bran, flax seed, or sunflower seed to finishing native Korean cattle (Hanwoo) on growth performances, carcass characteristics, fatty acid composition, free amino acid and peptide contents, and sensory evaluations of *Longissimus* muscle (LM). A total of 39 Hanwoo steers (average age of 22.2 mo and average body weight (BW) of 552.2 kg) were randomly divided into Control, rice bran (RB), flax seed (FS), or Sunflower seed (SS) groups. The steers were group fed for 273 d until they reached an average age of 31.2 mo. Final BW was 768.2, 785.8, 786.2, and 789.0 kg, and average daily gain was 0.79, 0.85, 0.82, and 0.84 kg for the Control, RS, FS, and SS groups, respectively ($p>0.05$). Fat thickness of the FS group (19.8 mm) was greater ($p<0.05$) than that of the other groups. Final yield grade converted into numerical values was 2.0 for the RB group, 1.7 for the Control and SS groups, and 1.4 for the FS group. Marbling degrees for the Control, SS, RB, and FS groups were 5.3, 5.1, 4.7, and 4.6, respectively. Percentages of palmitic acid ($C_{16:0}$), stearic acid ($C_{18:0}$), and arachidic acid ($C_{20:0}$) in the LM were not different among the groups. Palmitoleic ($C_{16:1}$) acid was higher ($p<0.05$) in the SS group. The concentration of oleic acid was highest ($p<0.05$) in the Control group (47.73%). The level of linolenic acid ($C_{18:3}$) was 2.3 times higher ($p<0.05$) in the FS group compared to the other groups. Methionine concentration was ($p<0.05$) higher in FS (1.7 mg/100 g) and SS (1.2 mg/100 g) steers than in the Control or RB groups. Glutamic acid and α-aminoadipic acid (α-AAA) contents were ($p<0.05$) higher in the FS group compared to the other groups. LM from the FS group had numerically higher ($p>0.05$) scores for flavor, umami, and overall palatability in sensory evaluations. In conclusion, supplementation of flax seed to diets of finishing Hanwoo steers improved sensory evaluations which might have been caused by increases in flavor related amino acids such as methionine, glutamic acid and α-AAA and peptides, anserine and carnosine, and their complex reactions. (**Key Words:** Flax Seed, Methionine, Glutamic Acid, α-Aminoadipic Acid, Sensory Evaluations, Hanwoo)

INTRODUCTION

For the last four decades, scientists have been trying to increase omega-3 polyunsaturated fatty acids (PUFAs) in meat because saturated fatty acids (SFAs) are associated with many human disorders although meat is an important food source. Feeding cattle with high-grain diets, supplementation of linseed or linseed oil into diets, or utilization of protected fat that is not degraded in the rumen could be included in these efforts. Recent reports indicate that monounsaturated fatty acid (MUFA) in human diets lowers blood cholesterol and reduces the risk for metabolic disorders (Whetsell et al., 2003), improves blood pressure, and increases insulin sensitivity (Gillingham et al., 2011). Fatty acids in beef are not only related with human health but also affect beef quality. Flavors attributed to volatile

* Corresponding Author: Chang Bon Choi.
E-mail: cbchoi@yu.ac.kr
[1] Department of Food and Nutrition, Yeungnam University, Gyeongsan 38541, Korea.
[2] Department of Animal Science, Gyeongbuk Provincial College, Yecheon 36830, Korea.
[3] Woosung Feed Co., Ltd., Daejeon 34379, Korea.

flavor components in cooked meat are the most important factors for the sensory quality of meat. Volatile components are derived from thermally induced reactions occurring during heating via four pathways: i) Maillard reaction of amino acid or peptides with reducing sugars, ii) lipid oxidation, iii) interaction between Maillard reaction products with lipid-oxidized products, and iv) vitamin degradation. Efforts to produce flavorful beef have been undertaken since the most important motivators influencing consumers' beef purchases are i) taste, ii) price, and iii) product consistency. To date, efforts to manage beef palatability attributes at the producer-level have focused specifically on pre-harvest management of beef tenderness. Lipids and produced volatile compounds during cooking are major contributors to the odor and flavor of meat. Thus, cattle production systems that encourage the deposition of intramuscular fat are fundamentally important for developing desirable beef flavor characteristics (Ritchie, 2005). Since the South Korean beef market was opened in 1992, scientists and farmers have worked to improve the competitiveness of native Korean cattle (Hanwoo) to imported beef. The main feeding strategy of Korean beef cattle industry practices include feeding high amount of concentrates and rice straw as the sole roughage source. This strategy is quite different from those of other countries such as the United States, Latin American nations (Brazil, Argentina, and Uruguay), Australia, and New Zealand that have implemented roughage-based feeding programs. The unique feeding program for Hanwoo in Korea has resulted in highly marbled beef with particularly high percentages of oleic acid (~50%), a major MUFA. Additionally, the MUFA/SFA ratios of Hanwoo, corn-fed Angus, and hay-fed Angus beef were found to be 1.28, 0.86, and 0.66, respectively (Smith et al., 2006). A high proportion of MUFAs including oleic acid can help to prevent human cardiovascular diseases (Kris-Etherton et al., 1999), elevated blood pressure (Rasmussen et al., 2006), and diabetes mellitus (Due et al., 2008). Furthermore, MUFAs are also acknowledged to have positive effects on beef palatability. High proportions of oleic acid in beef decrease the melting point of lipids extracted from sirloin produced by Hanwoo steers (Kim et al., 2013b). According to a panel test, Hanwoo beef with high MUFA percentages also has high palatability scores compared to imported beef with relatively low MUFA concentrations (Kim et al., 2011). In addition, information provided by previous studies implies that beef produced with a concentrate-based feeding system has high scores for tenderness, juiciness, umami, and overall palatability. Based on these data, this study was realized to determine whether feeding finishing diet of Hanwoo steers supplemented with rice bran, flax seed, or sunflower seed would affect animal performances and carcass characteristics. Additionally, the fatty acid composition, free amino acid and peptide contents, and sensory evaluation results of *longissimus* muscle (LM) were also assessed.

MATERIALS AND METHODS

Experimental animals and diets

A total of 39 Hanwoo steers (average age of 22.2±2.3 mo and average body weight [BW] of 552.2±32.8 kg) in local farm were divided into Control (10 animals), rice bran (RB, 10 animals), flax seed (FS, 9 animals), or sunflower seed (SS, 10 animals).

The steers were group fed and housed in groups of four or five animals in 5.0×10.0 m pens (a total of eight pens). The cattle were fed for 273 d until they reached an average age of 31.2±2.3 mo. Commercial concentrates (The Hanwoomaru Finishing, Woosung Feed Co., Ltd., Daejeon, Korea) for finishing Hanwoo steers were fed to the Control group (Table 1) and used as a basal concentrate for the RB, FS, and SS groups. For the concentrates given to the RB, FS, and SS group, 5% of the Control concentrate was replaced with RB, FS, and SS, respectively. Ten kg of concentrates per d were given to each animal until the end of the experiment. The steers had free access to rice to rice straw as the sole roughage source and water *ad libitum*. Chemical composition of the concentrates for each treatment group is shown in Table 2.

Table 1. Ingredients of commercial concentrate for finishing Hanwoo steers (DM basis)

Ingredients	%
Corn	25.0
Wheat	22.0
Barley	10.0
Wheat bran	5.5
Rice bran	5.0
Corn gluten feed	5.0
Coconut oil	5.0
Palm kernal	5.0
Canola	3.0
DDGS	3.0
Cottonseed whole	3.0
Cane molasses	3.0
Limestone	1.5
Salt	0.8
Sodium bicarbonate	0.7
Magnesium oxide	0.3
Mineral premix	0.2
Vitamin premix	0.1
Others*	1.9

DM, dry matter; DDGS, distiller's dried grains with solubles.
* Pre-/probiotics, organic minerals, buffers, etc.

Table 2. Chemical composition of concentrates fed to the finishing Hanwoo steers

Item	Control	RB	FS	SS
	------------------------- % ----------------------			
Dry matter	87.50	87.53	87.73	87.68
Crude protein	12.00	12.10	12.35	12.30
Ether extract	4.15	4.86	5.92	5.94
Crude fiber	6.21	6.32	6.62	6.85
Crude ash	7.12	7.20	6.95	6.96
NFE	58.02	57.06	55.89	55.62
NFC	43.23	42.17	41.07	41.07
Ca	0.90	0.86	0.86	0.87
P	0.45	0.53	0.46	0.46
ADF	9.00	9.09	9.80	10.28
NDF	21.00	21.21	21.44	21.40
TDN	74.00	74.51	75.70	75.65

RB, rice bran; FS, flax seed; SS, sunflower seed; NFE, nitrogen-free extract; NFC, non-fibrous carbohydrate; ADF, acid detergent fiber; NDF, neutral detergent fiber; TDN, total digestible nutrients.

Carcass evaluation

At the end of the experiment, all steers were transported to a local abattoir (Gyeongsan, Korea) and slaughtered by exsanguination after stunning. After chilling the carcasses for 24 h at 4°C, cold carcass weight, indices (marbling degree, LM area, fat thickness, meat color, fat color, texture, and maturity at the 13th rib) for yield grade and quality grade of the carcasses were measured according to the Korean Beef Carcass Grading Standard (2014).

Total lipid extraction

Samples of LM taken between the 12th and 13th ribs were removed from each carcass and transported to the laboratory while kept at 4°C. After trimming the subcutaneous fat, the samples were homogenized and stored in a freezer at –80°C until analysis. Total lipids were extracted from the beef samples using the modified method by Folch et al. (1957). Approximately 5 g of muscle tissue was homogenized (Polytron PT2100, Kinematica Inc., Bohemia, NY, USA) with 5 mL of chloroform:methanol (2:1, vol:vol) solution and incubated at room temperature for 30 min. The homogenate was filtered through Whatman GF/C filters (Whatman Ltd., Maidstone, UK) and rinsed with an additional 10 mL of chloroform:methanol solution. The extracted lipid was combined with 8 mL of 0.74% KCl and stirred for 1 min. After the phases separated, the lipid layer was transferred to 20-mL scintillation vials (Wheaton, Millville, NJ, USA) and the solvents were evaporated by heating (C-WBE, Changshin Science, Seoul, Korea) at 60°C under nitrogen.

Fatty acid composition analysis

The lipids extracted from LM were methylated according to the method by Lepage and Roy (1986) and the fatty acid composition was analyzed using gas chromatography. Approximately 200 μL of each lipid extract were converted into fatty acid methyl esters by adding 2 mL methanol:benzene (4:1, vol:vol) solution and 200 μL of acetyl chloride, and heated for 40 min at 100°C on a heating block (Lab & Tools, Anyang, Korea). One mL isooctane (Sigma-Aldrich, Saint Louis, MO, USA) and 8 mL of 6% potassium carbonate (Junsei, Tokyo, Japan) were added, and the solution was centrifuged (1580R, Labogene, Daejeon, Korea) for 10 min at 500×g. One (1) μL of the supernatant was subjected to analysis with a gas chromatography (Clarus 500; Perkin Elmer, Shelton, CT, USA) equipped with a fused column SP 2560 (100 m in length, 0.25 mm in diameter, 0.2 um film thickness) using nitrogen as the carrier gas (flow rate of 1.0 mL/min). The initial oven temperature was 210°C; the injector and detector temperatures were 240°C and 250°C, respectively.

Measurement of free amino acid contents

Free amino acid contents in the samples were analyzed as previously described by Henderson et al. (2000). Ten grams of LM sample were dissolved in 200 mL of 80% ethanol and stored at room temperature for 12 hours. The solution was filtered (Whatman filter paper No. 2, GE Healthcare Bio-Sciences Corp., Piscataway, NJ, USA). The filtrate was concentrated on an evaporator (Eyela Rotary evaporator N-1100, Tokyo Rikakikai, Co., Tokyo, Japan) at 45°C and dissolved in 40 mL of distilled water. Infranatant separated by funnel after adding ethyl ether (20 mL; Sigma-Aldrich, Saint Louis, MO, USA) was evaporated. The residue was dissolved in 20 mL of 0.2 M citrate buffer and filtered through 0.45 μm syringe filter (Toyo Roshi Kaisha, Ltd., Tokyo, Japan) and the free amino acids were analyzed with HPLC (L-8900, Hitachi, Tokyo, Japan). Cation exchange resin (4.6 mm×60 mm) was used as a column, Lithium Citrate Buffer solution (0.35 mL/min) was used as a mobile phase A, and Ninhydrin reagents (0.3 mL/min) solution was used as a mobile phase B. Free amino acid mixture standard solution (Wako Pure Chemical Co., Osaka, Japan) was used as derivatives.

Sensory evaluation

Ten panelists were trained according to the modified methods of the National Live Stock and Meat Board (1995). LM samples used for sensory evaluation had the same quality grade (1^{++} grade according to the Korean Beef Carcass Grading Standard). The samples were cut into portions 2×2.5×1 cm (L×W×D) in size and cooked at 220°C for 60 s on the front and 90 s on the back. The cooked samples were immediately served to the panelists. The tenderness, juiciness, flavor, umami, oiliness, and overall palatability (1; extremely tough, dry, not flavorful, bland, dry, or unacceptable to 8; extremely tender, juicy, flavorful, intense, oily, or acceptable) of the steaks were evaluated.

Statistical analysis

Data were analyzed with a one-way analysis of variance using SPSS (2011) Ver. 19.0 (SPSS Inc., Chicago, IL, USA). The steers were randomly assigned into pens and pens served as the experimental unit for growth performance, carcass characteristics, carcass grade, fatty acid composition, free amino acid contents, and sensory evaluation data. Significances were tested at 5% level and Tukey's multiple range test was used to compare results among the treatment groups.

RESULTS AND DISCUSSION

Growth performance

Final BW of the Control, RB, FS, and SS groups was 768.2 ±36.2, 785.8±48.2, 786.2±18.4, and 789.0±73.3 kg, respectively. The mean initial BW of all the experimental Hanwoo steers was 552.2±32.8 kg. The average daily gain (ADG) was 0.79±0.1, 0.85±0.1, 0.82±0.1, and 0.84±0.2 kg for the Control, RB, FS, and SS steers, respectively. No differences ($p>0.05$) were observed among groups (Table 3).

Forster et al. (1993) compared the growth performance of steers given diets supplemented with grazing tall fescue, clover, and Bermuda grass hay with either low or high levels of corn or rice bran, and found that the rate of BW gain for rice bran-supplemented calves was greater than the nonsupplemented controls (1.06 kg/d). Rice bran-supplemented calves were also found to have a greater rate of BW gain compared to nonsupplemented controls in a different investigation (Till et al., 1991).

The performance of beef cattle consuming diets supplemented with rice bran was not better compared to those given corn and soybean hulls (Gadberry et al., 2007). BW and ADG were increased and feed requirement was reduced to 14.5% when rice bran was fed to finishing Hanwoo steers (Kim et al., 2013a). In another investigation (Kronberg et al., 2011), Angus steers were grazed and fed ground flax seed (FLX, 0.20% BW) or ground corn and soybean meal (CSBM, 0.28% BW) and compared to those fed basal diet (Control). ADG values were 1.04, 1.09, and 0.89 kg for the FLX, CSBM, and Control groups, respectively ($p<0.05$). When grazing steers were given diets supplemented with flax seed, ADG was 0.28 kg higher than that of the Control group and feed efficiency was better

Table 3. Carcass characteristics of the Hanwoo steers after consuming rice bran, flax seed, and sunflower seed during the finishing period

	Control[1]	RB[2]	FS[3]	SS[4]
No. of heads	10	10	9	10
Initiation age (mo)	23.1±2.4[12]	21.5±1.7	21.8±2.6	22.3±2.6
Termination age (mo)	32.1±2.4	30.5±1.7	30.8±2.6	31.2±2.6
BW (kg)				
Initial	552.8±35.5	551.2±33.0	553.8±34.1	551.2±33.8
Final	768.2±36.2	785.8±48.2	786.2±18.4	789.0±73.3
ADG (kg)	0.79±1.00	0.85±0.94	0.82±0.11	0.84±0.19
Cold carcass wt (kg)	442.5±27.9	450.2±32.6	451.9±13.3	454.6±43.4
Yield traits				
Fat thickness (mm)	14.2±4.4^{b}	14.8±6.7^{b}	19.8±4.6^{a}	15.1±4.5^{ab}
Longissimus muscle area (cm^3)	91.2±10.5	96.4± 5.8	95.9±6.2	92.4±8.6
Yield index	63.8±2.8	63.9±5.0	60.7±3.4	63.1±3.8
Yield grade[5]	1.7±0.5	2.0±0.7	1.4±0.5	1.7±0.7
Quality grade				
Marbling score[6]	5.3±1.0	4.7±1.9	4.6±2.1	5.1±1.7
Meat color[7]	5.0± 0.0	4.9±0.3	5.0±0.0	5.0±0.5
Fat color[8]	3.0±0.0	3.0±0.0	3.0±0.0	3.0±0.0
Texture[9]	1.0±0.0	1.3±0.5	1.4±0.5	1.2±0.4
Maturity[10]	2.3±0.5	2.1±0.3	2.0±0.0	2.0±0.0
Quality grade[11]	2.4±0.5	2.1±1.0	2.0±1.1	2.2±0.9

RB, rice bran; FS, flax seed; SS, sunflower seed; BW, body weight; ADG, average daily gain, SD, standard deviation.
[1] Commercial finishing concentrate. [2] 95% commercial finishing concentrate+5% rice bran.
[3] 95% commercial finishing concentrate+5% flax seed. [4] 95% commercial finishing concentrate+5% sunflower seed.
[5] Converted into numeric values (grade A = 3, B = 2, and C = 1).
[6] 9 = the most abundant, 1 = devoid. [7] 1 = bright red, 7 = dark red. [8] 1 = white, 7 = yellowish. [9] 1 = fine, 3 = coarse. [10] 1 = immature, 9 = mature.
[11] Converted into numeric values (4 = grade 1^{++}, 3 = grade 1^{+}, 2 = grade 1, and 1 = grade 2).
[12] Mean±SD.
[ab] Mean values in the same row with different superscript letters are significantly ($p<0.05$) different.

compared to corn-supplemented steers without differences in ADG (Scholljegerdes and Kronberg, 2010). Feeding whole sunflower seeds to finishing steers (Gibb et al., 2004) increased dry matter intake ($p = 0.02$), ADG ($p = 0.01$), and gain:feed ($p = 0.01$). Other report has shown that finishing steers fed diet supplemented with sunflower seed did not affect final BW and feed efficiency with reducing dry matter intake ($p = 0.005$) and ADG ($p = 0.02$) (Shah et al., 2006).

Carcass characteristics

Supplementing RB, FS, and SS to finishing Hanwoo steers resulted in cold carcass weights of 450.2±32.6, 451.9±13.3, and 454.6±43.4 kg, respectively ($p>0.05$; Table 3). Fat thickness of the FS group (19.78±4.60 mm) was greater ($p<0.05$) than that of the other groups. LM area of the RB, FS, SS, and Control groups was 96.4±5.8, 95.9±6.2, 92.4±8.6, and 91.2±10.5 cm^2, respectively. Final yield grade converted into numerical values considering cold carcass weight, fat thickness, and LM area of the steers was 2.0±0.7 for the RB group, 1.7±0.5 for the Control animals, 1.7±0.7 for the SS group, and 1.4±0.5 for the FS cattle. Marbling degree values for the Control, SS, RB, and FS groups were 5.3±0.9, 5.1±1.7, 4.7±1.9, and 4.6±2.1, respectively ($p>0.05$). Meat color and fat color among the treatment groups were not affected by dietary treatment. Carcass texture (1 = fine, 3 = coarse) was in the order of FS, RB, SS, and Control group ($p<0.05$) and the Control group had the highest ($p<0.05$) scores in maturity (1 = immature, 9 = mature). Final carcass quality grades converted into numeric values (4 = grade 1^{++} , 3 = grade 1^{+}, 2 = grade 1, and 1 = grade 2) based on marbling degree, meat color, fat color, texture, and maturity were in the order of Control, SS, RB and FS group without differences observed ($p>0.05$).

In a previous study, fat thickness and LM area were not different when finishing steers were fed hay and 0.38% or 0.76% BW of rice bran for 84 d compared to the Control or corn-fed groups (Forster et al., 1993). Kim et al. (2013a) reported that supplementing the diets of finishing Hanwoo steers with rice bran results in cold carcass weight of 470.0 kg, which was heavier than that of Control animals. However, no differences in fat thickness, LM area, or marbling degree were observed. Kronberg et al. (2011) discovered that there were no changes in hot carcass weight, LM area, fat thickness, or marbling degree when Angus steers were fed a control diet, flax seed (0.2% BW), ground corn, and soybean meat (0.3% BW) for 85 d. The inclusion of flax in the diets of finishing beef heifers did not affect ($p = 0.32$) fat thickness over the 12th rib (Maddock et al., 2006a). Shah et al. (2006) reported that the fat depth of steers fed diets containing sunflower seed was reduced ($p = 0.02$) to 13.8 from 17.0 mm.

Fatty acid composition

When Hanwoo steers were fed diet supplemented with rice bran, flax seed, and sunflower seed during the finishing period, percentages of myristic acid ($C_{14:0}$) in the LM increased ($p<0.05$) compared to the Control group. However, palmitic acid ($C_{16:0}$), stearic acid ($C_{18:0}$), and arachidic acid ($C_{20:0}$) percentages were unaffected (Table 4). Among the MUFAs, palmitoleic acid ($C_{16:1}$) concentrations were higher ($p<0.05$) in the SS animals than the Control group. The level of oleic acid ($C_{18:1}$) was highest ($p<0.05$) in the Control group (47.7%±2.7%) compared to the RB, FS, and SS groups. Additionally, the concentration of linolenic acid ($C_{18:3}$), a PUFA belonging to the n-3 family, was 2.3 times higher ($p<0.05$) in the FS group compared to the other groups. The level of total SFAs in the RB group was 45.4%±1.9%, the highest ($p<0.05$) among the treatment groups. The concentration of total unsaturated fatty acids (UFAs) and total MUFAs in the Control group was 57.1%±2.1% and 53.7%±2.3%, respectively, and were the highest values among the groups. There were no differences in total PUFA levels among the treatment groups although LM from the FS group had a high concentration of linolenic acid ($C_{18:3}$).

Previously, Kim et al. (2011) reported that there were no changes in myristic acid ($C_{14:0}$), palmitic acid ($C_{16:0}$), stearic acid ($C_{18:0}$), or oleic acid ($C_{18:1}$) percentages in LM of Hanwoo steers fed rice bran (0.1% of concentrate) for 314 d. Additionally, feeding rice bran had no effects on the percentages of conjugated linoleic acid (CLA) or trans vaccenic acid in the LM of finishing Hanwoo steers. In a different study, no differences ($p>0.22$) were observed in the concentrations of myristic acid ($C_{14:0}$), palmitic acid ($C_{16:0}$), stearic acid ($C_{18:0}$), or total SFAs in the LM of flax seed-fed steers (Kronberg et al., 2011). Intake of flax seed can result in red meat with an increased concentration of beneficial n-3 fatty acids (Maddock et al., 2006b). Meat from cattle fed flax has been reported to contain higher levels of omega-3 fatty acids (Drouillard et al., 2004). However, feeding flax at a concentration of 8% of the finishing diet does not improve the fatty acid profile of steers (Maddock et al., 2003). Intake of sunflower seed decreased ($p<0.05$) palmitic acid ($C_{16:0}$) and linolenic acid ($C_{18:3}$) levels with increasing ($p = 0.05$) the prevalence of oleic acid ($C_{18:1}$), linoleic acid ($C_{18:2}$), cis-9, trans-11 CLA, and trans-10, cis-12 CLA in subcutaneous fat (Gibb et al., 2004).

Free amino acid and peptide contents

No change on levels of alanine (Ala), glycine (Gly), serine (Ser), threonine (Thr), lysine (Lys), arginine (Arg), histidine (His), isoleucine (Ile), leucine (Leu), phenylalanine (Phe), valine (Val), or aspartic acid (Asp) was observed in the LM of Hanwoo steers fed rice bran, flax

Table 4. Fatty acid composition of *longissimus* muscle of Hanwoo steers after consuming rice bran, flax seed, and sunflower seed during the finishing period

	Control[1]	RB[2]	FS[3]	SS[4]
	%			
$C_{10:0}$	0.11±0.04[5,b]	0.21±0.10[ab]	0.17±0.06[ab]	0.23±0.19[a]
$C_{12:0}$	0.18±0.06	0.20±0.04	0.22±0.07	0.22±0.05
$C_{14:0}$	4.00±0.58[b]	4.96±0.93[a]	4.89±0.65[a]	5.06±1.42[a]
$C_{14:1}$	0.24±0.05	0.27±0.05	0.30±0.06	0.30±0.10
$C_{16:0}$	28.27±1.74	29.92±1.35	28.86±1.79	29.08±1.97
$C_{16:1}$	5.55±0.66[b]	6.46±1.15[ab]	6.24±1.25[ab]	6.91±1.19[a]
$C_{18:0}$	10.30±0.97	10.21±1.45	10.19±1.24	10.14±1.25
$C_{18:1}$	47.73±2.66[a]	44.25±1.49[b]	45.52±1.82[b]	44.69±2.83[b]
$C_{18:2}$	2.76±0.50	2.67±0.57	2.46±0.56	2.58±0.31
$C_{18:3}$	0.34±0.07[b]	0.32±0.13[b]	0.75±0.16[a]	0.33±0.15[b]
$C_{20:0}$	0.08±0.03	0.08±0.03	0.09±0.06	0.09±0.04
$C_{20:1}$	0.15±0.09	0.16±0.07	0.12±0.08	0.12±0.09
$C_{20:4}$	0.29±0.07	0.30±0.14	0.20±0.14	0.27±0.11
SFA	42.83±2.09[b]	45.37±1.87[a]	44.25±2.42[ab]	44.58±2.33[ab]
UFA	57.06±2.11[a]	54.42±1.90[b]	55.58±2.41[ab]	55.19±2.23[ab]
MUFA	53.67±2.36[a]	51.14±1.46[b]	52.18±2.07[ab]	52.02±2.14[ab]
PUFA	3.39±0.55	3.28±0.65	3.40±0.77	3.17±0.44
U/S	1.34±0.12	1.20±0.09	1.26±0.13	1.24±0.11
M/S	1.26±0.12	1.13±0.08	1.18±0.11	1.17±0.11
P/S	0.08±0.01	0.07±0.02	0.08±0.02	0.07±0.01
FI[6]	1.17±0.12[a]	1.05±0.06[b]	1.10±0.09[ab]	1.09±0.09[ab]

RB, rice bran; FS, flax seed; SS, sunflower seed; SFA, saturated fatty acid; UFA, unsaturated fatty acid; MUFA, monounsaturated fatty acid; PUFA, polyunsaturated fatty acid; U/S, unsaturated fatty acid/saturated fatty acid; M/S, monounsaturated fatty acid/saturated fatty acid; P/S, polyunsaturated fatty acid/saturated fatty acid; FI, flavor index, SD, standard deviation.
[1] Commercial finishing concentrate. [2] 95% commercial finishing concentrate+5% rice bran.
[3] 95% commercial finishing concentrate+5% flax seed. [4] 95% commercial finishing concentrate+5% sunflower seed.
[5] Mean±SD.
[6] FI (flavor index) = $(C_{14:1}+C_{16:1}+C_{18:1})/(C_{14:0}+C_{16:0}+C_{18:0}+C_{18:2})$.
[abc]Mean values in the same row with different superscript letters are significantly ($p<0.05$) different.

seed, and sunflower seed during the finishing period (Table 5). The concentration of methionine (Met), a sulfur-containing amino acid, was higher ($p<0.05$) in the FS and SS groups than the Control and RB steers. The level of methionine in the FS group (1.67±0.41 mg/100 g) was even higher ($p<0.05$) than that found in the SS group (1.17±0.30 mg/100 g). Glutamic acid (Glu), known to have a positive relationship with beef taste, and tyrosine (Tyr) contents were remarkably elevated ($p<0.05$) in the FS group compared to the other steers. Alpha-aminoadipic acid (α-AAA) contents in the FS group were higher ($p<0.05$) than levels in the Control, RB, or SS group. Additionally, anserine (Ans) and carnosine (Car) concentrations was numerically higher ($p>0.05$) in the FS steers relative to the other groups.

Beef flavor, which develops during heating process, depends on the amount and proportion of precursor compounds present. Among various components of meat, proteins, lipids, and carbohydrates play key roles in the development of flavor since they are capable of developing into important flavor precursors when heat applied (Mottram, 1998).

Sensory evaluation results

According to sensory evaluation, tenderness (5.7±0.3) and juiciness (5.5±0.3) scores were the highest in RB group whereas flavor (5.6±0.6), umami (5.5±0.5), and overall palatability (5.5±0.4) scores were the highest in FS group without statistical significance ($p>0.05$) (Table 6).

Recent consumer research has shown that once tenderness is within an acceptable range or when variation in tenderness has been minimized, flavor becomes the most important determinant for beef consumers' assessments of overall satisfaction (Killinger et al., 2004). Flavor results from the combination of basic tastes (sweet, sour, bitter, salt, and umami) derived from water-soluble compounds and odors produced from a myriad of substances present in the food product from the onset or via various reactions (Farmer and Mottram, 1994). The flavors and aromas associated with beef are generally those that develop during heating. When water-soluble substances produced from precursor compounds dissolve in saliva, they bind to the

Table 5. Free amino acid and peptide contents in *longissimus* muscle of Hanwoo steers fed rice bran, flax seed, and sunflower seed during the finishing period

	Control[1]	RB[2]	FS[3]	SS[4]
	----------	mg/100 g	----------	
Ala	23.82±6.75[5]	23.46±4.40	26.29±3.54	23.12±3.89
Gly	5.91±1.57	5.39±1.21	6.04±1.09	5.86±0.89
Ser	2.99±1.06	3.03±0.84	3.08±1.33	2.61±0.39
Thr	3.14±1.18	3.19±0.98	2.80±0.59	2.45±0.41
Lys	4.29±1.65	4.22±1.92	4.27±1.04	4.17±0.63
Arg	4.33±1.56	4.74±1.11	4.55±1.15	3.92±0.67
His	2.53±0.78	2.46±0.44	2.66±0.41	2.42±0.47
Ile	2.66±1.02	2.67±0.66	2.79±0.51	2.42±0.45
Leu	4.57±1.92	4.68±1.43	5.06±1.10	4.01±0.84
Met	0.56±0.42[b]	0.45±0.71[b]	1.67±0.41[a]	1.17±0.30[a]
Phe	2.90±1.06	2.93±0.89	3.23±0.64	2.50±0.46
Val	3.39±1.24	3.35±0.70	3.62±0.71	3.28±0.60
Asp	0.30±0.15	0.30±0.09	0.13±0.20	0.26±0.23
Glu	2.39±1.09[b]	3.00±1.40[b]	4.58±1.96[a]	2.38±0.96[b]
Tau	21.40±4.95	26.55±5.27	26.96±7.47	25.18±8.54
Tyr	2.91±1.18[ab]	2.86±0.90[ab]	3.26±0.75[a]	2.39±0.47[b]
α-AAA	0.00±0.00[b]	0.10±0.22[b]	0.53±0.27[a]	0.07±0.12[b]
Anserine	78.31±21.07	71.94±7.35	88.18±15.51	82.65±20.98
Carnosine	291.40±77.25	292.71±61.81	329.74±24.44	322.04±52.01

RB, rice bran; FS, flax seed; SS, sunflower seed; α-AAA, α-amino aminoadipic acid; SD, standard deviation.
[1] Commercial finishing concentrate. [2] 95% commercial finishing concentrate+5% rice bran.
[3] 95% commercial finishing concentrate+5% flax seed. [4] 95% commercial finishing concentrate+5% sunflower seed.
[5] Mean±SD.
[abc] Mean values in the same row with different superscript letters are significantly ($p<0.05$) different.

taste buds and stimulate responses that are perceived in the brain. Odors are generated when volatile compounds bind to receptors in the olfactory bulb behind the nasal cavity and stimulate a response.

Intake of rice bran and rice oil for 126 d by feedlot-finished steers was found to improve tenderness ($p = 0.81$), juiciness ($p = 0.33$), and palatability ($p = 0.22$) of steaks extracted from the LM between the 10th and 12th ribs (Donicht et al., 2011). When finishing Hanwoo steers were fed rice bran for 314 d, juiciness and overall palatability of the LM muscle was improved ($p<0.05$) compared to the Control animals (Kim et al., 2011). An early study by Bowling et al. (1977) comparing forage-finished and grain-finished beef suggested that the consumer or retailer may discriminate against forage-finished beef due to the color (muscle or fat) and palatability (flavor, tenderness, or both) of the meat.

When yearling beef heifers were fed feedlot diets that

Table 6. Sensory evaluations for *longissimus* muscle of Hanwoo steers fed rice bran, flax seed, and sunflower seed during the finishing period

	Control[1]	RB[2]	FS[3]	SS[4]
Tenderness[5]	5.50±0.84[11]	5.67±0.32	5.25±0.49	5.10±0.85
Juiciness[6]	5.08±0.65	5.50±0.26	5.20±0.00	5.35±0.78
Flavor[7]	5.25±0.78	5.40±0.50	5.60±0.57	5.45±0.07
Umami[8]	5.00±0.53	5.07±0.31	5.55±0.49	5.10±0.42
Oiliness[9]	5.10±0.68	5.40±0.53	5.15±0.21	5.35±0.92
Overall palatability[10]	4.95±0.70	5.23±0.38	5.50±0.42	5.05±0.35

RB, rice bran; FS, flax seed; SS, sunflower seed; SD, standard deviation.
[1] Commercial finishing concentrate. [2] 95% commercial finishing concentrate+5% rice bran.
[3] 95% commercial finishing concentrate+5% flax seed. [4] 95% commercial finishing concentrate+5% sunflower seed.
[5] 1 = extremely tough, 8 = extremely tender. [6] 1 = extremely dry, 8 = extremely juicy.
[7] 1 = extremely unflavorful, 8 = extremely flavorful. [8] 1 = extremely absent, 8 = extremely present.
[9] 1 = extremely dry, 8 = extremely oily. [10] 1 = extremely bad, 8 = extremely good.
[11] Mean±SD.

contained no flax, whole flax, rolled flax, or ground flax, the sensory panel tenderness ratings ($p = 0.44$), flavor ratings ($p = 0.35$), and Warner-Bratzler shear force (WBSF; $p = 0.06$) were unaffected (Maddock et al., 2006b). Maddock et al. (2003) reported that steaks from flax-fed steers were less juicy and tender than steaks from steers finished on a corn-based control diet. Drouillard et al. (2004) did not note any differences in sensory traits or shear force among steers and heifers fed different concentrations of flax or Holstein steers fed 5% flax. Supplementation with ground flax seed did not change ($p>0.12$) WBSF measurements for steaks from the grazing steers, nor did this diet influence ($p>0.29$) the tenderness, juiciness, or flavor intensity of the meat. However, there appeared to be differences ($p = 0.04$) in sensory perception of the off-flavor intensity for steaks from flax seed-fed steers. Panelists sensed a slightly more intense off-flavor in steaks from steers fed flax seed compared to steaks from steers fed the control diet ($p = 0.07$). The off-flavor intensity ratings for steaks from steers fed flax seed may be related to greater concentrations of linolenic acid ($C_{18:3}$) (Kronberg et al., 2011).

Addition of SS that contain high oleic acid levels to corn diets increased ($p = 0.02$) juiciness whereas high-linoleic acid SS elevated initial and overall tenderness ($p = 0.02$) of LM steaks from steers (Shah et al., 2006). Mir et al. (2003) noted that tenderness scores tended to increase compared to those for vitamin E-supplemented steers when sunflower oil containing about 70% linoleic acid ($C_{18:2}$) is fed to finishing steers. Duckett et al. (2001) reported similar results for meat from high oil containing corn-fed steers and suggested that the cause of differences in flavor may be related to the fatty acid composition of the meat.

In conclusion, supplementation of flax seed to the concentrates of finishing Hanwoo steers tended to improve ($p>0.05$) flavor, umami, and overall palatability of LM. These results might be caused by increases in the concentration of free amino acids, glutamic acid, methionine, and α-AAA, and peptides, anserine and carnosine, and their complex reactions. Further detailed studies, however, to understand mechanisms for the improved beef flavors by feeding flax seed in the diet of finishing beef cattle are necessary.

ACKNOWLEDGMENTS

The authors gratefully acknowledge the financial support (2010) of Woosung Feed Co., Ltd. This study was also supported by a 2014 Yeungnam University research grant (214A061048).

REFERENCES

Bowling, R. A., G. C. Smith, Z. L. Carpenter, T. R. Dutson, and W. M. Oliver. 1977. Comparison of forage-finished and grain-finished beef carcasses. J. Anim. Sci. 45:209-215.

Donicht, P. A. M. M., J. Restle, L. Da S. Freitas, A. M. Callegaro, M. S. Weise, and I. L. Brondani. 2011. Fat sources in diets for feedlot-finished steers - carcass and meat characteristics. Ciência Anim. Bras. 12:487-496.

Drouillard, J. S., M. A. Seyfert, E. J. Good, E. R. Loe, B. Depenbusch, and R. Daubert. 2004. Flaxseed for finishing beef cattle: Effects on animal performance, carcass quality, and meat composition. Plant Sci. 55:149-159.

Duckett, S. K., B. A. Jardner, M. A. Winds, and F. N. Owen. 2001. Impact of high oil corn on beef steak quality. J. Anim. Sci.(Suppl.). 78:155.

Due, A., T. M. Larsen, K. Hermansen, S. Stender, J. J. Holst, S. Toubro, and A. Astrup. 2008. Comparison of the effects on insulin resistance and glucose tolerance of 6-mo high-monounsaturated-fat, low-fat, and control diets. Am. J. Clin. Nutr. 87:855-862.

Farmer, L. J. and D. S. Mottram. 1994. Lipid-Maillard interactions in the formation of volatile aroma compounds. J. Article 35: 313-326.

Folch, J., M. Lees, and G. H. Sloane Stanley. 1957. A simple method for the isolation and purification of total lipides from animal tissues. J. Biol. Chem. 226:497-509.

Forster, L. A. Jr., A. L. Goetsch, D. L. Sr. Galloway, and Z. B. Johnson. 1993. Feed intake, digestibility, and live weight gain by cattle consuming forage supplemented with rice brand and (or) corn. J. Anim. Sci. 71:3105-3014.

Gadberry, M. S., P. A. Beck, and S. A. Gunter. 2007. Review: Rice milling coproducts as feedstuffs for beef cattle. The Prof. Anim. Sci. 23:309-315.

Gibb, D. J., F. N. Owens, P. S. Mir, Z. Mir, M. Ivan, and T. A. McAllister. 2004. Value of sunflower seed in finishing diets of feedlot cattle. J. Anim. Sci. 82:2679-2692.

Gillingham, L. G., S. Harris-Janz, and P. J. Jones. 2011. Dietary monounsaturated fatty acids are protective against metabolic syndrome and cardiovascular disease risk factors. Lipids 46:209-228.

Henderson, J. W., R. D. Ricker, B. A. Bidlingmeyer, and C.Woodward. 2000. Rapid, accurate, sensitive, and reproducible HPLC analysis of amino acids. Agilent Technologies Technical Note 5980-1193E.

Killinger, K. M., C. R. Calkins, W. J. Umberger, D. M. Feuz, and K. M. Eskridge. 2004. A comparison of consumer sensory acceptance and value of domestic beef steaks and steaks from a branded, Argentine beef program. J. Anim. Sci. 82:3302-3307.

Kim, S. I., K. K. Jung, D.Y. Kim, J. Y. Kim, and C. B. Choi. 2011. Effects of supplementation of rice bran and roasted soybean in the diet on physico-chemical and sensory characteristics of M. longissimus dorsi of Hanwoo steers. Korean J. Food Sci. Anim. Resour. 31:451-459.

Kim, S. I., G. H. Lee, and C. B. Choi. 2013a. Effects of supplementary rice bran and roasted soybean in the diets on carcass characteristics and composition of CLA in Hanwoo steers. J. Anim. Sci. Technol. 55:435-442.

Kim, S. I., B. R. Cho, and C. B. Choi. 2013b. Effects of sesame meal on growth performances and fatty acid composition, free amino acid contents, and panel tests of loin of Hanwoo steers. J. Anim. Sci. Tech. 55:451-460.

Kris-Etherton, P. M., T. A. Pearson, Y. Wan, R. L. Hargrove, K. Moriarty, V. Fishell, and T. D. Etherton. 1999. High-monounsaturated fatty acid diets lower both plasma cholesterol and triacylglycerol concentrations. Am. J. Clin. Nutr. 70:1009-1015.

Kronberg, S. L., E. J. Scholljegerdes, A. N. Lepper, and E. P. Berg. 2011. The effect of flaxseed supplementation on growth, carcass characteristics, fatty acid profile, retail shelf life, and sensory characteristics of beef from steers finished on grasslands of the northern Great Plains. J. Anim. Sci. 89:2892-2903.

Korean Beef Carcass Grading Standard. 2014. Korean Standards for Grading Animal Products. Ministry of Agriculture, Food and Rural Affairs, Jeonju, Korea.

Lepage, G. and C. C. Roy. 1986. Direct transesterification of all classes of lipids in a one-step reaction. J. Lipid Res. 27:114-120.

Maddock, T. D., V. L. Anderson, P. T. Berg, R. J. Maddock, and M. J. Marchello. 2003. Influence of level of flaxseed addition and time fed flaxseed on carcass characteristics, sensory panel evaluation and fatty acid content of fresh beef. Proc. 56th Recip. Meats Conf. 56:110 (Abstr.)

Maddock, T. D., M. L. Bauer, K. B. Koch, V. L. Anderson, R. J. Maddock, G. Barceló-Coblijn, E. J. Murphy, and G. P. Lardy. 2006a. Effect of processing flax in beef feedlot diets on performance, carcass characteristics, and trained sensory panel ratings. J. Anim. Sci. 84:1544-1551.

Maddock, T. D., B. Kreft, R. J. Maddock , V. L. Anderson, and G. P. Lardy. 2006b. Effect of including flax in beef creep feed on performance and subsequent carcass characteristics. J. Anim. Vet. Adv. 5:156-160.

Mir, P. S., T. A. McAllister, S. Zaman, S. D. Morgan Jones, M. L. He, J. L. Aalhus, L. E. Jeremiah, L. A. Goonewardene, R. J. Weselake, Z. Mir. 2003. Effect of dietary sunflower oil and vitamin E on Beef cattle performance, carcass characteristics and meat quality. Can. J. Anim. Sci. 83:53-66.

Mottram, D. S. 1998. Flavour formation in meat and meat products: A review. Food Chem. 62:415-424.

National Live Stock and Meat Board. 1995. Research guidelines for cookery, sensory evaluation, and instrumental tenderness measurements of meat. American Meat Science Association, Champaign, IL, USA.

Rasmussen, B. M., B. Vessby, M. Uusitupa, L. Berglund, E. Pedersen, G. Riccardi, A. A. Rivellese, L. Tapsell, and K. Hermansen. 2006. Effects of dietary saturated, monounsaturated, and n-3 fatty acids on blood pressure in healthy subjects. Am. J. Clin. Nutr. 83:221-226.

Ritchie, H. 2005. Strategies for managing marbling in beef cattle. Michigan State University. East Lansing, MI, USA.

Scholljegerdes, E. J. and S. L. Kronberg. 2010. Effect of supplemental ground flaxseed fed to beef cattle grazing summer native range on the northern Great Plains. J. Anim. Sci. 71:3199-3205.

Shah, M. A., P. S. Mir, J. L. Aalhus, J. Basarab, and E. K. Okine. 2006. Effects of sunflower seed inclusion in finishing diets for steers on performance, carcass characteristics, muscle and adipose fatty acid composition and meat quality. Can. J. Anim. Sci. 86:37-48.

Smith, S. B., D. K. Lunt, K. Y. Chung, C. B. Choi, R. K. Tume, and M. Zembayashi. 2006. Adiposity, fatty acid composition, and delta-9 desaturase activity during growth in beef cattle. J. Anim. Sci. J. 77:478-486.

SPSS. 2011. SPSS Release Ver. 19.0. SPSS Inc., Chicago, IL, USA.

Till, A. R., M. R. Hunt, T. Panggabean, D. Bulo, and G. J. Blair. 1991. The live weight gain of cattle at pasture in South Sulawesi supplemented with locally available by-products. Asian Australas. J. Anim. Sci. 4:85-90

Whetsell, M. S., E. B. Rayburn, and J. D. Lozier. 2003. Human health effects of fatty acids in beef. Fact Sheet: West Virgina University & USDA Agriculture. http://www.wvu.edu/~agexten/forglvst/humanhealth.pdf Accessed November 13, 2014.

3

^{1}H-Nuclear Magnetic Resonance-Based Plasma Metabolic Profiling of Dairy Cows with Fatty Liver

Chuang Xu[a],*, Ling-wei Sun[1,a], Cheng Xia, Hong-you Zhang, Jia-san Zheng, and Jun-song Wang[2]
Department of Clinical Veterinary Medicine, Animal Science and Technology College,
Heilongjiang August First Land Reclamation University, Daqing 163319, China

ABSTRACT: Fatty liver is a common metabolic disorder of dairy cows during the transition period. Historically, the diagnosis of fatty liver has involved liver biopsy, biochemical or histological examination of liver specimens, and ultrasonographic imaging of the liver. However, more convenient and noninvasive methods would be beneficial for the diagnosis of fatty liver in dairy cows. The plasma metabolic profiles of dairy cows with fatty liver and normal (control) cows were investigated to identify new biomarkers using ^{1}H nuclear magnetic resonance. Compared with the control group, the primary differences in the fatty liver group included increases in β-hydroxybutyric acid, acetone, glycine, valine, trimethylamine-*N*-oxide, citrulline, and isobutyrate, and decreases in alanine, asparagine, glucose, γ-aminobutyric acid glycerol, and creatinine. This analysis revealed a global profile of endogenous metabolites, which may present potential biomarkers for the diagnosis of fatty liver in dairy cows. (**Key Words:** Dairy Cow, Fatty Liver, ^{1}H-Nuclear Magnetic Resonance, Metabolomics, Plasma)

INTRODUCTION

Fatty liver is a common metabolic disorder in dairy cows during the transition period. Up to 65% of dairy cows are affected by moderate (triacylglycerol [TAG] 50 to 100 mg/g of wet liver) or severe (TAG ≥100 mg/g of wet liver) fatty liver during early lactation (Jorritsma et al., 2000). The primary pathological feature of fatty liver is the excessive deposition of fat, which predominantly consists of TAG, in the liver. Most transition dairy cows are in a state of negative energy balance (NEB) due to increased energy demands after parturition, coupled with lagging dry matter intake (DMI) (Hayirli et al., 2002). The most direct factor in the induction of fatty liver is a NEB (Oikawa et al., 2010). The economic results of fatty liver are reduced milk yield (MY), poor fertility, high risks of other periparturient diseases, and early culling of affected animals. As fatty liver can lead to substantial economic losses in the dairy industry, its prevention is of the utmost importance.

Several risk factors have been suggested, which include a high body condition score caused by prolonged lactation because of reproductive failure and overfeeding in late lactation and the dry period, a high rate of body lipid mobilization around calving, a low feed intake, and a low protein content in the diet (Sejersen et al., 2012). However, there is no consensus regarding the overall metabolic state for fatty liver in dairy cows. Many new holistic metabolomic platforms have been used for the identification of new predictive markers for the early diagnosis of this disease.

At present, the diagnosis of fatty liver has been restricted to biochemical or histological examinations of liver biopsy specimens (Gonzalez et al., 2011). Biopsy of the liver is the most reliable method for accurate estimation

* Corresponding Author: Chuang Xu.
E-mail: xuchuang7175@gmail.com
[1] Jiangsu Engineering Technology Research Center of Meat Sheep & Goat Industry, Nanjing Agricultural University, Nanjing, 210095, China.
[2] Center for Molecular Metabolism, School of Environmental and Biological Engineering, Nanjing University of Science Technology, Nanjing, 210095, China.
[a] These two authors contributed equally to this work.

of the degree of fatty infiltration. It can be used to determine the concentration of triglycerides (TGs) and the severity of the fatty liver (Herdt et al., 1983). For on-farm testing, liver biopsies are impractical due to the injury to the animals and the time needed to acquire and analyze the samples. Biopsy causes decreased feed intake and increased danger of infection and hemorrhage in cows (Smith et al., 1997). Furthermore, liver biopsy has been found to induce behavioral changes for up to 19 h afterwards; particularly for behavior previously associated with pain. Ultrasonographic imaging of the liver is a recently developed, noninvasive method; however, it requires skilled operating experience and is costly.

Due to the rapid development of metabolomics in recent years, the use of this approach for disease biomarker assessment has become popular. As the preferred platform of the metabolomics technologies, ^{1}H nuclear magnetic resonance (NMR) can provide a comprehensive metabolic profile of proton-containing, low-molecular-weight metabolites, and requires a minimal amount of sample (Song et al., 2013). The objective of the current study was to use ^{1}H NMR-based plasma metabolomics to examine fatty liver in dairy cattle to obtain information for understanding the metabolic pathways, metabolic networks, and pathogenesis of the fatty liver.

MATERIALS AND METHODS

Ethics statement

This study was carried out in strict accordance with the recommendations in the Guide for the Care and Use of Laboratory Animals of the National Institutes of Health. The protocol was approved by the Committee on the Ethics of Animal Experiments of the Heilongjiang Bayi Agricultural University (Permit Number: 20120319-1). All surgery was performed under sodium pentobarbital anesthesia, and all efforts were made to minimize suffering.

Animals

After obtaining the owner's permission, blood and tissue samples from cows were collected from a commercial dairy farm located in Heilongjiang Province, China. All samples—from 171 Holstein cows (14 to 21 days after calving)—were collected prior to feeding in the morning. The samples were collected from March 2014 to September 2014.

For this experiment, the cows were fed a total mixed ration (TMR) at 0070, 1300, and 1900 h. The TMR consisted of 55.60% dry matter (DM), 16.40% crude protein, 5.60% fat, 34.30% neutral detergent fiber, 22.00% acid detergent fiber, 1.07% calcium, 0.49% phosphorus, 0.32% magnesium, 0.13% sodium, 1.40% potassium, 0.39% chloride, 0.22% sulfur, and 1.75 mcal/kg DM net energy for lactation (Zhang et al., 2013).

The data collected during examination of the cows included age, parity, postpartum days, MY, body condition score (BCS), and DMI. The BCS (5-point scale) of 1 (thin) to 5 (obese) points with 0.25 intervals is used to assess body fat stores, and describes the appearance of seven body regions in Holstein dairy cows (Wildman et al., 1982). The MY of cows in this experiment was recorded using the calibrated weigh jars in the milking parlor on the same day *as* the collection of the blood samples. Individual feed intake was measured daily. Samples of feeds and TMR were collected weekly and dried at 60°C for 72 h. The DM percentages of the feed ingredients were used to adjust ration components each week, and DM percentages of TMR were used to calculate daily DMI.

Sample collection

The tissue and blood liver samples obtained from the experimental cows were collected on the same day. The blood samples were obtained from the jugular vein, stabilized in sodium heparin, and immediately centrifuged at 1,400×*g* for 20 min at 4°C. The plasma samples obtained were subsequently stored at –80°C until further analysis. Liver transfixion pins were used for the collection of liver tissue samples from the 11th or 12th right intercostal space. Ten milliliters of procaine 2% (CDM Lavoisier, Paris, France) was used to anaesthetize the skin around the 12th intercostal space. Liver tissue samples were collected with tailor-made biopsy needles (Berlin Model, 2.5 mm×25 cm; Eickemeyer Medizintechnik für Tierärzte, Tuttlingen, Germany), and biopsy specimens (150 to 350 mg of liver tissue) were stored at –20°C until determination of total lipid and TG concentrations.

Plasma biochemistry

Plasma alanine transaminase/glutamate pyruvate transaminase (ALT/GPT), TG, glycerol, creatine kinase (CK), non-esterified fatty acids (NEFA), glucagon, acetoacetate, fibroblast growth factor 21 (FGF21), glucose (Glc), β-hydroxybutyric acid (BHBA), and aspartate transaminase (AST) were photometrically analyzed (Abx Pentra 400; Horiba, Kyoto, Japan). Insulin (INS) and growth hormone (GH) concentrations in the blood samples were measured using radioimmunoassay kits (Beckman Coulter, Miami, FL, USA and Medilab, Malmö, Sweden, respectively), which have been validated for use in bovine plasma. The mean intra-assay coefficients of variation (CV) for duplicate samples were 3.9% and 3.5% for INS and GH, respectively. All inter-assay CVs were <10%.

Hepatic triglyceride content test

The liver tissue samples were tested in copper sulfate

solution and water with specific gravities of 1.025 and 1.055 (Herdt et al., 1983), respectively. Based on the buoyancy of the liver tissue in these liquids, the samples were classified as containing >35% lipid, >25% lipid, or <15% lipid (Herdt et al., 1983). Cows with TG contents >35% in the liver were grouped into the fatty liver group, and cows with TG contents <13% in the liver and with no clinical symptoms were grouped into the control group.

Sample preparation

Prior to ^{1}H NMR analysis, the plasma samples were thawed at room temperature. Deuterium oxide was added to each plasma sample (300 μL), which consisted of 150 μL buffer solution (pH 7.4, 0.2 mol/L Na_2HPO_4, and 0.2 mol/L NaH_2PO_4) and 150 μL sodium 3-trimethylsilyl-(2,2,3,3-D4) propionate (TSP; 1 mg/mL; Sigma-Aldrich, St. Louis, MO, USA). The plasma samples were centrifuged at 4°C (12,000 ×*g*) for 10 min. The aliquot of the resulting mixture (550 μL) was subsequently transferred to a 5 mm NMR tube.

Acquisition of ^{1}H NMR plasma spectra

Conventional ^{1}H NMR of the plasma samples were performed at 500 MHz on a Bruker Avance-500 spectrometer (Billerica, MA, USA) at 25°C. The spectra of the samples were recorded using the water-suppressed Carr–Purcell–Meiboom–Gill sequence with a spin–spin relaxation delay of 40 ms to suppress the broad signals of micromolecules. To provide sufficient data points for each resonance prior to Fourier transformation, the free induction decays were zero-filled to 64 K, and an exponential line broadening factor of 0.5 Hz was applied. All plasma ^{1}H NMR spectra were corrected for phase and baseline distortions using Bruker Topspin 3.0 software (Bruker GmbH, Karlsruhe, Germany) and were referenced to TSP (CH_3, δ 0.0).

Metabolite identification

The metabolites were assigned based on chemical shift and identified from a library of in-house pure compounds, Chenomx NMR suite (Version 7.5; Chenomx, Inc., Edmonton, Alberta, Canada), and database query (Madison: http://mmcd.nmrfam.wisc.edu/, HMDB: http://www.hmdb.ca/, etc.), and were identified by ^{1}H-^{1}H correlation spectroscopy and ^{1}H-^{13}C heteronuclear single quantum correlation. In order to enhance the information obtained from global metabonomic profiling of fatty liver, the Kyoto Encyclopedia of Genes and Genomes (http://www.genome.jp/kegg/) was utilized to map the marker metabolites with regards to the dairy cow metabolic pathways.

Data reduction of ^{1}H NMR data

Using the MestReNova software (Version 8.0.1; Mestrelab Research SL, A Coruña, Spain), all ^{1}H NMR spectra were automatically reduced to American Standard Code for Information Interchange (ASCII) files. The ASCII files were imported into "R" (version 2.7.2; http://www.r-project.org) to eliminate phase and baseline variations. To reduce variability in peak positions, peak alignment scripts were built into the R software. The spectra (range of δ 0.5 to 4.3) were binned into integrated segments of equal width (δ 0.003) to assess differences in the concentrations between the samples. The aligned spectra were then normalized using probabilistic quotient normalization (Dieterle et al., 2006).

Multivariate analysis

Multivariate analysis was conducted on the ^{1}H NMR data, which included unsupervised principal component analysis (PCA) and supervised orthogonal projections to latent structures discriminant analysis (OPLS-DA).

First, an initial overview of the PCA analysis was used to decrease the dimensionality of the data and display the internal structure of the datasets in an unbiased way. Then, the OPLS-DA models were constructed to identify the marker metabolites between the different groups. The OPLS-DA model was generated using t[1]P and t[2]O, which represent the first principal component and the second orthogonal component, respectively. In the OPLS-DA model, the X variable and the Y variable represent the peak intensities in the ^{1}H NMR spectra and the predictive classifications, respectively. With a 10-fold cross-validation in the OPLS-DA models, Q^2 and R^2Y values were obtained, which represent the predictive ability of the model and the explained variance, respectively. Score plots were used to identify differential metabolites between the two groups and to combine the reliability and correlation from the OPLS-DA models.

Each point, the center of each ellipse, and the margin in the OPLS-DA score plots represent an individual sample, mean, and standard deviation (SD), respectively. Based on the cross-validated residuals (CV-ANOVA), the S-plot from the OPLS-DA analysis was used to indicate the contributions to clustering and to identify the significant metabolites between the two groups (Song et al., 2013). The p(corr)[1] axis and p[1] axis towards the predictive variation shown in the S-plot represent the correlation and magnitude of the spectral bins, respectively. The points in the S-plot represent the bins of the OPLS-DA. In the S-plot, the top or bottom points represent a stronger contribution to the class separation and more change in the variables compared with the middle points. The color-coded loading plots from the OPLS-DA analysis indicate changes in the metabolite between the two groups. In the color-coded loading plots, signals with a positive direction indicate a decreased metabolite level in the fatty liver group; in contrast, signals with a negative direction indicate an increased metabolite level in the fatty liver group (Sun et al., 2014). The color of the coefficient plots represent the importance of the significant metabolites in explaining the scores. In the coefficient plots, red indicates a more significant contribution to the separation between the two groups, and the opposite is true of blue.

The biplot (correlation circle) obtained from the partial least squares discriminant analysis (PLS-DA) represents the biochemical parameters and the correlation of metabolites between the two groups. The *x*-axis, *y*-axis, and the concentric circles in the biplot represent the first and second components of the PLS-DA model and the explained variance, respectively (Song et al., 2013).

Univariate analysis

Statistical differences among groups in terms of clinical data, plasma biochemistry, and plasma enzymes data, and ^{1}H NMR spectra of potential biomarkers from the differential metabolites were calculated by one way of analysis of variance (ANOVA) using SPSS statistical software (Version 11.0; SPSS, Inc., Chicago, IL, USA). A p values of <0.05 were considered to be statistically significant. Data were expressed as the mean±SD.

Receiver operating characteristic (ROC) curves were constructed for the ^{1}H NMR spectra data of the identified metabolites using SPSS statistical software. In this study, ROC curves were used to test the diagnostic value of the potential biomarkers from the differential metabolites of the fatty liver. Typically, area under the curve values greater than 0.8 and larger positive likelihood ratio values, which were calculated from the ROC analysis, indicate excellent predictive ability (Zhang et al., 2013).

RESULTS

Liver evaluation and clinical information

The liver tissues from the 171 cows were tested with the liver TG content test method described in previous studies (Herdt et al., 1983). Thirty-two cows with fatty liver and 32 healthy cows were selected for this study, as shown in Table 1. The clinical data for the control and fatty liver groups revealed biochemical differences between the groups (Table 1). There were no significant differences between the two groups in terms of age, parity, and days in milk (p>0.05). However, MY (p<0.05) and BCS (p<0.01) were significantly different between the two groups.

Table 1. Clinical data for the control (C) and fatty liver (F) groups

Parameters	C	F	p-values
Number	32	32	
Age (yr)	3.46±1.06	3.50±0.85	0.860
Parity	1.78±0.83	1.65±0.87	0.555
Days in milk	15.50±6.02	16.31±4.30	0.557
MY (kg/d)	30.75±3.70	28.32±4.73	0.030
BCS	3.16±0.32	2.88±0.34	0.001

MY, milk yield; BCS, body condition score; SD, standard deviation.
Data are expressed as the mean±SD.

Plasma biochemistry analysis

Compared with the control group, the main differences in the fatty liver group included increased NEFA (p<0.001) and BHBA (p<0.001), and decreased CK (p<0.001), FGF21 (p<0.001), INS (p<0.01), Glc (p<0.01), TG (p<0.05) and AST (p<0.05) (Table 2).

^{1}H NMR spectra analysis

Figure 1 shows the representative ^{1}H NMR spectra (δ 0.5 to 4.3) of the plasma samples from the control and fatty liver groups, with the major metabolites labeled. The main differences in the fatty liver group compared with the control group were increased isobutyrate (IB), BHBA, acetone (ACTN), citrulline (Citn), and trimethylamine-*N*-oxide (TMAO) (p<0.001), valine (Val) and glycine (Gly) (p<0.01), and decreased alanine (Ala), glycerol, α-Glc, and β-Glc (p<0.001), asparagine (Asn) and γ-aminobutyric acid (GABA) (p<0.01), and creatinine (Cr) (p<0.05) (Table 3). A supervised OPLS-DA analysis was conducted to identify significant metabolite changes and to ascertain differences between the control and fatty liver groups (Table 3).

A PCA with pareto-scaled data was first conducted; however, the two groups were not well clustered (data not shown). Thus, an OPLS-DA model was constructed to identify the differences in the metabolites between the fatty

Table 2. Results of the identified plasma metabolites and plasma enzymes obtained from the photometric analysis

No.	Metabolites	C	F	p-values
1	ALT/GPT (U/L)	4.18±3.11	3.50±2.34	0.226
2	AST (U/L)	58.77±31.63	23.00±5.42	0.028
3	TG (mmol/L)	0.10±0.02	0.09±0.01	0.028
4	Glycerol (mg/dL)	33.14±15.58	34.27±15.32	0.428
5	CK (U/L)	1.48±0.76	0.97±0.86	<0.000
6	NEFA (mmol/L)	0.55±0.29	12.24±0.62	<0.000
7	GC (ng/L)	50.10±9.47	48.46±8.83	0.677
8	INS (μIU/mL)	6.97±1.00	6.19±1.06	0.004
9	ACAC (mg/dL)	2.88±0.46	3.00±1.13	0.934
10	GH (ng/mL)	5.63±1.20	5.33±1.25	0.678
11	FGF21 (pg/mL)	322.47±88.16	501.00±94.69	<0.000
12	Glc (mmol/L)	3.14±0.71	2.70±1.31	0.003
13	BHBA (mmol/L)	0.87±0.30	3.14±1.21	<0.000

C, control; F, fatty liver; ALT/GPT, alanine transaminase/glutamate pyruvate transaminase; TG, triglyceride; CK, creatine kinase; NEFA, non-esterified fatty acid; GC, glucagon; INS, insulin; ACAC, acetoacetate; GH, growth hormone; FGF21, fibroblast growth factor 21; Glc, glucose; BHBA, β-hydroxybutyric acid; AST, aspartate transaminase; SD, standard deviation.
Data are expressed as the mean±SD.

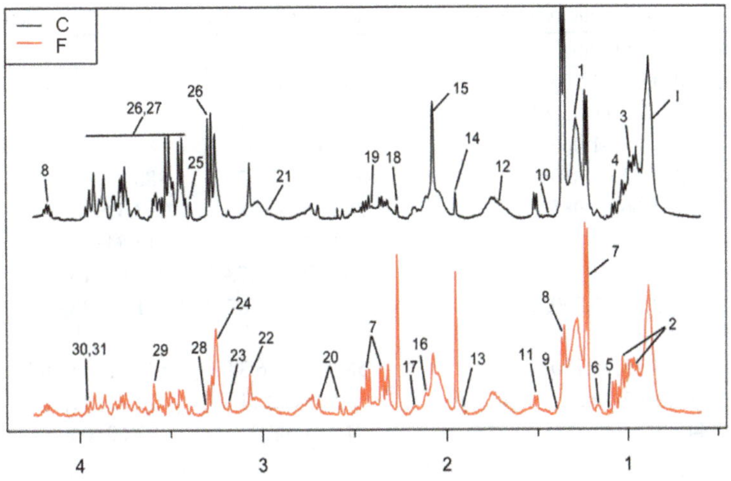

Figure 1. Typical ^{1}H nuclear magnetic resonance–metabolomics spectra of the plasma extracted from samples; the identified metabolites are labeled. "C" and "F" indicate the control and fatty liver groups, respectively. Metabolites: 1 low-density lipoprotein/very-low-density lipoprotein, 2 isoleucine, 3 leucine, 4 valine, 5 isobutyrate, 6 ethanol, 7 β-hydroxybutyric acid, 8 lactate, 9 threonine, 10 α-hydroxy-n-valerate, 11 alanine, 12 lysine, 13 gamma-aminobutyric acid, 14 acetate, 15 N-acetyl glycoprotein, 16 O-acetyl glycoprotein, 17 glutamate/glutamine, 18 acetone, 19 succinate, 20 citrate, 21 asparagine, 22 creatinine, 23 citrulline, 24 choline, 25 glycerol, 26 β-glucose, 27 α-glucose, 28 trimethylamine-*N*-oxide, 29 glycine, 30 betaine, and 31 creatine.

liver and control groups (Figure 2). The OPLS-DA score plot (Figure 2A) demonstrates an obvious separation between the fatty liver and control groups ($p<0.05$), with an R^2Y of 91.3% and a Q^2 of 80.1%.

The loadings S-plot was used to identify the variant spectral bins for the inter-class differences (Figure 2B). This plot is typically applied to identify metabolites, and to reduce the risk of false positives in the metabolite selection. It is also often used to visualize covariance p(1) (*x*-axis) and correlation p(corr) (*y*-axis), which are obtained from the OPLS-DA model (Song et al., 2013). In the S-plot, the lower left and the upper right quadrants exhibit positive and negative correlation and covariance, respectively. The significant metabolites in the lower left quadrant were increased in the fatty liver group; in contrast, the significant metabolites in the upper right quadrant were decreased in the fatty liver group. The further away from the center of the S-plot (Figure 2B), the more significant the contribution of the endogenous metabolites for clustering in the score plot. The S-plot illustrated that the metabolites that separated the two groups were primarily BHBA (δ 1.220 to 1.237), lipoprotein (δ 1.244 to 1.271), glutamine (Gln) and glutamate (Glu) (δ 2.255 to 2.266), and Glc (δ 3.296 to 3.422).

The color-coded corresponding coefficient plots of the OPLS-DA revealed additional detailed information regarding the metabolic differences between the fatty liver and healthy control groups (Figure 2C). The value of $|r^2|$, which represents the absolute correlation coefficient of each variable, displays the class separation of the two groups. In the color-coded loadings plot, red indicates a more significant contribution to the separation between the two groups compared with blue (Zhang et al., 2013). The color-coded loading plots (Figure 2C) show a decrease in the levels of Ala, α-Glc, β-Glc, and GABA, and an increase in the levels of low-density lipoprotein/very-low-density lipoprotein, Val, BHBA, acetate, glycerol, and lactate in the fatty liver group compared with the control group.

Based on the results of the OPLS-DA S-plot (Figure 2B) and the color-coded loadings plot (Figure 2C), 31 metabolites were identified as potential metabolite biomarkers of fatty liver in cows. *T*-tests were used to calculate the p-values of the concentrations of the 31 metabolites between the two groups; the results are summarized in Table 3.

Correlations between the endogenous metabolites and the biochemical parameters

In Figure 3, the biplot of the PLS-DA model was based on all ^{1}H NMR and biochemical data, and was used to further investigate the correlations between metabolites and biochemical responses in the fatty liver and control groups. This plot was generated using the biochemical levels as *Y* variables and the metabolite contents as *X* variables (R^2 = 0.81, Q^2 = 0.64). The metabolites in the right quadrants are positively correlated with the fatty liver group, while the metabolites in the left quadrants are negatively correlated with the fatty liver group. As shown in Figure 3, the metabolites glycerol and Glc, which were detected by biochemical methods or NMR, are both positively correlated with the fatty liver group, which demonstrates the reliability of our results. In addition, Figure 3 shows that Glc and INS are negatively correlated with BHBA, acetate, ACTN, and citrate; Glu and Gln are negatively correlated

Table 3. Results from the ^{1}H nuclear magnetic resonance spectra analysis and the classification of the identified metabolites

No.	Metabolite	Chemical shift (ppm)	C	F	p-value	LR+	AUC
1	LDL/VLDL	0.866-0.920, 1.250-1.275, 1.30-1.35, 2.0-2.05, 3.225-3.262	3,690.40±541.84	3744.73±756.84	0.8374	0.9990	0.5010
2	Isoleucine	0.925-0.960, 1.035-1.060	492.13±62.93	487.24±60.05	0.8374	1.2903	0.5020
3	Leucine	0.965-0.993, 1.825-1.850	377.51±47.67	380.06±66.36	0.8804	4.1290	0.4829
4	Valine	1.010-1.031, 1.065-1.085	267.26±35.88	303.49±45.74	0.0025	8.2581	0.7077
5	Isobutyrate	1.090-1.115	4.94±0.81	14.84±6.12	<0.0000	12.3871	0.8942
6	Ethanol	1.180-1.220,	302.09±260.78	232.01±156.45	0.4483	1.8433	0.5685
7	BHBA	1.225-1.238, 2.305-2.360, 2.398-2.430, 4.164-4.205	657.14±131.87	1,351.49±383.50	<0.0000	27.8710	0.9808
8	Lactate	1.350-1.375, 4.155-4.180	580.43±276.34	549.51±157.87	0.9613	1.1470	0.4929
9	Threonine	1.378-1.415, 3.63-3.65	57.05±22.71	77.72±41.70	0.0787	5.6774	0.6492
10	α-Hydroxy-n-valerate	1.420-1.445	11.15±0.27	7.60±0.87	0.0965	0.9677	0.3589
11	Alanine	1.500-1.525	101.68±18.63	70.67±15.15	<0.0000	0.9677	0.1008
12	Lysine	1.72-1.76	186.61±17.31	195.30±22.26	0.1499	5.1613	0.6472
13	GABA	1.905-1.942, 2.278-2.302	107.12±18.03	103.67±26.57	0.0035	4.1290	0.4103
14	Acetate	1.945-1.952	159.05±76.30	194.80±74.44	0.0733	3.6129	0.6532
15	NAGP	2.068-2.076	135.90±29.03	149.40±41.86	0.2656	5.1613	0.5978
16	OAGP	2.100-2.120	103.99±17.85	103.87±15.67	0.9773	1.5484	0.5101
17	Glutamate/ Glutamine	2.150-2.200, 2.475-2.525	131.39±17.85	120.51±34.17	0.2642	3.0968	0.3861
18	Acetone	2.255-2.275	53.94±27.96	452.42±194.79	<0.0000	28.9032	0.9798
19	Succinate	2.375-2.4	67.37±4.07	70.57±10.90	0.2090	13.4194	0.6361
20	Citrate	2.550-2.554, 2.690-2.720	65.22±12.78	57.85±12.92	0.0565	1.0323	0.3478
21	Asparagine	2.975-3.020	134.27±12.90	120.02±19.44	0.0031	1.0323	0.2692
22	Creatinine	3.065-3.075	93.80±11.94	87.32±13.43	0.0362	2.0645	0.3236
23	Citrulline	3.14-3.195	83.69±20.13	104.15±226.59	<0.0000	2.0645	0.1421
24	Choline	3.225-3.265	594.69±86.51	568.84±10.28	0.3987	2.0645	0.4073
25	Glycerol	3.59-3.70	33.70±7.91	22.62±7.02	<0.0000	0.9677	0.1512
26	β-Glucose	3.275-3.300, 3.415-3.47, 3.478-3.530, 3.54-3.585, 3.73-3.825	1,387.69±205.07	1,026.91±398.55	<0.0000	1.0323	0.0857
27	α-Glucose	3.85-3.890, 3.927-3.975, 3.415-3.47, 3.478-3.530, 3.54-3.585, 3.73-3.825	1,363.10±206.03	1,019.59±374.66	<0.0000	0.9677	0.0837
28	TMAO	3.305-3.325	5.09±0.27	14.84±7.83	<0.0000	14.4516	0.8266
29	Glycine	3.585-3.60	58.10±11.59	67.84±8.96	0.0011	3.0968	0.7560
30	Betaine	3.950-3.970	35.09±6.11	33.86±4.28	0.4482	1.1561	0.4103
31	Creatine	3.970-3.985	8.02±3.55	7.66±2.39	0.7589	1.2496	0.5071

C, control; F, fatty liver; LR+, a positive likelihood ratio; AUC, area under the curve; LDL/VLDL, low-density lipoprotein/very-low-density lipoprotein; BHBA, β-hydroxybutyric acid; GABA, γ-aminobutyric acid; NAGP, N-acetyl glycoprotein; OAGP, O-acetyl glycoprotein; TMAO, trimethylamine-*N*-oxide; SD, standard deviation.
Data of peak intensity are expressed as the mean±SD.

with AST, ALT/GPT, and GH.

DISCUSSION

Many studies have demonstrated that metabolomics techniques are powerful tools for identifying metabolites from multiple biological samples with biochemical variation. In addition, we have used gas chromatography/mass spectrometry and NMR techniques to assess clinical and subclinical ketosis in dairy cattle and identified significant metabolites (Zhang et al., 2013; Sun et al., 2014). In the present study, ^{1}H NMR-based metabolomics was used to identify plasma metabolites in dairy cows with fatty liver and controls. The results clearly indicate significant differences in the metabolite concentrations between the control cows and the cows with fatty liver. In this study, BHBA and ACTN were higher in the fatty liver group. Conversely, the plasma Glc levels in the fatty liver group were lower compared with the control group. These results (Table 3) are consistent with the experimental data shown in Table 2.

Previous studies have showed that the lipid content in the liver of dairy cows is closely related to liver specific gravity (Herdt et al., 1983). To estimate the liver lipid

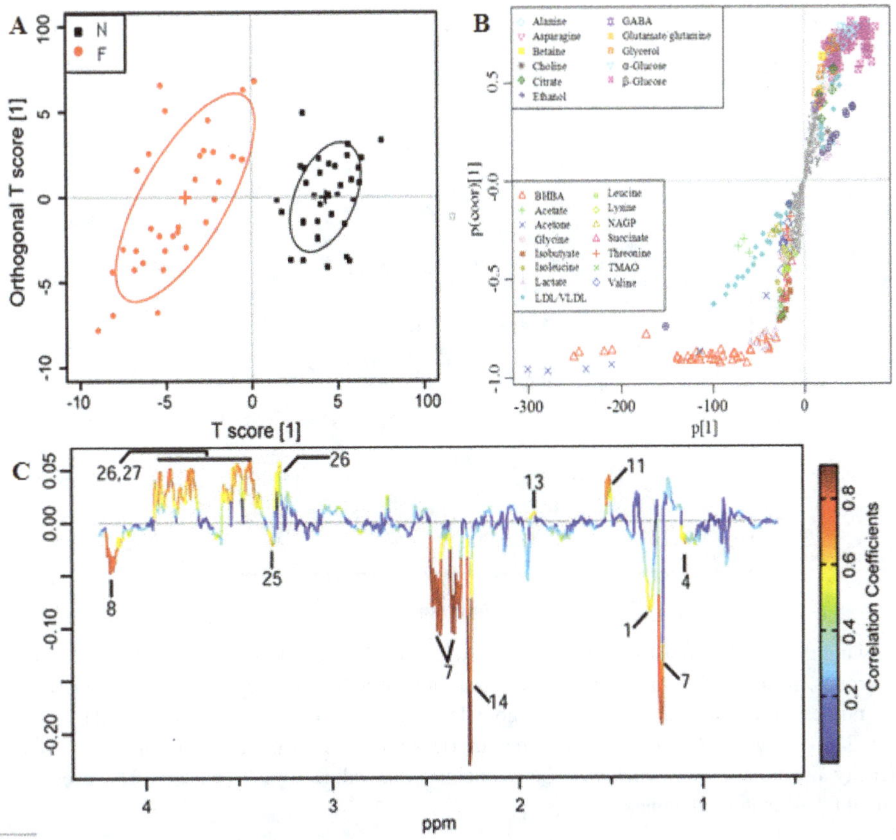

Figure 2. Orthogonal projections to latent structures discriminant analysis (OPLS-DA) model with the metabolites labeled in Figure 1. (A) Score plot obtained from the OPLS-DA analysis of the control group (■) and the fatty liver group (•) (n = 32); (B) S-plot obtained from the OPLS-DA analysis of the control and fatty liver groups corresponding to the score plot (Figure 2A); (C) color-coded loadings plot from the OPLS-DA analysis illustrating the metabolite components that differ between the two groups. Red indicates the greatest difference, with statistical significance; blue indicates the smallest difference, with no significant difference between the groups. Metabolites: 1 low-density lipoprotein/very-low-density lipoprotein, 4 valine, 7 β-hydroxybutyric acid, 8 lactate, 11 alanine, 13 γ-aminobutyric acid, 14 acetate, 25 glycerol, 26 β-glucose, and 27 α-glucose.

content of dairy cows, this observation was used as the basis for a clinical test. An increase in liver fat concentration during the peripartum period is extremely common in dairy cows and, to some degree, is likely normal. To obtain accurate biomarkers, cows with TG contents between 13% and 34% in the liver were not included in the study. In this study, cows with TG contents >35% in the liver were included in the fatty liver group, and cows with no clinical signs and TG contents <13% in the liver were included in the control group.

As a result of high milk production, high-yielding dairy cows typically experience NEB during early lactation. Dairy cows with NEB exhibit increased gluconeogenesis and mobilization of body TG and protein (Sun et al., 2014). Compared with monogastric animals, Glc regulation is very complicated in ruminant animals. Due to the lack of carbohydrates in the body and/or increasing energy requirements, ruminant animals predominately depend on non-carbohydrate compounds to meet their energy demands via gluconeogenesis (Laffel, 1999). Therefore, the primary changes include high ketone bodies, glucopenia, and high TG in the blood (Xu et al., 2008). In this study, the high levels of IB, BHBA, and ACTN and the low levels of Ala, Asn, Glc, and Cr in the fatty liver group were similar to those found in previous research (Xu et al., 2008).

During metabolism, chain amino acids such as Val, leucine (Leu), and isoleucine (Ile) produce Ala. Blood Ala is one of the primary materials required for the synthesis of Glc through the gluconeogenesis pathway. The concentration of Ala in blood has been found to be highly correlated with many metabolic pathways, including gluconeogenesis, glycolysis, and the tricarboxylic acid cycle (TCA cycle) (Mukherjee et al., 2010). In energy-deficient states, Ala enters the TCA cycle to generate Glc for energy, and can also give rise to Glc via the Ala cycle. In the current experiment, Ala decreased in the fatty liver group compared with the control group. This result implies that low concentrations of blood Ala are closely associated with fatty liver; a similar result was reported in previous research (Mukherjee et al., 2010).

In the current experiment, the levels of blood Asn were lower in the cows with fatty liver than in the control cows,

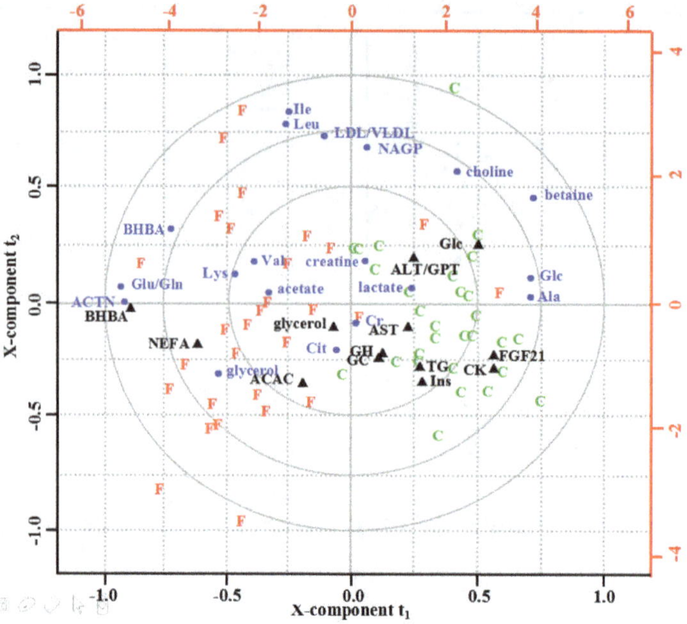

Figure 3. The biplot showing the correlation between the endogenous metabolites and biochemical parameters. "C" and "F" indicate the control and fatty liver groups, respectively. ALT/GPT, alanine transaminase/glutamate pyruvate transaminase; TG, 'triglyceride; CK, creatine kinase; NEFA, non-esterified fatty acid; GC, glucagon; INS, insulin; ACAC, acetoacetate; GH, growth hormone; FGF21, fibroblast growth factor 21; Glc, glucose; BHBA, β-hydroxybutyric acid; AST, aspartate transa-minase; LDL/VLDL, low-density lipoprotein/very-low-density lipoprotein; Ile, isoleucine; Leu, leucine; Val, valine; Lys, lysine; ACTN, acetone; Cr, creatinine; Ala, alanine; Cit, citrate; Glu/Gln, glutamate/glutamine.

which implies that, as a glucogenic amino acid, Asn can be metabolized by gluconeogenesis or the TCA cycle. Moreover, it also plays an important role in glycosylation, which is closely related to protein structure and function (Patterson, 2005).

In this study, the concentration of blood Cr was significantly lower in the fatty liver group than in the control group. Previous research has demonstrated that the concentration of Cr in the blood or urine provides an accurate measure of renal function (Shemesh et al., 1985). Another study demonstrated a relationship between low levels of serum Cr and increased risk for the development of type 2 diabetes (Harita et al., 2009). These findings imply that the levels of Cr typically decrease due to weakened aerobic metabolism during NEB, which is an important mesostate of the TCA cycle.

As an important component of bile acids, Gly is necessary for dietary fat digestion and long-chain fatty acid absorption. Furthermore, Gly has been used to treat metabolic disorders such as obesity, ischemia–reperfusion injuries, cardiovascular disease, and diabetes (Wu et al., 2012). In the current study, the concentrations of Gly were higher in the fatty liver group than in the control group, which indicates that Gly was used to correct lipid metabolic disorder during the functional compensatory period.

A previous study has demonstrated that Val can regulate lipid metabolism and increase fat loss via increased energy expenditure (Du et al., 2012). In animal models, Val induces non-alcoholic fatty liver disease via oxidative stress. Another study indicated that a low Val supply could represent a limiting factor for milk protein synthesis (Haque et al., 2013). The results of the present study indicate that the levels of Asn were higher in the cows with fatty liver than in the control group, suggesting that Val can reverse the liver damage caused by lipid metabolic disturbance.

Our results suggest an increased demand for energy via a potential increase in the precursors for the TCA cycle and gluconeogenesis compared with dairy cows in lactation (Moyes et al., 2013). A series of metabolic abnormalities were associated with fatty liver, which could be considered as diagnostic biomarkers.

Liver disease is typically caused by metabolic abnormalities. Liver damage can also result from characteristic changes in plasma amino acid levels. Some plasma amino acids are highly related to liver damage, with clear concentration changes after liver damage (Ning et al., 1967). As an inhibitory neurotransmitter, deficiency in GABA may cause decreased attention span, memory alterations, mood changes, and drowsiness, and these symptoms are consistent with the clinical symptoms of cows with fatty liver (Xu et al., 2008; Mukherjee et al., 2010). Another study indicated that GABA in cows can promote gastric juice production, thus improving growth rate and feed intake (Mukherjee et al., 2008). The fatty liver cow group showed significant deficiency of plasma GABA compared with the control group. The results of the present study indicate that GABA acts as an inhibitor of fatty liver and obesity by activating hepatic function and avoiding NEB. We suggest that the downstream events of GABA deficiency may be important in the initiation of fatty liver disease.

Previous studies have shown increased level of TMAO, which acts as a chemical toxin, inducing liver damage in a mouse model accompanied by the suppression of glycolysis, stimulation of fatty acid oxidation, and increased levels of oxidative stress (Zhao et al., 2011). A study published in 2013 indicated that TMAO is related to cholesterol metabolism in artery walls, intestines, and the liver. In artery walls, TMAO can reduce cholesterol removal and increase cholesterol deposition (Koeth et al., 2013).

In the urea cycle, Citn is produced from ornithine and carbamoyl phosphate. A high plasma Citn level is related to several human liver-damage diseases, such as liver fibrosis and type I/II citrullinemia (Faghfoury et al., 2011; Takahashi et al., 2012). In addition, the circulating Citn concentration is, in humans, a biomarker of intestinal functionality (Crenn et al., 2003). In the current study, the level of Citn was lower in the plasma of cows with fatty liver than in healthy cows, which suggests that fatty liver is related to liver-damage diseases in cows.

Fatty liver in mice is a multi-factorial disease in which abnormal TG accumulation in hepatocytes can result from a number of distinct alterations. It has recently been *reported* that synthesis of TGs requires esterification of one glycerol molecule with three fatty acids, and aquaporin-9 has been found to play a major role in glycerol import by mouse hepatocytes (Gena et al., 2013).

Glycerol, TMAO, and Citn have important effects on liver damage and lipid imbalance. To our knowledge, these metabolites have not been studied in fatty liver in cows. Additional research is required to explain the specific mechanisms involved, and disease diagnostic assays require more complex procedures.

Routine plasma biochemistry was performed for all dairy cows in this study. Several compounds, such as Glc, BHBA, CK, and NEFA, confirmed the hepatic TG content-based grouping. Most of the parameters, especially glycerol, TG, ALT/GPT, and AST, which are supposed to be closely related to fatty liver, failed to be effective biomarkers.

Metabolomics provides a new viewpoint and technology platform to assist in the early diagnosis and efficacy assessment of many diseases. The testing of entire datasets for metabolic features is critical for assessing the processes of fatty liver. ^{1}H NMR-based metabolomics is a novel method for detecting cows with fatty liver. In this study, plasma metabolic profiling and data analysis methods were used to predict the diagnosis of cows with and without fatty liver, with high specificity and sensitivity. The development of fatty liver involves disturbances in the metabolism of amino acids and fatty acids, as well as in gluconeogenesis (Figure 4). In the present study, the levels of 13 metabolites were significantly different between dairy cows with fatty liver and control cows. Our study demonstrates that these 13 metabolites can be used in future

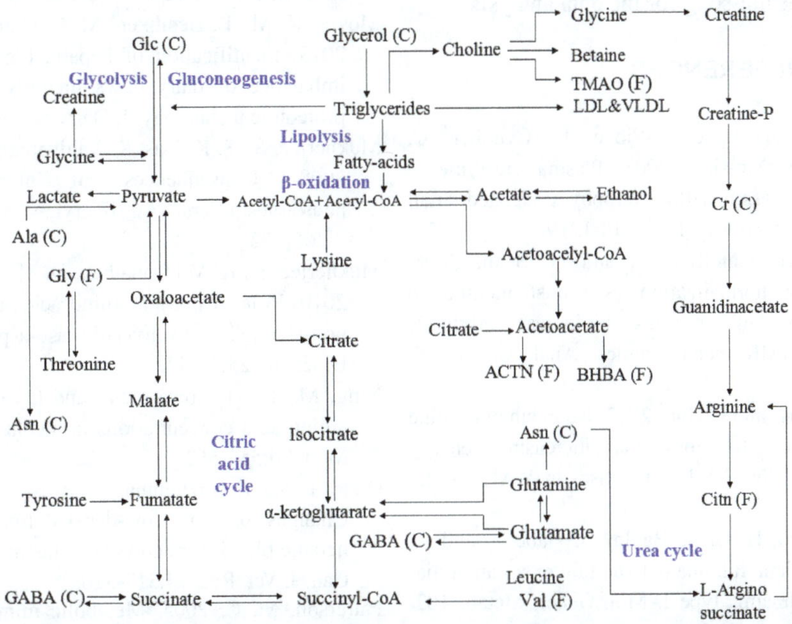

Figure 4. Schematic representation of the most relevant metabolic differences between the two groups. "F" and "C" indicate the metabolites that were higher or lower in the fatty liver group compared with the control group, respectively. Metabolites: LDL/VLDL, low-density lipoprotein/very-low-density lipoprotein; Ala, alanine; Asn, asparagine; Glc, glucose; BHBA, β-hydroxybutyric acid; Cr, creatinine; GABA, γ-aminobutyric acid; glycerol; ACTN, acetone; Citn, citrulline; Gly, glycine; IB, isobutyrate; TMAO, trimethylamine-*N*-oxide; Val, valine.

as potential diagnostic biomarkers for dairy cows with fatty liver. Previous studies have also identified the concentrations of BHBA, Citn, Ala, Ile, Leu, Gly, and Glc in cows blood as potential metabolite biomarkers (Zhang et al., 2013; Sun et al., 2014). Future studies should evaluate and optimize the diagnostic abilities of the 13 metabolites found in this study for fatty liver.

CONCLUSION

Through ^{1}H NMR detection and data analysis, 31 metabolites were identified between the fatty liver and control groups. The results demonstrate that the ^{1}H NMR technique combined with multivariate statistical analysis can be used to access the changes and progression of fatty liver and to discover potential biomarkers for this disease in cows. In future, changes in the potential metabolites and metabolic pathways could present new strategies for the diagnosis and prevention of fatty liver in dairy cows.

ACKNOWLEDGMENTS

This study was sponsored by the Chinese National Natural Science Foundation (Grant numbers 31001094 and 31272625), the Chinese National Science and Technology Support Program (Grant numbers 2013BAD21B01), the China Spark Program (Grant numbers 2012GA670001), and the Hei Long Jiang Province Plan (Grant numbers 1252-NCET-003). We appreciated the technical support provided by Prof. Wang Jun-song for the data analysis.

REFERENCES

Crenn, P., K. Vahedi, A. Lavergne-Slove, L. Cynober, C. Matuchansky, and B. Messing. 2003. Plasma citrulline: a marker of enterocyte mass in villous atrophy-associated small bowel disease. Gastroenterology 124:1210-1219.

Dieterle F., A. Ross, G. Schlotterbeck, and H. Senn. 2006. Probabilistic quotient normalization as robust method to account for dilution of complex biological mixtures. Application in 1H NMR metabonomics. Anal. Chem. 78: 4281-4290.

Du Y., Q. Meng, Q. Zhang, and F. Guo. 2012. Isoleucine or valine deprivation stimulates fat loss via increasing energy expenditure and regulating lipid metabolism in WAT. Amino Acids 43:725-734.

Faghfoury H., J. Baruteau, H. O. de Baulny, J. Häberle, and A. Schulze. 2011. Transient fulminant liver failure as an initial presentation in citrullinemia type I. Mol. Genet. Metab. 102: 413-417.

Gena, P., M. Mastrodonato, P. Portincasa, E. Fanelli, D. Mentino, A. Rodríguez, R. A. Marinelli, C. Brenner, G. Frühbeck, M. Svelto, and G. Calamita. 2013. Liver glycerol permeability and aquaporin-9 are dysregulated in a murine model of non-alcoholic fatty liver disease. PLoS One. 8:e78139.

Gonzalez, F. D., R. Muino, V. Pereira, R. Campos, and J. L. Benedito. 2011. Relationship among blood indicators of lipomobilization and hepatic function during early lactation in high-yielding dairy cows. J. Vet .Sci. 12:251-255.

Haque, M. N., H. Rulquin, and S. Lemosquet. 2013. Milk protein responses in dairy cows to changes in postruminal supplies of arginine, isoleucine, and valine. J. Dairy Sci. 96:420-430.

Harita, N., T. Hayashi, K. K. Sato, Y. Nakamura, T. Yoneda, G. Endo, and H. Kambe. 2009. Lower serum creatinine is a new risk factor of type 2 diabetes: the Kansai healthcare study. Diabetes Care 32:424-426.

Herdt, T. H., L. Goeders, J. S. Liesman, and R. S. Emery. 1983. Test for estimation of bovine hepatic lipid content. J. Am. Vet. Med. Assoc. 182:953-955.

Laffel, L. 1999. Ketone bodies: A review of physiology, pathophysiology and application of monitoring to diabetes. Diabetes Metab. Res. Rev. 15: 412–426.

Jorritsma R., H. Jorritsma, Y. H. Schukken, and G. H. Wentink. 2000. Relationships between fatty liver and fertility and some periparturient diseases in commercial Dutch dairy herds. Theriogenology 54:1065-1074.

Hayirli, A., R. R. Grummer, E. V. Nordheim, and P. M. Crump. 2002. Animal and dietary factors affecting feed intake during the prefresh transition period in Holsteins. J. Dairy Sci. 85:3430-3443.

Koeth, R. A., Z. Wang, B. S. Levison, J. A. Buffa, E. Org, B. T. Sheehy, E. B. Britt, X. Fu, Y. Wu, L. Li, J. D. Smith, J. A. DiDonato, J. Chen, H. Li, G. D. Wu, J. D. Lewis, M. Warrier, J. M. Brown, R. M. Krauss, W. H. Tang, F. D. Bushman, A. J. Lusis, and S. L. Hazen. 2013. Intestinal microbiota metabolism of L-carnitine, a nutrient in red meat, promotes atherosclerosis. Nat. Med. 19:576-585.

Moyes, K. M., E. Bendixen, M. C. Codrea, and K. L. Ingvartsen. 2013. Identification of hepatic biomarkers for physiological imbalance of dairy cows in early and mid lactation using proteomic technology. J. Dairy Sci. 96:3599-3610.

Mukherjee, S., S. K. Das, K. Vaidyanathan, and D. M. Vasudevan. 2008. Consequences of alcohol consumption on neurotransmitters -An overview. Curr. Neurovascular Res. 5:266-272.

Mukherjee, S., K. Vaidyanathan, D. M. Vasudevan, and S. K. Das. 2010. Role of plasma amino acids and gaba in alcoholic and non-alcoholic fatty liver disease-a pilot study. Indian J. Clin. Biochem. 25:37-42.

Ning, M., L. M. Lowenstein, and C. S. Davidson. 1967. Serum amino acid concentrations in alcoholic hepatitis. J. Lab. Clin. Med. 70:554-562.

Oikawa, S., Y. Mizunuma, Y. Iwasaki, and M. Tharwat. 2010. Changes of very low-density lipoprotein concentration in hepatic blood from cows with fasting-induced hepatic lipidosis. Can. J. Vet. Res. 74:317-320.

Patterson, M. C. 2005. Metabolic mimics: The disorders of N-linked glycosylation. Semin. Pediatr. Neurol. 12:144-151.

Sejersen, H., M. T. Sørensen, T. Larsen, E. Bendixen, and K. L. Ingvartsen. 2012. Liver protein expression in dairy cows with high liver triglycerides in early lactation. J. Dairy Sci. 95: 2409-2421.

Shemesh, O., H. Golbetz, J. P. Kriss, and B. D. Myers. 1985. Limitations of creatinine as a filtration marker in glomerulopathic patients. Kidney Int. 28:830-838.

Smith, T. R., A. R. Hippen, D. C. Beitz, and J. W. Young. 1997. Metabolic characteristics of induced ketosis in normal and obese dairy cows. J. Dairy Sci. 80:1569-1581.

Song, X. F., J. S. Wang, P. R. Wang, N. Tian, M. H. Yang, and L. Y. Kong. 2013. 1H NMR-based metabolomics approach to evaluate the effect of Xue-Fu-Zhu-Yu decoction on hyperlipidemia rats induced by high-fat diet. J. Pharm. Biomed. Anal. 78:202-210.

Sun, L. W., H. Y. Zhang, L. Wu, S. Shu, C. Xia, C. Xu, and J. S. Zheng. 2014. 1H-Nuclear magnetic resonance-based plasma metabolic profiling of dairy cows with clinical and subclinical ketosis. J. Dairy Sci. 97:1552-1562.

Takahashi, Y., S. Koyama, H. Tanaka, S. Arawaka, M. Wada, T. Kawanami, H. Haga, H. Watanabe, K. Toyota, C. Numakura, K. Hayasaka, and T. Kato. 2012. An elderly Japanese patient with adult-onset type II citrullinemia with a novel D493G mutation in the SLC25A13 gene. Intern. Med. 51:2131-2134.

Wildman, E. E., G. M. Jones, P. E. Wagner, R. L. Boman, H. F. Troutt Jr, and T. N. Lesch. 1982. A dairy cow body condition scoring system and its relationship to selected production characteristics. J. Dairy Sci. 65:495-501.

Wu, G., B. Imhoff-Kunsch, and A. W. Girard. 2012. Biological mechanisms for nutritional regulation of maternal health and fetal development. Paediatr. Perinat. Epidemiol. 26:4-26.

Xu, C., Z. Wang, G. W. Liu, X. B. Li, G. H. Xie, C. Xia, and H. Y. Zhang. 2008. Metabolic characteristic of the liver of dairy cows during ketosis based on comparative proteomics. Asian Australas. J. Anim. Sci. 21:1003-1010.

Zhang, H. Y., L. Wu, C. Xu, C. Xia, L. W. Sun, and S. Shu. 2013. Plasma metabolomic profiling of dairy cows affected with ketosis using gas chromatography/mass spectrometry. BMC Vet. Res. 9:186.

Zhao, X. J., C. Huang, H. Lei, X. Nie, H. Tang, and Y. Wang. 2011. Dynamic metabolic response of mice to acute mequindox exposure. J. Proteome Res. 10:5183-5190.

4

Effects of Varying Levels of Fungal (*Arachniotus* sp.) Treated Wheat Straw as an Ingredient of Total Mixed Ration on Growth Performance and Nutrient Digestibility in Nili Ravi Buffalo Calves

F. Shahzad*, M. Abdullah[1], A. S. Chaudhry[2], J. A. Bhatti[1], M. A. Jabbar[3], F. Ahmed[4], T. Mehmood[4], M. Asim[4], S. Ahmed[5], Z. Kamran, I. Irshad[6], and M. N. Tahir

University College of Veterinary and Animal Sciences, The Islamia University of Bahawalpur, Bahawalpur 63100, Pakistan

ABSTRACT: The study was carried out to explore the effects of replacing wheat straw with fungal treated wheat straw as an ingredient of total mixed ration (TMR) on the growth performance and nutrient digestibility in Nili Ravi buffalo male calves. Fungal treated wheat straw was prepared using *Arachniotus* sp. Four TMRs were formulated where wheat straw was replaced with 0 (TMR1), 33 (TMR2), 67 (TMR3), and 100% (TMR4) fungal treated wheat straw in TMR. All TMRs were iso-caloric and iso-nitrogenous. The experimental TMRs were randomly assigned to four groups of male calves (n = 6) according to completely randomized design and the experiment continued for four months. The calves fed TMR2 exhibited a significant improve in dry matter intake, average daily weight gain, feed conversion ratio and feed economics compared to other groups. The same group also showed higher digestibility of dry matter, crude protein, neutral-, and acid detergent fibers than those fed on other TMRs. It is concluded that TMR with 33% fungal-treated wheat straw replacement has a potential to give an enhanced growth performance and nutrient digestibility in male Nili Ravi buffalo calves. (**Key Words:** *Arachniotus* sp., Average Daily Gain, Fungal Treated Wheat Straw, Nili Ravi Buffalo Calves, Total Mixed Ration)

INTRODUCTION

The livestock sector is an imperative part of agriculture in Pakistan. Livestock accounted for approximately 55.9% of agriculture value and 11.8% in country's gross domestic product during 2013 through 2014 (Pakistan Economics Survey, 2014). The country has a huge population of livestock including cattle, buffalo, goat and sheep totaling approximately 172 million heads (Pakistan Economics Survey, 2014). The livestock are well adapted to sub-tropical environment, tolerant to endemic diseases and efficient convertors of poor quality forages into valuable products like milk, meat, skin, hides, bone and blood (Younas and Yaqoob, 2005). There is an utmost need to increase the livestock production potential to fulfill the growing demands of ever increasing population for livestock products. Amongst other factors, shortage of high quality feedstuffs and feeding of abundantly available poor quality dry roughages are the major constraints in the productivity of livestock sector (Sarwar et al., 2002).

Traditionally, crop residues contribute substantially nationwide to livestock feeding year around (Sarwar et al., 2002) but their proportion in the ration increases during periods of fodder scarcity. Wheat straw (WS), a crop

* Corresponding Author: F. Shahzad.
E-mail: faisalshehzad76@yahoo.com
[1] Department of Livestock Production, University of Veterinary and Animal Sciences, Lahore 42000, Pakistan.
[2] School of Agriculture, Food & Rural Development, Newcastle University, Newcastle, NE1 7RU, UK.
[3] Pakistan Agriculture Research Board, Lahore 42000, Pakistan.
[4] Buffalo Research Institute, Pattoki 55300, Pakistan.
[5] Department of Animal Nutrition, University of Veterinary and Animal Sciences, Lahore 42000, Pakistan.
[6] Department of Pathology, University of Veterinary and Animal Sciences, Lahore 42000, Pakistan.

residue that is emerging as a dominant feed source for a sustainable crop-livestock production system. It is also well known that WS is low in crude protein (CP) and energy, while its high fibre content limits voluntary intake to low levels in ruminants (Nisa et al., 2004). However, the low feeding value of WS could be improved through biotechnological methods (Selim et al., 2004) by carrying out fungal fermentation of WS in the presence of non-protein nitrogen (urea) and soluble carbohydrates (e.g. molasses). Different kinds of fungi have been used to improve the nutritive value of industrial by-products or highly fibrous agricultural wastes. Growth of fungal mycelium was previously found to enhance total protein contents in fermented feed (Fazaeli and Mirhadi, 2007). However, long incubation periods and slow degradation of fiber components during fermentation are main drawbacks of fungal application for nutritive improvement of WS.

Among the family members, *Arachniotus* fungi could be used to upgrade the nutritional value of fibrous feeds, especially in combination with solid state fermentation technology. *Arachniotus* sp. is a white rot fungus and has been used for the economic utilization of many waste products (Shaukat et al., 2006). The higher production of microbial biomass protein by fermentation of corn stover with *Arachniotus* sp. was already observed by Ahmed et al. (2010). And the potential utilization of corn stover can minimize the cost for growth of these microorganisms and enhance microbial biomass protein production by fermentation. The microbial biomass protein increases by utilizing the cellulose and hemi-cellulose of the substrate as an energy source for synthesis of protein. These methods are preferable due to their simplicity, specificity and circumventing the disadvantages of physical and chemical treatments (Misra et al., 2007). But the application of these methods at farm level in a livestock production system is very limited especially in developing countries like Pakistan. Undoubtedly, research is being carried out since the last decade to exploit the nutritive potential of fungal treated crop residues but is restricted to the laboratory level in Pakistan and there is not even a single feeding trial reported at the farm level in ruminants.

The present study gives the opportunity to evaluate varying levels of fungal treated WS as an ingredient of total mixed rations (TMR) on growth performance and nutrient digestibility in Nili Ravi buffalo calves. The purpose was to elucidate the potential of replacing the conventional feeding management practice with novel biotechnological methods and its effects on palatability and economic value of the diet.

MATERIALS AND METHODS

Animals and experimental design

The present study was conducted according to the guidelines of the Committee on Use of Animals in Research and Experimentation. Twenty four Nili Ravi buffalo male calves with the average age 9 to 12 month and a live body weight 112±7 kg were selected at Buffalo Research Institute, Pattoki, Kasur, Pakistan. The animals were tied and maintained in separate pens with individual feeding. Fresh clean drinking water was provided *ad libitum* during the whole experimental period. Animals were allocated in four groups in a completely randomized design and the experiment continued for four months.

Preparation of fungal treated wheat straw

Fungal treated WS was prepared according to the method described (Faisal Shahzad, personal communication). Briefly, WS with at a chop length of 2.5 to 3.0 cm was bedded on polythene cemented plotted covered room (4×4 m) with an adjustment of 24 electric rods (500 watt each) in criss-cross arrangement to control the temperature. Wheat straw (about 80 kg) was placed in a pre-fumed room and subjected to spraying of 105.6 L water solution containing ($MgSO_4.7H_2O$ [40 g], $CaCl_2$ [60 g], KH_2PO_4 [120 g] urea [120 g]) and inoculated with *Arachniotus* sp. (5 L inoculums). At 48 h of start of the experiment, molasses (6.4 L), rice polishing (12.8 kg) and corn steep liquor (48.0 L) were added to enhance the fermentation process. The incubation period continued for 4 days at approximately 28°C.

Treatments and experimental design

Four experimental TMR were formulated where WS was replaced with 0% (TMR1), 33% (TMR2), 67% (TMR3), and 100% (TMR4) fungal treated WS using the ingredients presented in Table 1. The experimental TMRs were randomly assigned to four groups of male calves (n = 6) according to completely randomized design. The TMRs were prepared according to National Research Council recommendations (NRC, 2001) for CP and energy for the growing animals and contained similar concentrations of CP (13.5%) and energy (2.4 Mcal/kg) on dry matter basis. The diets were offered *ad libitum* once daily and daily feed intake was recorded for individual animals during the sampling period. The residual feed from the feeding troughs was removed once per day. The ingredient and chemical compositions of all TMR are presented in Table 2.

Sampling and chemical analysis

Samples of TMR and feces were collected as per guidelines and stored at –20°C until analysis. TMR and fecal samples were analyzed for dry matter (DM), CP, and ash according to the procedures of Association of Official Analytical Chemists (AOAC, 2006). Neutral detergent fiber (NDF) and acid detergent fiber (ADF) and acid detergent lignin were determined according to the methods of Van

Table 1. Chemical composition of the ingredients

Ingredients	Maize silage	Fungal treated wheat straw	Wheat straw	Maize Broken	Wheat bran	Cotton seed cake	Maize gluten meal 30%	Rapeseed cake	Corn steep liquor	Cane molasses	Mineral mix	Urea
DM	30	92	92	89	89	90	91	90	50	74	100	100
CP	8	13	1.9	9	15	20	29	35	35	2	0	288
NDF	46	37	76	10	43	50	36	30	0	0.4	0	0
ADF	25	22	56	3	16	10	10	21	0	0	0	0
ADL	2.25	4.3	8.2	0.9	3	12	1.5	7.4	0	0	0	0
Ash	7	19	9	7	6	4	5	7	10	13	100	0
ME	2.58	2.1	1.55	3.1	2.5	2.29	3.03	2.75	1.8	2.78	0.0	0.0

The values are expressed as % age of dry matter unless otherwise stated.
DM, dry matter; CP, crude protein; NDF, neutral detergent fiber; ADF, acid detergent fiber; ADL, acid detergent lignin; ME, metabolizable energy (MCal/kg DM).

Soest et al. (1991).

Data recording and statistical analysis

Feed intake was recorded daily and live body weight fortnightly with digital weighing bridge. Feed conversion ratio (FCR; kg feed/kg gain) and feed economics (return from body gain/total feed cost) were calculated. The digestibility trial was carried out by total collection method during which total feces were collected for 5 consecutive days. Digestibility for DM, CP, and NDF which is hereby denoted by X was calculated according to the following equation:

$$Digestibilty\ (X) = \left(\frac{Dietary\ concentrarion\ of\ X - Fecal\ concentration\ of\ X}{Dietary\ concentration\ of\ X}\right) \times 100$$

Table 2. Ingredient and chemical composition of the experimental total mixed rations

Items	Experimental total mixed rations			
	TMR1	TMR2	TMR3	TMR4
Ingredients				
Maize silage	22.0	22.0	22.0	22.0
Fungal treated wheat straw	0.0	7.3	14.7	22.0
Wheat straw	22.0	14.7	7.3	0.0
Maize broken	23.5	10	14.5	0.5
Wheat bran	0.5	8.0	16.0	34.0
Cotton seed cake	13.0	9.0	4.0	0.5
Maize gluten meal 30%	2.0	14.0	14.5	9.0
Rapeseed cake	10.0	6.5	0.5	1.0
Corn steep liquor	3.5	1.0	3.0	2.0
Cane molasses	2.0	6.5	2.5	8.0
Mineral mix	1.0	1.0	1.0	1.0
Urea	0.5	0.0	0.0	0.0
Chemical composition				
Dry matter	75.5	75.9	75.5	75.0
Crude protein	13.7	13.6	13.7	13.6
Neutral detergent fiber	39.5	39.8	36.6	36.5
Acid detergent fiber	22.2	20.5	17.6	16.7
Ash	8.2	8.9	9.5	10.7
Metabolizable energy (Mcal/kg DM)	2.40	2.47	2.51	2.47
Price/kg feed	19.8	19.5	19.5	19.4

The values are expressed as % age of dry matter unless otherwise stated.
TMR, total mixed ration.

Data from the feeding trial were analyzed according to Completely Randomized Design using general linear model procedure of SAS 9.2 software (SAS, 2008; SAS User Manual, Version 9.2. SAS Inst. Inc., Cary, NC, USA). Each animal on a specific TMR was considered as the experimental unit. The linear and quadratic effects of increasing level of fungal treated WS in the diet were examined by replacing the qualitative variable diet in the model with the quantitative variable proportion of fungal treated WS using Fitted Line Polynomial Regression Analysis in MINITAB (version 16.1.1.0). Data are presented as mean±standard error of mean. The level of significance was set at $p<0.05$.

Following mathematical model was applied:

$$Y_{ij} = \mu + \tau_i + \varepsilon_{ij}$$

Where, Y_{ij} = each observation on j^{th} animal due to i^{th} treatment

μ = overall mean

τ_i = effect of ith treatment ($\sum\tau_i = 0$ and i = 1, 2, 3, 4)

ε_{ij} = random error associated with i^{th} treatment with the restriction that variance σ^2 and mean zero.

RESULTS

Fungal treatment of WS increased CP and ash contents while cell wall contents were reduced (Table 1). The results showed that TMR2 improved ($p<0.001$) average daily gain (ADG) of the calves and feed economics ($p<0.001$) and FCR ($p<0.001$) (Table 3). The increasing level of fungal treated WS produced quadratic effects on all these

Table 3. Effects of fungal treated wheat straw supplementation as a part of total mixed ration on growth performance, feed conversion ratio and feed economics of Nili Ravi buffalo calves

Parameters	Experimental total mixed rations[1]				SEM	p-value		
	TMR1	TMR2	TMR3	TMR4		TMR	L	Q
ADG (kg/d)	0.52^{ab}	0.56^{a}	0.51^{b}	0.48^{b}	0.011	<0.001	0.010	0.019
FCR (kg feed/kg gain)	5.49^{a}	5.27^{b}	5.49^{a}	5.57^{a}	0.033	<0.001	0.050	0.004
Cost/kg gain	108.8^{a}	103.2^{b}	107.0^{a}	107.9^{a}	0.61	<0.001	0.808	0.001
FE (return from body gain/total feed cost)	1.38^{b}	1.44^{a}	1.38^{b}	1.36^{b}	0.008	<0.001	0.057	0.001

All values are mean±SEM. Data represent the average of 6 animals in each group.
TMR, total mixed ration; SEM, standard error of mean; L, linear effect of increasing fungal treated wheat straw in TMR; Q, quadratic effect of increasing fungal treated wheat straw in TMR; ADG, average daily gain; FCR, feed conversion ratio; FE, feed economics.
[1] TMR1, TMR2, TMR3, and TMR4 represent 0%, 33%, 67%, and 100% replacement of wheat straw with fungal treated wheat straw respectively.

parameters.

Replacing 33% WS with fungal treated WS in TMR2 showed the greatest intakes of DM, CP, and NDF by calves as depicted in Table 4. The intake of ADF was the highest in the control group (TMR1). Data also showed that the increasing level of fungal treated WS had linear effects on DM, CP, NDF, and ADF intakes (Table 4). The digestibilities of CP, NDF, and ADF were the highest for calves fed the TMR2 diets and linearly influenced by the increasing levels of fungal treated WS.

DISCUSSION

The lower NDF and ADF contents might be an indication of cell wall breakdown due to *Arachniotus* sp. The increased ash may be due to the added minerals that had to be used in the medium to grow *Arachniotus* sp.

The increases in ADG and feed economics on TMR2 diets might be due to favorable increase in DM intake, improvement in the digestibility of all nutrients and may have been related to improved nitrogen status of animals. Further increase in fungal treated WS in TMR beyond this level led to decrease in ADG that corresponds well to decrease in DM intake. The reasons for this phenomenon were not clear. The relatively high ash content in TMRs with 67% and 100% fungal treated WS replacements might have resulted in reduced palatability of the diet; this could be one explanation to the underlying phenomenon. Fazaeli and Shafeyi (2003) concluded that WS could be supplemented with *Agaricus bisporus* up to 15% for finishing lambs beyond which nutrient balance is reduced due to high mineral contents. The results of the current study correspond well with those of several authors previously reporting the effectiveness of fungal treatments in different diets. *Pleurotus ostreatus* treated maize straw led to a 17.5% increase in ADG in Pelibuey lambs compared to control diet (Ramírez-Bribiesca et al., 2010). Similarly, significant increases in DM intake and growth rate were observed in West African dwarf lambs fed biologically treated maize cobs replacing wheat offal in guinea grass (*Panicum maximum*) based diets (Akinfemi and Ladipo, 2011). Abdel-Azim and co-authors (2011)

Table 4. Effects of fungal treated wheat straw supplementation as a part of total mixed ration on nutrients intake and digestibility in Nili Ravi buffalo calves

Parameters	Experimental total mixed rations[1]				SEM	p-value		
	TMR1	TMR2	TMR3	TMR4		TMR	L	Q
Intake (kg/d)								
Dry matter	2.88^{ab}	2.95^{a}	2.80^{ab}	2.71^{b}	0.049	<0.01	0.009	0.125
Crude protein	0.39^{ab}	0.40^{a}	0.38^{ab}	0.37^{b}	0.007	<0.01	0.006	0.076
NDF	1.14^{a}	1.17^{a}	1.02^{b}	0.99^{b}	0.018	<0.001	0.001	0.136
ADF	0.64^{a}	0.60^{a}	0.49^{b}	0.45^{c}	0.010	<0.001	0.001	0.784
Digestibility (%)								
Dry matter	64.5^{ab}	69.3^{a}	64.0^{ab}	60.9^{b}	1.29	<0.01	0.093	0.030
Crude protein	74.1^{ab}	77.2^{a}	72.9^{bc}	69.7^{c}	0.92	<0.001	0.020	0.016
NDF	54.1^{ab}	59.2^{a}	50.7^{ab}	44.0^{b}	2.49	<0.01	0.019	0.049
ADF	44.6^{ab}	48.6^{a}	40.7^{ab}	38.8^{b}	1.89	<0.01	0.026	0.250

All values are mean±SEM. Data for intake represent the average of 6 and those for digestibility represent the average of 3 animals in each group, respectively.
TMR, total mixed ration; SEM, standard error of mean; L, linear effect of increasing fungal treated wheat straw in TMR; Q, quadratic effect of increasing fungal treated wheat straw in TMR; NDF, neutral detergent fiber; ADF, acid detergent fiber.
[1] TMR1, TMR2, TMR3 and TMR4 represent 0%, 33%, 67%, and 100% replacement of wheat straw with fungal treated wheat straw respectively.

found a greater DM intake and an improved nitrogen balance in cross-bred lambs fed rice straw and corn stalks treated with *Trichoderma viride*. Kim et al. (2011) found that supplementing Oyster mushroom (*Pleurotus ostreatus*) spent substrate with selective lactic acid producing bacteria at the rate of 10% in calf starter diet improved the ADG in post weaning calves.

The reduced FCR exhibited by calves fed TMR2 diets is mainly attributable to improved nutritive value and better palatability as reflected by improved growth performance of the experimental animals; this is in agreement with the results of others (El-Kady et al., 2006). Similar results for FCR were found by Salman et al. (2008) in goats fed fungal treated sugar beet pulp. Fouda (2008) reported a lower feed cost/kg gain and a higher net profit values/kg gain in lambs fed fungal treated diets when compared with control groups. The experiments conducted by Abdel-Aziz (2002) and El Shafie et al. (2007) also showed a reduction in feed cost by replacing 40% to 50% concentrate feed mixture with biologically treated rice or WS.

The digestibility of DM in our study was increased by 7% units on TMR2 diets, which is closer to that (10%) found by Fazaeli et al. (2002) in cattle consuming fungal treated WS diets and that found by Kafilzadeh et al. (2009) in sheep consuming *Pleurotus florida* treated palm leaves, respectively. Similarly, Mahesh and Mohini (2013) found that feeding WS supplemented with *Crinipellis* sp. to Sahiwal calves significantly increased the digestibility of various nutrients and concluded that solid state fermentation with *Crinipellis* sp. holds the potential to upgrade the nutritive worth of WS.

Arora and Sharma (2009) conducted the solid state fermentation of WS collected from different regions of India with four different white rot fungi viz. *Phlebia brevispora*, *P. fascicularia*, *P. floridensis*, and *P. radiate*. They observed that *P. brevispora* was found to be the best organism that significantly enhanced the *in vitro* feed digestibility. In another experiment, 50% increase in *in vitro* digestibility of WS was observed with *P. floridensis* (Sharma and Arora, 2010). Okano et al. (2009) observed an increased *in vitro* digestibility of Madake bamboo (*Phyllostachys bambusoides*) when treated with white rot *Ceriporiopsis subvermispora* for 10 weeks in solid state fermentation chamber. Akinfemi (2010) reported that CP concentration was improved by treating peanut husk with *P. ostreatus*.

CONCLUSIONS

It is concluded that fungal (*Arachniotus* sp.) treatment of WS increases CP and ash contents while reducing the cell wall contents. Feeding rations with a 33% replacement of WS with fungal treated WS in the TMR enhances growth performance in Nili Ravi buffalo calves by improving feed intake and nutrient digestibility. Beyond this level, losses in performance due to reduced dry matter intake owing to reduced digestibility and palatability may occur. It is also concluded that the biological treatments are preferable than physical and chemical methods for environmental conservation. We recommend the further extension of our findings in other livestock species to improve the overall livestock production potential.

ACKNOWLEDGMENTS

This work was fully funded by the Higher Education Commission of Pakistan.

REFERENCES

Abdel-Azim, S. N., M. A. Ahmed, F. Abo-Donia, and H. Soliman. 2011. Evaluation of fungal treatment of some agricultural residues. Egypt J. Sheep Goat Sci. 6:1-13.

Abdel-Aziz, M. Y. 2002. Nutritional Studies on Biological Treatment of Agricultural by Product on Ruminants. M.Sc. Thesis, Zagazig University, Zagazig, Egypt.

Ahmed, S., F. Ahmad, and A. S. Hashmi. 2010. Production of microbial biomass protein by sequential culture fermentation of *Arachniotus* species and *Candida utilis*. Pak. J. Bot. 42:1225-1234.

Akinfemi, A. 2010. Bioconversion of peanut husk with white rot fungi: Pleurotus ostreatus and Pleurotus pulmonarius. Livest. Res. Rural Dev. 22:Article 49.

Akinfemi, A. and M. K. Ladipo. 2011. Effect of fungal treated maize cob on the performance of West African Dwarf Rams. In: International Conference on Research on Food Security, Natural Resource Management and Rural Development, University of Bonn, Bonn, Germany.

AOAC. 2006. Official Methods of Analysis, 18th ed., Association of Official Analytical Chemists, Washington, DC, USA.

Arora, D. S. and R. K. Sharma. 2009. Enhancement in *in vitro* digestibility of wheat straw obtained from different geographical regions during solid state fermentation by white rot fungi. BioResourc 4:909-920.

El-Kady, R. I., I. M. Awadalla, M. I. Mohamed, M. Fadel, and H. H. Abd El-Rahman. 2006. Effect of exogenous enzymes on the growth performance and digestibility of growing buffalo calves. Int. J. Agric. Biol. 8:354-359.

El-Shafie, M. H., A. A. Mahrous, and T. M. M. Abdel-Khalek. 2007. Effect of biological treatments for wheat straw on performance of small ruminants. Egypt J. Nutr. Feeds 10:635-648.

Fazaeli, H., Z. A. Jelan, H. Mahmodzadeh, J. B. Liang, A. Azizi, and A. Osman. 2002. Effect of fungal treated wheat straw on the diet of lactating cows. Asian Australas. J. Anim. Sci. 15:1573-1578.

Fazaeli, H. and A. Shafeyi. 2003. Use of mushroom spent wheat straw compost as animal feed In: 5th International Conference on Mushroom Biology and Mushroom Products, 8-12 April 2005, Shanghai, China.

Fazaeli, H. and S. A. Mirhadi. 2007. Nutritive value index of treated wheat straw with *Pleurotus* fungi fed to cow. Proc. Br. Soc. Anim. Sci. Southport, United Kingdom, April, 2007. p. 197.

Fouda, S. M. I. 2008. Studies of Nutrition on Sugar Cane Bagasse Previous Treated of Some Chemical and Biological Treatment. Ph.D. Thesis, Faculty of Agriculture, Al-Azhar University, Cairo, Egypt. pp. 115-116.

Kafilzadeh, F., F. Hozhabri, and A. Kabirifard. 2009. Effect of Pleurotus florida on *in vitro* gas production of wheat stubble and date palm leaf. Res. J. Biol. Sci. 4:37-41.

Kim, M.-K., H.-G. Lee, J.-A. Park, S.-K. Kang, and Y.-J. Choi. 2011. Recycling of fermented sawdust-based oyster mushroom spent substrate as a feed supplement for postweaning calves. Asian Australas. J. Anim. Sci. 24:493-499.

Mahesh, M. S. and M. Mohini. 2013. Nutritional evaluation of wheat straw treated with white-rot fungus Crinipellis sp. RCK-SC in Sahiwal calves. Livest. Sci. http://dx.doi.org/10.1016/j.livsci.2013.11.021

Misra, A. K., A. S. Mishra, M. K. Tripathi, R. Prasad, S. Vaithiyanathan, and R. C. Jakhmola. 2007. Optimization of Solid State Fermentation of Mustard (*Brassica campestris*) Straw for Production of Animal Feed by White Rot Fungi (*Ganoderma lucidum*). Asian Australas. J. Anim. Sci. 20:208-213.

Nisa, M., M. Sarwar, and M. A. Khan. 2004. Influence of *ad libitum* feeding of urea treated wheat straw with or without corn steep liquor on intake, *in situ* digestion kinetics, nitrogen metabolism, and nutrient digestion in Nili Ravi buffalo bull. Aust. J. Agric. Res. 55:229-236.

NRC. 2001. Nutrient Requirements of Dairy Cattle: Seventh Revised Edition. National Academy Press, Washington, DC, USA.

Okano, K., N. Ohkoshi, A. Nishiyama, T. Usagawa, and M. Kitagawa. 2009. Improving the nutritive value of madake bamboo, Phyllostachys bambusoides, for ruminants by culturing with the white-rot fungus Ceriporiopsis subvermispora. Anim. Feed Sci. Technol. 152:278-285.

Pakistan Economic Survey. 2013-2014. Government of Pakistan, Ministry of Food and Agriculture, Economic Adviser's Wing, Islamabad, Pakistan. pp. 13-15.

Ramírez-Bribiesca, J. E., A. S. Sánchez, L. M. Hernández-Calva, J. S. Alinas-Chavira, J. R. Galaviz-Rodriguez, R. G. Cruz-Monterrosa, and S. Vargas-López. 2010. Influence of Pleurotus ostreatus spent corn straw on performance and carcass characteristics of feedlot Pelibuey lambs. Ind. J. Anim. Sci. 80: 754-757.

Salman, F. M., R. I. El-Kadi, H. Abdel-Rahman, S. M. Ahmed, M. I. Mohamed, and M. M. Shoukry. 2008. Biologically treated sugar beet pulp as a supplement in goat rations. Int. J. Agric. Biol. 10:412-416.

Sarwar, M., M. A. Khan, and Z. Iqbal. 2002. Feed Resources for Livestock in Pakistan. Int. J. Agric. Biol. 4:186-192.

Selim, A. S. M., J. Pan, T. Takano, T. Suzuki, S. Koike, Y. Kobayashi, and K. Tanaka. 2004. Effect of ammonia treatment on physical strength of rice straw, distribution of straw particles and particle-associated bacteria in sheep rumen. Anim. Feed Sci. Technol. 115:117-128.

Shaukat., S., M. A. Bajwa, M. A. Waqar, and T. Sohail. 2006. Production, optimization, purification and characterization of glucomalyase from *Arachniotus* sp. J. Chem. Soc. Pak. 28:368-373.

Sharma, R. K. and D. S. Arora. 2010. Production of lignocellulolytic enzymes and enhancement of *in vitro* digestibility during solid state fermentation of wheat straw by *Phlebia floridensis*. Bioresour. Technol. 101:9248-9253.

Van Soest, P. J., J. B. Robertson, and B. A. Lewis. 1991. Methods for dietary fiber, neutral detergent fiber, and nonstarch polysaccharides in relation to animal nutrition. J. Dairy Sci. 74:3583-3597.

Younas, M. and M. Yaqoob. 2005. Feed resources of livestock in the Punjab, Pakistan. Livest. Res. Rural Dev. 17:Article 18.

5

Effects of Chromium Methionine Supplementation on Blood Metabolites and Fatty Acid Profile of Beef during Late Fattening Period in Holstein Steers

Jalil Ghassemi Nejad, Bae-Hun Lee, Byong-Wan Kim, Sang-Jip Ohh, and Kyung Il Sung*

College of Animal Life Sciences, Kangwon National University, Chuncheon 200-701, Korea

ABSTRACT: The objective of this study was to determine the effects of chromium methionine (Cr-Met) chelate supplementation on blood metabolites and fatty acid profile of beef from Holstein steers during late fattening period. Fifteen Holstein steers were allotted randomly into two groups including the control (non Cr-Met feeding, NCM, ave. body weight [BW] = 483±25.7 kg) and the treatment (Cr-Met feeding for 4 months, 4CM, ave. BW = 486±27.5 kg) group. The feeding amount of Cr-Met to animals was limited to 400 ppb/cow/d and was supplemented to total mixed ration. No difference in blood albumin, alkaline phosphatase, urea-nitrogen, calcium, creatine, glucose, total protein, triglyceride, and cholesterol were observed between the treatment groups ($p>0.05$). The level of high density lipoprotein was higher in the 4CM group than the NCM group, whereas low density lipoprotein was lower in the 4CM group ($p<0.05$). The fatty acid composition (caprate, laurate, myristate, pentadecanoate, palmitate, palmitoleate, margarate, cis-11 heptadodecanoate, stearate, oleate, trans-vaccenate, linoleate, cis-11 eicosenoate, docosa hexaenoic acid, and docosa pentaenoic acid) of the beef showed no difference between the two groups ($p>0.05$). The arachidonic acid level tended to be higher in the 4CM than the NCM group ($p = 0.07$). Cr-Met had no influence ($p>0.05$) on the ratio of saturated, unsaturated, unsaturated/saturated, monounsaturated/saturated and polyunsaturated/saturated fatty acids whereas the ratio of polyunsaturated fatty acids (PUFA) in the 4CM group was comparatively higher than the NCM group ($p<0.05$). This study concluded that feeding Cr-Met supplementation in 400 ppb/d to Holstein steers for 4 months during late fattening period can improve some blood metabolites and beef quality by increasing PUFA and gamma-linoleate compositions of beef. (**Key Words:** Chromium Methionine Chelate, Blood Metabolites, Fatty Acids, Beef Steers, Beef Quality)

INTRODUCTION

A recent study has reported that Cr-Met in the form of chelate could improve carcass characteristics including marbling score in Korean native steers (Sung et al., 2015). Another follow up study in Holstein steers during raising and late fattening period showed that 4 months is an optimum period for feeding Cr-Met chelate to improve daily gain and carcass characteristics of Holstein steers during the fattening period; however, fatty acids profile of beef was not measured (Song et al., 2013). Therefore, the apparent effects of Cr-Met chelate supplementation on blood metabolites and profile of fatty acids of beef in Holstein beef steers during late fattening period has not yet been investigated.

Chromium (Cr) is one of the essential micronutrients for ruminants and is considered to be a metabolic modifier. The organic source of Cr is a promising form due to higher bioavailability than inorganic sources (NRC, 1997). Cr supplementation has been reported to show an influence on some blood metabolites such as glucose (Wang et al., 2007) and could improve carcass quality such as intermuscular fat and percentage of muscle (Boleman et al., 1995) in pigs. Furthermore, Cr is also known to be a key constituent of glucose tolerance factor (regulates blood glucose level) and maintains glucose homeostasis (Sung et al., 2015). In addition, supplemental Cr was also found to play an important role in serum cholesterol homeostasis (Ohh et al., 2004). Cr supplementation decreased the level of total cholesterol, low density lipoprotein (LDL) cholesterol, and

* Corresponding Author: Kyung Il Sung.
E-mail: kisung@kangwon.ac.kr

triglyceride in the blood, but increased the level of high density lipoprotein (HDL) cholesterol (Anderson, 1995; Sung et al., 2015). Schwarz and Merts (1959) and Bunting et al. (1994) reported that Cr supplementation can alter glucose metabolism in rats and in calves. Positive responses were reported in pigs by Page et al. (1993), Lindemann et al. (1995), Boleman et al. (1995), and Mooney and Cromwell (1995, 1997) in swine for carcass leanness. Moreover, according to Ohh et al. (2004), the positive effect of Cr supplementation can be associated with its obvious influence on the systematic division of energy between adipose and lean tissue. Reports of lessened fat over the 10th rib and decreased yield grades have been reported in lambs supplemented with Cr tripicolinate (Kitchalong et al., 1995). However, others reported no responses in carcass leanness to supplemental Cr (Harris et al., 1995; Ward et al., 1995). Likewise, Cr research has been conducted in dairy (Al-Saiady et al., 2004) and beef cattle (Swanson et al., 2000; Pollard et al., 2002; Stahlhut et al., 2006). Treatments were control with no Cr-Met supplementation and the different level of Cr-Met supplementation in diet with or without yeast resulted in improving carcass quality including marbling score and glucose tolerance rate. The aim of the present study was to evaluate the effects of Cr-Met chelate on blood metabolites and fatty acids profile of beef from Holstein beef steers.

MATERIALS AND METHODS

Treatments, feeding and experimental procedure

To assess the effect of Cr-Met on performance and beef quality, fifteen Holstein steers were randomly assigned into two dietary treatment groups; Non Cr-Met feeding (NCM, 7 head), Cr-Met feeding over a 4 months (4CM, 8 head). Average body weights of groups were 483±25.7 kg, 486±27.5 kg for NCM and 4CM, respectively at the beginning of the experiment. The feeding amount of Cr-Met (Innobio Co., Ltd., Shiheung, Korea) to animals was limited to 400 ppb/cow/d. The duration for the study was 4 months. The rate of forage to concentrate was 20:80 and forages included alfalfa cube and perennial ryegrass hay according to NRC requirements of beef cattle (2000). Concentrate, alfalfa cubes and Perennial ryegrass hay were fed twice daily, in the morning at 0900 am and at 0500 pm in the afternoon to achieve controlled feeding of fixed ratio of F:C as 20:80. The intake of concentrate was 8±0.4 kg (dry matter [DM]) and the forage intake was 2 kg (1±0.2 kg of alfalfa cube +1±0.2 kg of Perennial ryegrass hay, DM) for both groups. The composition of the experimental diets is presented in Table 1.

Feed, blood and beef quality analysis

Common nutrients were analyzed according to AOAC (1990); neutral detergent fiber and the level of acid detergent fiber were analyzed using the method of Goering and Van Soest (1991). A total digestible nutrient (Table 1) was calculated using the regression equation Wardeh (1981). Steers had *ad libitum* access to water. Blood samples from each steer at the beginning and the end of the experiment was collected from the jugular vein at 1300 h by using a vacutainer (no additive: BD, Franklin Lakes, NJ, USA) for serum separation. The serums samples were collected at the beginning, on a monthly basis and at the end of the experiment by centrifuging it at 2,500×g for 15 minutes and the samples were stored at –20°C for further processing. The serum samples were later analyzed for total protein, albumen, alkaline phosphatase (ALP), calcium, creatine, triglyceride, cholesterol, glucose, LDL, and HDL, using kits abiding with the manufacturer's protocol (Modular analytic E170, Roche, Germany). Upon completion of the field experiment, all steers were slaughtered in order to measure the fatty acids composition of beef from loin side. Samples from each steer were frozen at –20°C for 12 hours, and were thawed prior to analysis. According to the method of lipid extraction (Folch et al., 1957), 6 g of sample and chloroform/methanol (2:1) solution were homogenized in a 25 mL homogenizer (Diax 6000, Heidolph, Germany) at 1,100×g for 30 seconds. Next 6 mL of 0.88% KCl solution was added to the homogenate, followed by centrifugation at 2,500×g (GS-6R Centrifuge, Beckman, Ramsey, MN, USA) for 10 minutes. The fluid was filtered through filter paper and lipid was concentrated using a nitrogen gas concentrator (MGS-2200, Eyelaa Tokyo Rikakikai Co., Ltd, Tokyo, Japan) following the method of AOAC (1990). Each of the fatty acid methyl ester standards (Sigma-Aldrich Co., Saint Louis, MO, USA) was qualitatively compared with retention time, and the analytic conditions used for Gas Chromatography (Agilent 6890N, Agilent Technologies, Santa Clara, CA, USA). For this, a sample of beef and tissue was kept for the split ratio of 1:10. The oven injector

Table 1. Chemical composition of experimental feed

	DM	CP	CF	EE	Ash	NDF	ADF	NFE	TDN
	% DM								
Concentrate	89.0	19.0	6.0	2.2	7.5	19.4	10.6	62.2	72.1
Alfalfa cube	89.7	14.4	20.4	1.5	12.1	46.6	38.2	51.6	56.2
Perennial ryegrass hay	92.6	6.7	29.8	2.5	5.3	62.3	35.7	55.7	56.8

DM, dry matter; CP, crude protein; CF, crude fiber; EE, ether extract; NDF, neutral detergent fiber; ADF, acid detergent fiber; NFE, nitrogen free extract; TDN, total digestible nutrient.

was heated with 220°C. A carrier of gas 1mL/min was heated with 150°C for one min. A column HP-Innowax (30 m length×0.32 id×0.25 μm thicknesses) maintained for a detector temperature of 275°C. The oven maintained the temperature of 200°C to 250°C at 3°C/min and 250°C for 5 min.

Statistical analysis

All the data are reported as the sample mean±the standard deviation. Pairwise comparisons between means of different groups were performed using a t-test. The difference between two subsets of data is considered statistically significant if the t-test gives a significance level p (p value) less than 0.05.

RESULTS AND DISCUSIONS

Blood metabolites

In assessing the effect of Cr-Met supplementation on normal body functions (renal function, clinical enzymology etc.) we observed that Cr-Met supplementation did not affect ($p>0.05$) albumen, ALP, calcium and creatine in blood (Table 2). Al-Saiady et al. (2004) also did not find any change in the albumin level in Holstein cows supplemented with chelated Cr. Similar results for creatine levels in pigs was observed by Lindemann et al. (2008). Kaneko et al. (1989) reported that an increase in ALP activity may occur in response to bone or liver damage. However, in the current study no increase ($p>0.05$) in ALP activity was observed. Furthermore, no effect on urea nitrogen and total protein was observed in any of the other trials (Ohh and Lee, 2005; Sung et al., 2015). Similar results were reported by Kitchalong et al. (1995) in lambs and Bunting et al. (1994) in steers with Cr-picolinate.

Table 2. Effect of Cr-Met on blood levels of Holstein steers during late fattening period

	NCM (n = 7)	4CM (n = 8)	p value
Albumin (g/dL)	2.9±0.7	3.3±0.2	0.45
Alkaline phosphatase (U/L)	262.3±137.6	196.5±24.6	0.27
Urea-nitrogen (mg/dL)	9.0±4.9	10.2±2.6	0.58
Calcium (mg/dL)	9.6±0.6	9.7±0.7	0.73
Creatine (mg/dL)	1.2±0.1	1.3±0.2	0.81
Glucose (mg/dL)	93.3±4.0	87.1±5.0	0.37
Total protein (g/dL)	8.1±1.7	6.9±0.6	0.29
Triglyceride (mg/dL)	24.0±6.0	26.7±5.5	0.78
Cholesterol (mg/dL)	164.0±7.5	141.5±12.1	0.43
HDL (mg/dL)	72.3±5.3^{b}	101.3±8.3^{a}	0.01
LDL (mg/dL)	51.0±5.3^{a}	35.1±7.5^{b}	0.04

NCM, non Cr-Met feeding; CM, Cr-Met feeding; HDL, high density lipoprotein; LDL, low density lipoprotein, ±, standard deviation.

$^{a-b}$ Means with different superscript in same row differs significantly ($p<0.05$).

However, Dominguez-vara (2009) showed inconsistent results in urea nitrogen levels with the addition of Cr-yeast. In this study, supplementation with Cr-Met chelate did not affect ($p>0.05$) serum glucose levels in Holstein steers. Previous studies (Kegley et al., 1997) also found no effect of Cr supplementation on plasma glucose concentration in cattle; however, reports by Chang et al. (1995) and Stahlhut et al. (2006) demonstrated that Cr supplementation lowered plasma glucose concentration in growing and finishing steers. Blood cholesterol has a diagnostic value for situations like hypothyroidism (Kaenko, 1989). However, no difference ($p>0.05$) was observed between triglyceride, and cholesterol in the current study (Table 2). Lindemann et al. (2008) also found no deviation in cholesterol and triglyceride values from that of the control group. Their findings support the results of this study. Bunting et al. (1994) reported decreased cholesterol levels in unstressed calves fed with Cr-picolonate. Supplemental Cr also plays an important role in serum cholesterol homeostasis. Cr supplementation decreased the level of the blood's total cholesterol, LDL cholesterol, and triglyceride but increased the level of HDL cholesterol (Anderson, 1995). Dietary saturated fat and cholesterol lead to elevated levels of cholesterol in the blood. A possible mechanism of Cr on amino acid synthesis has been predicted due to the role of insulin in amino acid uptake. However, other activities of Cr in protein metabolism have not been reported (Roginski and Mertz, 1969). In this study, HDL was remarkably increased whereas LDL was decreased ($p<0.05$, Table 2). This result is in agreement with Sung et al. (2015) who reported higher HDL and lower LDL levels in Korean native steers fed with supplemented Cr.-Met chelate. Al-Saiady (2004), however, reported increased cholesterol levels in dairy cows with Cr supplementation. HDL helps to prevent narrowing of the artery walls by removing excess cholesterol and transporting it to the liver for excretion. LDL carries cholesterol for cell building needs, but leaves behind any excess on artery walls and in tissues (Wang et al., 2007; Sung et al., 2015). High LDL and low HDL levels indicate diets high in refined carbohydrates and/or carbohydrate sensitivity. Low levels of HDL are strong indicators of insulin resistance, but in this experiment we observed adverse amounts due to using Cr-Met in diets.

Fatty acid composition of beef

In the current study, no significant differences ($p>0.05$) in composition of fatty acids (caprate, laurate, myristate, pentadecanoate, palmitate, palmitoleate, margarate, cis-11 heptadodecanoate, stearate, oleate, trans-vaccenate, linoleate, cis-11 eicosenoate, docosa hexaenoic acid, and docosa pentaenoic acid) were observed. Whereas, gamma linolenate (C18:3n6, $p<0.05$) and arachidonic acid (C20:4n6, $p = 0.07$) were higher in the 4CM group than the

NCM group. There was no difference ($p>0.05$) in gamma linolenate level between the two treatment groups. The level of arachidonic acid in 4CM was higher ($p<0.05$) than other fatty acids (Table 3). The fatty acid composition of beef plays an important role in the quality of beef for consumers (Sung et al., 2015). Cr supplementation has been employed to manipulate the quality of beef due to its biological function on body fat and muscle metabolism (Page et al., 1993; Sung et al., 2015). Plasma non-esterified fatty acids (NEFA) reflect body fat mobilization in response to a negative energy balance or stress conditions. During an energy deficit, animals break down triglycerides (fat) stored in adipose tissue. NEFA enters the blood stream to be transported to organs and tissue throughout the body. The concentration of NEFA measured in blood has been shown to reflect fat mobilized from body fat reserves. Plasma NEFA concentrations were lower in beef cows receiving supplemental Cr shortly after calving (days 97 and 155), especially young cows (Stahlhut et al., 2006). Matthews et al. (2001) reported that swine supplemented with Cr picolinate had lower plasma NEFA concentration vs. control. Similarly, studies on sheep (Kitchalong et al., 1995) and dairy cattle (Hayirli et al., 2001) found that Cr supplementation lessened plasma NEFA concentration. Conversely, supplemental Cr had no effect on plasma NEFA concentration in growing steers (Chang and Mowat, 1992; Bunting et al., 1994). In a study by Stahlhut et al. (2006), NEFA concentration responses to Cr were affected by the amount of copper. Cr supplementation decreased plasma NEFA concentration in non- copper supplemented cows. Beef quality could have been different at the beginning of the experiment but we were unable to slaughter the steers at both the beginning and the end of the experiment. However, the body weight of steers at the beginning of the experiment showed no significant difference while cows were receiving specific amounts of feed through controlled feeding. Therefore, any changes to the fatty acids profile of beef may be due to Cr-Met supplementation since the intake of DM was similar between the two groups and blood parameters were not different at the beginning of the experiment. The response to organic Cr supplementation in beef quality of farm animals is varied among animal species, form of dietary Cr, and the level of Cr supplementation. However, the results have varied presumably due to other extrinsic factors such as onset and level of supplementation, nutrients and level of basal diet, breed and species (Page et al., 1993).

Cr-Met had no effect on the ratio of saturated, unsaturated, unsaturated/saturated, monounsaturated/saturated, and polyunsaturated/saturated fatty acids while the ratio of polyunsaturated fatty acids (in the 4CM group was considerably higher ($p<0.05$) than the control group (Table 4). Cr can affect fat mobilization from body stores to meet nutrient demands. In this regard, alteration in some fatty acid compositions can be explained. Cr in the form of chelate especially with methionine has a greater potential to be absorbed through increased digestion in the gastro-intestinal tract (Mertz, 1993; Ohh et al., 2004; Sung et al., 2015). Since cattle tend to produce less intramuscular fat, it is expected that the polyunsaturated fatty acid (PUFA) level of the fat will be higher. In this study higher ($p<0.05$) PUFA in 4CM was observed. The finishing diet strongly

Table 3. Effect of Cr-Met supplementation on fatty acid profile of beef (loin side) from Holstein steers during late fattening period

Fatty acid profile (%)	NCM (n = 7)	4CM (n = 8)	p value
C10:0 (caprate)	0.06±0.01	0.05±0.02	0.57
C12:0 (laurate)	0.09±0.01	0.08±0.01	0.73
C14:0 (myristate)	3.26±0.13	3.11±0.66	0.69
C15:0 (pentadecanoate)	0.27±0.03	0.24±0.02	0.49
C16:0 (palmitate)	29.01±0.62	28.72±2.75	0.47
C16:1n7 (palmitoleate)	4.74±0.95	4.58±0.77	0.83
C17:0 (margarate)	0.71±0.11	0.69±0.10	0.75
C17:1n6 (cis-11-heptadecanoate)	0.67±0.15	0.65±0.04	0.57
C18:0 (stearate)	11.31±0.86	12.29±1.24	0.43
C18:1n9 (oleate)	47.00±0.29	46.49±2.42	0.39
C18:1n7 (trans-vaccenate)	1.91±0.10	1.76±0.36	0.37
C18:2n6 (linoleate)	0.24±0.04	0.24±0.03	0.81
C18:3n6 (gamma-Linoleate)	0.07±0.06^{b}	0.23±0.03^{a}	0.04
C20:1n9 (cis-11-Eicosenoate)	0.26±0.10	0.29±0.04	0.68
C20:4n6 (arachidonate)	0.32±0.07	0.48±0.10	0.07
C22:4n6 (DHA)	0.04±0.07	0.05±0.09	0.73
C22:5n3 (DPA)	0.04±0.08	0.05±0.08	0.39
Total	100.00	100.00	

NCM, non Cr-Met feeding; CM, Cr-Met feeding; DPA, docosapentaenoic acid; DHA, docosahexaenoic acid; ±, standard deviation.

$^{a-b}$ Means with different superscript in same row differs significantly ($p<0.05$).

Table 4. Effect of Cr-Met on fatty acid ratio of Holstein steers beef (loin side) during late fattening period

Fatty acids (%)	NCM (n = 7)	4CM (n = 8)	p value
SFA	44.72±1.35	44.55±2.10	0.47
UFA	55.28±1.35	54.45±2.10	0.59
Total	100.00	100.00	
MUFA	54.58±1.34	53.71±2.06	0.73
PUFA	0.70±0.08^{b}	0.87±0.07^{a}	0.04
UFA/SFA	1.24±0.07	1.15±0.10	0.31
MUFA/SFA	1.22±0.07	1.19±0.10	0.83
PUFA/SFA	0.02±0.01	0.02±0.01	0.83

NCM, non Cr-Met feeding; CM, Cr-Met feeding; SFU, saturated fatty acid; UFA, unsaturated fatty acid; MUFA, monounsaturated fatty acid; PUFA, polyunsaturated fatty acid; ±, standard deviation.

$^{a-b}$ Means with different superscript in same row differs significantly ($p<0.05$).

influences the fatty acid composition of beef (Smith et al., 2009). Zea et al. (2007) reported that saturated fatty acids (SFA) were higher in animals fed with concentrates, while animals fed with silage had higher levels of PUFA and a higher PUFA/SFA ratio. Furthermore, Smith et al. (2009) reported that grain feeding arouses the activity of adipose tissue stearoyl-CoA desaturase in marbling adipose tissue and lowers ruminal isomerization/hydrogenation of dietary PUFA, resulting in a noticeable increase in monounsaturated fatty acid (MUFA) in beef over time. The current study found no differences ($p>0.05$) in the values of SFA, unsaturated fatty acid (UFA), MUFA, UFA/SFA, MUFA/SFA, and PUFA/SFA.

CONCLUSIONS

It can be concluded that feeding Cr-Met supplementation in 400 ppb/d to Holstein steers for 4 months during late fattening period can improve some blood metabolites and beef quality regarding fatty acid composition by increasing PUFA and gamma-linoleate compositions of beef. However, further research is warranted to validate the present results.

ACKNOWLEDGMENTS

Authors are thankful to Austin Thelen (Department of Communication, Oregon State University, USA) and Lindsay M. Loehden (Linfield College, USA) for their help in editing the manuscript.

REFERENCES

Al-Saiady, M. Y., M. A. Al-Shaikh, S. Y. Al-Mufarrej, T. A. Al-Showeimi, H. H. Mogawer, and A. Dirrar. 2004. Effect of chelated chromium supplementation on lactation performance and blood parameters of Holstein cows under heat stress. Anim. Feed Sci. Technol. 117:223-233.

Anderson, R. A. 1995. Chromium, glucose tolerance, and lipid metabolism. J. Adv. Med. 8:37.

AOAC. 1990. Official Methods of Analysis. 15th edition. Association of Official Analytical Chemists, Washington, DC, USA.

Boleman, S. L., S. J. Boleman, T. D. Bidner, L. L. Southern, T. L. Ward, J. E. Pontif, and M. M. Pike. 1995. Effect of chromium picolinate on growth, body composition, and tissue accretion in pigs. J. Anim. Sci. 73:2033-2042.

Bunting, L. D., J. M. Fernandez, Jr. D. L. Thompson, and L. L. Southern. 1994. Influence of chromium picolinate on glucose usage and metabolic criteria in growing Holstein calves. J. Anim. Sci. 72:1591-1599.

Chang, H., D. N. Mowat, and B. A. Mallard. 1995. Supplemental chromium and niacin for stressed feeder calves. Can. J. Anim. Sci. 75:351-358.

Chang, X. and D. N. Mowat. 1992. Supplemental chromium for stressed and growing feeder calves. J. Anim. Sci. 70:559-565.

Domínguez-Vara, I. A., S. S. González-Muñoz, J. M. Pinos-Rodríguez, J. L. Bórquez-Gastelum, R. Bárcena-Gama, G. Mendoza-Martínez, L. E. Zapata, and L. L. Landois-Palencia. 2009. Effects of feeding selenium-yeast and chromium-yeast to finishing lambs on growth, carcass characteristics, and blood hormones and metabolites Anim. Feed Sci. Technol. 152:42-49.

Folch, J., M. Lee, and G. H. Sloane-Stanley. 1957. A simple method for the isolation and purification of total lipides from animal tissues. J. Boil. Chem. 226:497-509.

Goering, H. K. and P. J. Van Soest. 1991. Forage fiber analyses (apparatus, reagents, procedures, and some applications). Agric. Handbook. No.379. ARS, USDA, Washington, DC, USA.

Harris, J. E., S. D. Crow, and M. D. Newcomb. 1995. Effect of chromium picolinate on growth performance and carcass characteristics on pigs fed adequate and low protein diets. J. Anim. Sci. 73 (Suppl. 1):194 (abstract).

Hayirli, A., D. R. Bremmer, S. J. Bertics, M. T. Socha, and R. R. Grummer. 2001. Effect of chromium supplementation on production and metabolic parameters in periparturient dairy cows. J. Dairy Sci. 84:1218-1230.

Kaneko, J. J. 1989. Reference values for blood gas and electrolyte determinations. In: Clinical Biochemistry of Domestic Animals. 4th ed. (Ed. J. J. Kaneko). Academic Press, San Diego, CA, USA. 564 p.

Kegley, E. B., J. W. Spears, and T. T. Brown. 1997. Effect of shipping and chromium supplementation on performance, immune response, and disease resistance of steers. J. Anim. Sci. 75:1956-1964.

Kitchalong, L., J. M. Fernandez, L. D. Bunting, L. L. Southern, and T. D. Bidner. 1995. Influence of chromium tripicolinate on glucose metabolism and nutrient partitioning in growing lambs. J. Anim. Sci. 73:2694-2705.

Lindemann, M. D., G. L. Cromwell, H. J. Monegue, and K. W. Pursur. 2008. Effect of chromium source on tissue concentration of chromium in pigs. J. Anim. Sci. 86:2971-2978.

Lindemann, M. D., C. M. Wood, A. F. Harper, E. T. Kornegay, and R. A. Anderson. 1995. Dietary chromium picolinate additions improve gain:feed and carcass characteristics in growing-finishing pigs and increase litter size in reproducing sows. J. Anim. Sci. 73:457-465.

Matthews, J. O., L. L. Southern, J. M. Fernandez, J. E. Pontif, T. D. Bidner, and R. L. Odgaard. 2001. Effect of chromium picolinate and chromium propionate on glucose and insulin kinetics of growing barrows and on growth and carcass traits of growing-finishing barrows. J. Anim. Sci. 79:2172-2178.

Mertz, W. 1993. Chromium in human nutrition: A review. J. Nutr. 123:626-633.

Mooney, K. W. and G. L. Cromwell. 1995. Effects of dietary chromium picolinate supplementation on growth, carcass characteristics, and accretion rates of carcass tissues in growing-finishing swine. J. Anim. Sci. 73:3351- 3357.

Mooney, K. W. and G. L. Cromwell. 1997. Efficacy of chromium picolinate and chromium chloride as potential carcass modifiers in swine. J. Anim. Sci. 75:2661-2671.

NRC. 1997. The role of chromium in animal nutrition. National academy of sciences. National Academy Press, Washington, DC, USA.

Nutrient Requirements of Beef Cattle. 2000. 7th rev. ed. National Academy Press, Washington, DC, USA.

Ohh, S. J., C. H. Kim, J. S. Shin, K. I. Sung, and H. S. Kim. 2004. Effects of different forms of Chromium supplements on serum glucose, insulin and lipids in rats. J. Feed Sci. Nutr. 9:342-345.

Ohh, S. J. and J. Y. Lee. 2005. Dietary chromium-methionine chelate supplementation and animal performance. Asian Australas. J. Anim. Sci. 18:898-907.

Page, T. G., L. L. Southern, T. L. Ward, and D. L. Thompson Jr. 1993. Effect of chromium picolinate on growth and serum and carcass traits of growing-finishing pigs. J. Anim. Sci. 71:656-662.

Pollard, G. V., C. R. Richardson, and T. P. Karnezos. 2002. Effects of supplemental organic chromium on growth, feed efficiency and carcass characteristics of feedlot steers. Anim. Feed Sci. Technol. 98:121-128.

Roginski, E. E. and W. Mertz. 1969. Effect of chromium supplementation on glucose and amino acid metabolism in rats fed a low protein diet. J. Nutr. 97:525-30.

Schwarz, K. and W. Mertz. 1959. Chromium (III) and the glucose tolerance factor. Arch. Biochem. Biophys. 85:292 (Letters to the Editors).

Smith, S. B., C. A. Gill, D. K. Lunt, and M. A. Brooks. 2009. Regulation of fat and fatty acid composition in beef cattle. Asian Australas. J. Anim. Sci. 22:1225-1233.

Song, S. Y., J. Ghassemi Nejad, S. J. Ohh, B. H. Lee, H. S. Kim, and K. I. Sung. 2013. Effects of chromium-methionine chelate feeding for different duration on growth and carcass characteristics of Holstein steers in the late fattening stage. Ann. Anim. Resour. Sci. 24:38-43.

Stahlhut, H. S., C. S. Whisnant, K. E. Lloyd, E. J. Baird, L. R. Legleiter, S. L. Hansen, and J. W. Spears. 2006. Effect of chromium supplementation and copper status on glucose and lipid metabolism in Angus and Simmental beef cows. Anim. Feed Sci. Technol. 128:263-265.

Sung, K. I., J. Ghassemi Nejad, S. M. Hong, S. J. Ohh, B. H. Lee, J. L. Peng, D. H. Ji, and B. W. Kim. 2015. Effects of forage level and chromium-methionine chelate supplementation on performance, carcass characteristics and blood metabolites in Korean native (Hanwoo) steers. J. Anim. Sci. Technol. 57:14-20.

Swanson, K. C., D. L. Harmon, K. A. Jacques, B. T. Larson, C. J. Richards, D. W. Bohnert, and S. J. Paton. 2000. Efficacy of chromium-yeast supplementation for growing beef steers. Anim. Feed Sci. Technol. 86:95-105.

Wang, M. Q., Z. R. Xu, L. Y. Zha, and M. D. Lindemann. 2007. Effects of chromium nanocomposite supplementation on blood metabolites, endocrine parameters and immune traits in finishing pigs. Anim. Feed Sci. Technol. 139:69-80.

Ward, T. L., L. L. Southern, and R. A. Anderson. 1995. Effect of dietary chromium source on growth, carcass characteristics, and plasma metabolite and hormone concentrations in growing-finishing swine. J. Anim. Sci. 73 (Suppl.):189 (abstract).

Wardeh, M. F. 1981. Models for Estimating Energy and Protein Utilization for Feeds. PhD Thesis, Utah State University, Logan, UT, USA.

Zea, S. J., M. D. D. Diaz, and J. A. C. Santaolalla. 2007. Sex and beef production system on meat and fat quality. Archivos de Zootecnia 56:817-828.

6

Nocturnal Light Pulses Lower Carbon Dioxide Production Rate without Affecting Feed Intake in Geese

De-Jia Huang and Shyi-Kuen Yang*

Department of Animal Science and Biotechnology, Tunghai University, Taichung 40704, Taiwan

ABSTRACT: This study was conducted to investigate the effect of nocturnal light pulses (NLPs) on the feed intake and metabolic rate in geese. Fourteen adult Chinese geese were penned individually, and randomly assigned to either the C (control) or NLP group. The C group was exposed to a 12L:12D photoperiod (12 h light and 12 h darkness per day), whereas the NLP group was exposed to a 12L:12D photoperiod inserted by 15-min lighting at 2-h intervals in the scotophase. The weight of the feed was automatically recorded at 1-min intervals for 1 wk. The fasting carbon dioxide production rate (CO_2 PR) was recorded at 1-min intervals for 1 d. The results revealed that neither the daily feed intake nor the feed intakes during both the daytime and nighttime were affected by photoperiodic regimen, and the feed intake during the daytime did not differ from that during the nighttime. The photoperiodic treatment did not affect the time distribution of feed intake. However, NLPs lowered ($p<0.05$) the mean and minimal CO_2 PR during both the daytime and nighttime. Both the mean and minimal CO_2 PR during the daytime were significantly higher ($p<0.05$) than those during the nighttime. We concluded that NLPs lowered metabolic rate of the geese, but did not affect the feed intake; both the mean and minimal CO_2 PR were higher during the daytime than during the nighttime. (**Key Words:** Feed Intake, Goose, Intermittent Lighting, Metabolic Rate, Photoperiod)

INTRODUCTION

Geese have much lower feed efficiency than broilers, especially in the fattening period (Chen et al., 2003). Thus it is desirable to increase the feed efficiencies of geese during both growing and fattening periods. Within a reasonable range, a high feed intake results in a high weight gain and feed efficiency, because tissue accretion occurs only when the ingested nutrients exceed the requirements for maintenance. Therefore, increasing feed intake is a potential method for increasing both weight gain and feed efficiency in geese. To maximize feed intake and growth rate broiler chickens are usually kept on a continuous or nearly continuous lighting schedule. However, intermittent lighting has been shown to result in some benefits, including increased feed efficiency (Weaver et al., 1982; Ketelaars et al., 1986; Apeldoorn et al., 1999) and increased weight gain (Ohtani and Leeson, 2000). Geese ingest more feed per hour during the daytime than during the nighttime (Chu, 2012). The geese which were subjected to a 1-h light pulse inserted in the scotophase (i.e. skeletal long photoperiod) concentrated their nighttime feed intake at the particular hour (Chu, 2012). In the present study, we aimed to increase feed intake in geese by using several short-time light pulses during the scotophase.

Although birds are quiet during the dark period, we assumed that short light pulses only shortly increase their activity, and do not substantially raise their heat production during the night. Historically, indirect calorimetry has mainly relied upon measurement of oxygen consumption. However, in recent years, increased attention has been placed on measuring carbon dioxide (CO_2) production, because the current CO_2 analyzers have higher resolution than the typical O_2 analyzers for detecting the changes in air composition. In birds, the thermal equivalents of carbohydrates, fats, and proteins for CO_2 production are 5.047, 6.694, and 6.597 kcal/L, respectively (Robbins, 1993). When little is known about the catabolized substrates,

* Corresponding Author: Shyi-Kuen Yang.
E-mail: skyang@thu.edu.tw

large errors can occur when estimating energy expenditure based on CO_2 production; however, errors are <1.5% when the catabolized substrates are limited to fats and proteins (i.e. in the fasting status). Therefore, the carbon dioxide production rate (CO_2 PR) theoretically is an acceptable index of energy expenditure rate for birds in fasting status. In this study, we also determined the fasting CO_2 PR to evaluate the effect of the photoperiod regimen on the energy expenditure rate. Briefly, the purpose of this study was to determine the effect of nocturnal light pulses on the feed intake and carbon dioxide production rate in geese.

MATERIALS AND METHODS

Fourteen female adult Chinese geese (a breed of domesticated geese) before laying period were used in this study. The ages of the geese were >2 yr, and the body weights of geese ranged from 4.0 to 4.5 kg. The experimental protocols used in this study were approved by the Experimental Animal Care and Use Committee of Tunghai University. Before the experiment, the geese were raised in a 230-m^2 enclosed paddock with other geese, mallards, Muscovy ducks, and peacocks. The paddock contained a 50-m^2 shelter and a 40-m^2 pond. The birds were exposed to the natural photoperiod, and were fed a commercial feed (fattening diet for duck; Fuso, Taichung, Taiwan) twice per day. Just before the beginning of the experiment, the natural day length was about 12 h (from civil dawn to civil dusk) and was increasing. The geese used in this study were transferred to a light-tight barn, and were randomly assigned to either the control (C) or the nocturnal light pulse (NLP) group. Each group was kept at a separated room, which was divided into individual pens (1.5×1.0 m). During the experimental period, all geese were penned individually, and allowed free access to feed and water. The control group was exposed to a 12L:12D photoperiod (12 h light and 12 h darkness per day; light on at 06:15 h and off at 18:15 h). The NLP group was exposed to a 12L:12D photoperiod inserted by 15-min lighting at 2-h intervals in the scotophase. The light was offered with fluorescent tubes. The light intensity on floor during the lighting period ranged from 60 to120 lx. A small red bulb, which generated a light intensity <1.0 lx at night, was lighted all day for each group.

This study included a 2-wk adaptation period, a 1-wk feed intake recording period, and a 1-d CO_2 PR recording period. During the experimental period, geese did not enter their laying period. After the 2-wk adaptation period, the geese were weighed. The mean body weights for C and NLP groups were 4.28±0.21 (mean±standard deviation) and 4.30±0.14 kg, respectively. When the feed intake was recorded, the weight of feed was automatically recorded at 1-min intervals for 1 wk. The feed was placed on a digital platform balance (ML4001, Mettler Toledo, Columbus, OH, USA) connected to a PC for recording the weight data. Both readability and repeatability of the balance were 0.1 g. After the 1-wk feed intake recording, all geese were kept in the experimental room, and were sequentially (2 geese once) transferred into metabolic chambers within 10 d. Before transferring to the metabolic chamber the geese were weighed after a 24-h fasting. The mean body weights of geese in C and NLP groups were 4.11±0.14 and 4.20±0.16 kg, respectively. The CO_2 PR values were recorded at 1-min intervals for 1 d. The room of metabolic chamber was light-tight, and the lighting schedule was adjusted to the experimental photoperiod for each group. Within the metabolic chamber, water was supplied in a sink, but feed was deprived.

The air-tight metabolic chamber was made of clear acrylic plastic (72×72×82 cm), and had a removable lid to allow the addition and removal of a stainless cage. Ambient air was supplied from the outdoors by an exhaust fan through an adjustable valve. The air in the chamber was mixed well by 2 small fans. The CO_2 concentration, temperature, and linear velocity of air flow were detected at the outlet of the chamber. The CO_2 was detected using a non-dispersive infrared dual wavelength type CO_2 sensor (KCD-HP100x, Korea Digital Co., Seoul, Korea), with an accuracy of ±3% full scale +2% reading. The temperature was detected using a thermocouple sensor (Type K, Jetec, Taichung, Taiwan), with a detection error of ±0.2°C at room temperature. The linear velocity of the air flow was detected using a transmitter based on a hot-film anemometer (EE576, E+E Elektronik, Engerwitzdorf, Austria), with an accuracy of ±0.05 m/s +2% reading. The CO_2 concentration, temperature, and air flow velocity data were transmitted to and recorded on a multiple-channel recorder (TRM2006A000T, Toho Electronics Inc., Kanagawa, Japan) at 1-min intervals. The goose was placed into a stainless cage (50×54×70 cm) fitted with a sink to supply water. During the experimental period, the air flow was maintained at approximately 60 L/min. The room temperature were maintained at 23°C to 25°C, and the actual temperature of the chamber outlet was maintained at 22°C to 25°C. An empty identical metabolic chamber without goose was used as a blank for calculating the CO_2 PR.

The data of individual goose was considered as an experimental unit for statistical analysis. The CO_2 PR data of one goose in NLP group were discarded due to the fault of facility. The CO_2 PR was expressed on the basis of metabolic size (kg$^{0.75}$). The feed intake of each goose used for drawing and statistical analysis was the mean calculated from the 7-d records. The mean and the minimal CO_2 PR of each goose during daytime or nighttime for statistical analysis were the mean and the minimal values during the time phase, respectively. Data were analyzed by repeated measures analysis of variance (ANOVA) using the general linear model (PROC GLM) of SAS statistical software (SAS Inst. Inc., Cary, NC, USA). The statistical model was:

$$y_{ij} = \mu + Trt_i + Tm_j + Trt_i \times Tm_j + e_{ij}$$

where y_{ij} is the observation, μ is the overall population mean, Trt_i is the fixed effect of treatment (i = 1, 2), Tm_j is the fixed effect of measuring time (j = 1, 2), $Trt_i \times Tm_j$ is the interaction between treatment and time, and e_{ij} is the residual term associated with the measure.

The statistical significances of differences between means were determined by the least squares means procedure with the significance set at $p<0.05$. Because the carbon dioxide production rates rhythmically fluctuated with an approximately 3.5-h period, the curves were also smoothed by the simple moving averages to find the diurnal trends. In smoothing, the value for a given time point was the mean which was calculated by using Microsoft Excel from the 105-min consecutive records both before and after the given time point. For example, the value at 10:00 was the mean of the records from 08:15 to 11:45.

RESULTS

The feeding patterns varied among geese. Some geese frequently nibbled, whereas others fed infrequently but ingested a great amount of feed per bout (data not shown). The ingestion bouts were neither synchronized among geese, nor synchronized among dates for a given goose (data not shown). The mean accumulative feed intakes during a day in both groups are shown in Figure 1a. The accumulative feed intake curves of both groups almost overlap, and resemble a straight line. The feeding rate (g/min) in both groups fluctuated throughout a day (Figure 1b and 1c). The

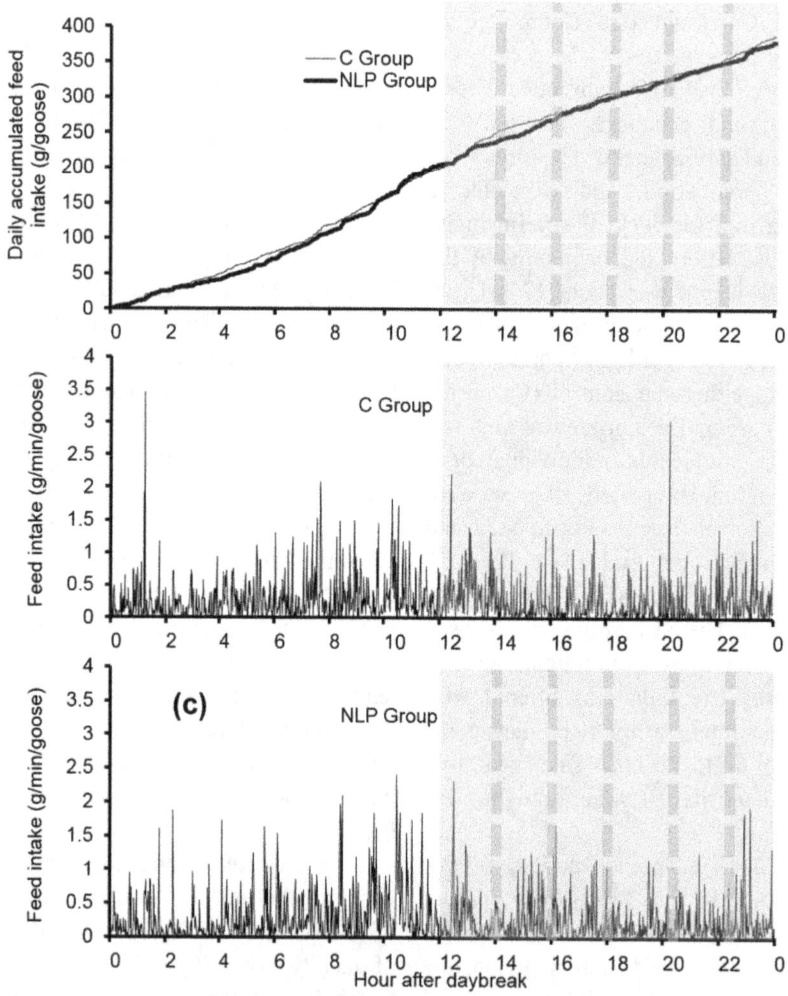

Figure 1. The effects of nocturnal light pulses (NLP) on the daily accumulated feed intake (a) and feeding rate (b and c) during a day in Chinese geese. C group was exposed to a 12L:12D photoperiod (12 h light and 12 h darkness per day); NLP group was exposed to a 12L:12D photoperiod inserted by 15-min lighting at 2-h intervals in scotophase. The shaded areas indicate the scotophase, and the vertical broken lines indicate the 15-min nocturnal light pulses in NLP group. Both the daily accumulated feed intake and feeding rate were the means of 7-day records. n = 7 for both groups.

Table 1. The effects of nocturnal light pulses[1] and time phase on the feed intake and carbon dioxide (CO_2) production rate in geese

Trait[2]	C group		NLP group		RMSE	p-value		
	Day	Night	Day	Night		Trt	Tm	Trt×Tm
Feed intake[2,3] (g/phase/goose)	194.3	185.5	206.4	176.9	40.9	0.911	0.239	0.519
Mean CO_2 production rate (mL/min/kg$^{0.75}$)[3]	19.2^a	16.1^b	16.3^b	12.8^c	1.9	0.002	0.001	0.825
Minimal CO_2 production rate (mL/min/kg$^{0.75}$)[3]	13.9^a	12.8^{ab}	11.6^{bc}	10.1^c	1.3	<0.001	0.022	0.716

NLP, nocturnal light pulse; RMSE, root mean square error; Trt, photoperiodic treatment; Tm, time phase (day vs night).

[1] C group was exposed to a 12L:12D photoperiod (12 h light and 12 h darkness per day); NLP group was exposed to a 12L:12D photoperiod inserted by 15-min lighting at 2-h intervals in scotophase. This study included a 2-wk adaptation period, a 1-wk feed intake recording, and a 1-d CO_2 production rate recording.

[2] The feed intake for each goose was the average of 7-day records, and was separated into two phase (day and night).

[3] n = 7 for each mean except for the mean and minimal CO_2 production rates of NLP group, where n = 6.

[a,b,c] Means in a same row without a common superscript letters differ significantly ($p<0.05$).

feed intake did not exhibit obvious relationship with nocturnal light pulses. Summarized data showed that feed intake did not differ between photoperiod regimens or between day and night, and there was no significant interaction between treatment and time phase (Table 1).

The CO_2 PR of each goose exhibited several fluctuations per day (data not shown). The curve of group mean of CO_2 PR for C group exhibited ultradian rhythmicity which period was about 3.5 h (Figure 2). The amplitudes of fluctuations were high in the daytime, and dampened in the nighttime. When the curve was smoothed by the moving averages, it exhibited an obvious circadian rhythm with a nadir at the middle night and acrophase at the middle day. The curve for NLP group almost parallels and is always below that for C group (Figure 2). The smoothed curves for C and NLP group clearly exhibit the parallelism and the same phase shift. In NLP group, the CO_2 PR did not increase during the nocturnal intermittent lighting periods. Summarized data showed that both the mean and minimal CO_2 PRs during the daytime were significantly higher ($p< 0.05$) than those during the nighttime (Table 1). Both the mean and minimal CO_2 PR were significantly lowered ($p<0.05$) by NLPs during both the daytime and the nighttime, compared with C group (Table 1).

DISCUSSION

In the present study, the feeding behavior of geese in both groups was evenly distributed throughout a day, and the feed intake during the day was not different from that during the night. The result did not prove the circadian rhythm in ingestion. Originally we hypothesized that NLPs

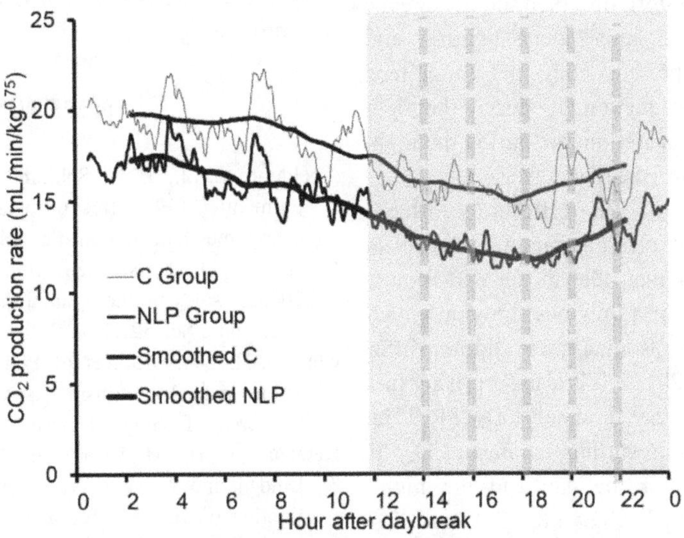

Figure 2. The effect of nocturnal light pulses (NLP) on the carbon dioxide production rate during a day in Chinese geese. C group was exposed to a 12L:12D photoperiod (12 h light and 12 h darkness per day); NLP group was exposed to a 12L:12D photoperiod inserted by 15-min lighting at 2-h intervals in scotophase. The shaded areas indicate the scotophase, and the vertical broken lines indicate the 15-min nocturnal light pulses in NLP group. The smoothed curves are drawn on the basis of the simple moving averages, which are the means of 105-min consecutive records before and after the given time points. The numbers of observations for the C and NLP groups are 7 and 6, respectively.

stimulate geese to feed during the lighting periods. However, the results in the present study did not agree with this hypothesis. Feeding was neither entrained nor stimulated by the NLPs. In a previous study, the White Roman geese subjected to a skeletal photoperiod, in which 1-h lighting was inserted at the 8th hour of 16-h darkness, had a feed intake per hour during the continuous lighting hours similar to that of the short day controls (exposed to 8L:16D); however, their feed intake during the inserted lighting hour was substantially higher than those during the dark hours and the continuous lighting hours (Chu, 2012). The inconsistency between studies may be attributable to the differences in the length of the continuous lighting and the time, number, and length of light pulses.

Unexpectedly, the mean CO_2 PR did not increase during the intermittent lighting periods. In addition, the NLPs lowered the mean CO_2 PR. These results imply that the NLPs do not disturb the resting of geese, at least after a 2-wk adaptation period. The reasons for the decrease in heat production caused by NLPs are not clear. However, a similar phenomenon was found in voles. The mean energy expenditure levels of the voles experienced NLPs were lower than those of the voles exposed to a short day photoperiod (Zubidat et al., 2007). In addition, although the heat production during the light phase is higher than that during the dark phase in chickens (Apeldoorn et al., 1999; Ohtani and Leeson, 2000), the differences in total daily heat production between groups subjected to intermittent lighting and continuous lighting varied among studies. Ohtani and Leeson (2000) reported that intermittent lighting increased heat production, in contrast with Ketelaars et al. (1986); Apeldoorn et al. (1999) observed no difference between groups subjected to intermittent lighting and continuous lighting. In the present study, the equal feed intake and the reduced heat production rate during the night suggested that the net energy accretion rate during the night might be higher than that during the day in geese.

The fasting minimal CO_2 PR reflects the basal metabolic rate. In this study, the fasting minimal CO_2 PR in the lighting hours was faster than that in the dark hours. This result differed from that of a previous study, in which the minimal CO_2 PR of White Roman geese did not differ between day and night (Chu, 2012). It is also surprising that NLPs significantly lowered the minimal CO_2 PR. The differences between treatments regarding the mean CO_2 PR (2.9 and 3.3 mL/min/kg$^{0.75}$ for the day and the night, respectively) and the minimal CO_2 PR (2.3 and 2.7 mL/min/kg$^{0.75}$ for the day and the night, respectively) suggested that the decrease in heat production caused by the NLPs mainly resulted from a lowered basal metabolism. The differences between day and night regarding the mean CO_2 PR (3.1 and 3.5 mL/min/kg$^{0.75}$ for the control and the treated groups, respectively) and the minimal CO_2 PR (1.1 and 1.5 mL/min/kg$^{0.75}$ for the control and the treated groups, respectively) suggested that the lowered basal metabolism contributed to only small portion of the difference in the mean CO_2 PR between day and night.

In conclusion, under a 12L:12D photoperiod, the feed intakes did not differ between day and night, but both the mean and minimal carbon dioxide production rates during the daytime were higher than those during the nighttime. Under a 12L:12D photoperiod, nocturnal light pulses affect neither the amount of feed ingested nor the time distribution of feed intake, but lowered both the mean and minimal fasting carbon dioxide production rates.

IMPLICATIONS

Ingestion behavior and activity in broilers are controlled by lighting; they ingest and are active in the light, and rest in the dark. In this study, we found that geese ingest equal amount of feed during the day and night, but their heat production rates in the daytime are higher than those in the nighttime. The nocturnal light pulses lower the fasting metabolic rate of geese during both day and night without affecting feed intake. The lowered fasting metabolic rate implies that the nocturnal light pulses may lower the energy requirement for maintenance and improve the efficiency of energy accumulation.

ACKNOWLEDGMENTS

We thank Miss Pei-Wen Chen for her assistance in the data processing.

REFERENCES

Apeldoorn, E. J., J. W. Schrama, M. M. Mashaly, and H. K. Parmentier. 1999. Effect of melatonin and lighting schedule on energy metabolism in broiler chickens. Poult. Sci. 78:223-229.

Chen, Y. H., J. C. Hsu, B. L. Shih, D. C. Lio, and M. T. Chen. 2003. A study on the optimal marketing age of geese. J. Chin. Soc. Anim. Sci. 32:111-121.

Chu, H.-H. 2012. Effects of Photoperiod on Feeding, Growth Traits and Metabolism in Geese. MS Thesis, Tunghai University, Taichung, Taiwan.

Ketelaars, E. H., M. Verbrugge, W. van der Hel, J. M. van de Linden, and W. M. A. Verstegen. 1986. Effect of intermittent lighting on performance and energy metabolism of broilers. Poult. Sci. 65:2208-2213.

Ohtani, S. and S. Leeson. 2000. The effect of intermittent lighting on metabolizable energy intake and heat production of male broilers. Poult. Sci. 79:167-171.

Robbins, C. T. 1993. Wildlife Feeding and Nutrition. Academic Press, San Diego, CA, USA.

7

Effects of Rice Straw Supplemented with Urea and Molasses on Intermediary Metabolism of Plasma Glucose and Leucine in Sheep

Mohammad Khairul Alam, Yasumichi Ogata, Yukari Sato, and Hiroaki Sano*
Department of Animal Sciences, Faculty of Agriculture, Iwate University, Morioka 020-8550, Japan

ABSTRACT: An isotope dilution method using [U-^{13}C]glucose and [1-^{13}C]leucine (Leu) was conducted to evaluate the effects of rice straw supplemented with urea and molasses (RSUM-diet) on plasma glucose and Leu turnover rates in sheep. Nitrogen (N) balance, rumen fermentation characteristics and blood metabolite concentrations were also determined. Four sheep were fed either mixed hay (MH-diet), or a RSUM-diet with a crossover design for two 21 days period. Feed allowance was computed on the basis of metabolizable energy at maintenance level. The isotope dilution method was performed as the primed-continuous infusion on day 21 of each dietary period. Nitrogen intake was lower (p = 0.01) for the RSUM-diet and N digestibility did not differ (p = 0.57) between diets. Concentrations of rumen total volatile fatty acids tended to be higher (p = 0.09) for the RSUM-diet than the MH-diet. Acetate concentration in the rumen did not differ (p = 0.38) between diets, whereas propionate concentration was higher (p = 0.01) for the RSUM-diet compared to the MH-diet. Turnover rates as well as concentrations of plasma glucose and Leu did not differ between diets. It can be concluded that kinetics of plasma glucose and Leu metabolism were comparable between the RSUM-diet and the MH-diet, and rumen fermentation characteristics were improved in sheep fed the RSUM-diet compared to the MH-diet. (**Key Words:** Intermediary Metabolism, Molasses, Rice Straw, Sheep, Stable Isotope, Urea)

INTRODUCTION

The economic point of view and sustainable environmental concerns encourage investigating the possibility of using the crop residue as animal feed. Rice straw is abundantly available crop residues in most tropical and sub-tropical countries and commonly used as a diet for ruminants, although it is low crude protein (CP) content and fermented poorly in the rumen (Alam et. al., 2010; Sarnklong et al., 2010). Supplementation of nitrogenous substrates to rice straw is reported to recover its CP deficiency and improve its digestibility through providing necessary ammonia (NH_3) for rumen microbial activities which is essential for better rumen fermentation characteristics (Wanapat et al., 2009). It was also reported that supplementation of nitrogen (N) sources in combination with energy substrates to straw diets improved feed intake, digestive function along with ruminal characteristics in ruminants through influencing rumen microbial growth (Rooke and Armstrong, 1989; Can et al., 2004; Wu et al., 2005). In most of the tropical countries it is a common practice to use urea and molasses as the sources of N and soluble carbohydrate, which provide required NH_3 and energy substrates for rumen microbial activities (Toppo et al., 1997; Tedeschi et al., 2002; Zinn et al., 2003). However, the effect of rice straw supplemented with urea and molasses (RSUM-diet) on intermediary metabolism of plasma nutrient kinetics in ruminant is scanty.

Glucose is an important energy source for the brain and body tissues and it is particularly important for growth and lactation. Leucine (Leu) is an essential amino acid and requirements of it in ruminants are met from microbes grown in the rumen. It could be expected that the RSUM-diet might influence plasma glucose and Leu metabolism in sheep through providing required NH_3 and energy for rumen microbial activities to increase the dietary carbohydrate fermentation. Therefore, the present study was designed to evaluate the effect of the RSUM-diet on turnover rates (TR)

* Corresponding Author: Hiroaki Sano.
E-mail: sano@iwate-u.ac.jp

of plasma glucose and Leu using the isotope dilution methods along with the determination of N balances rumen characteristics and blood metabolite concentrations in sheep.

MATERIALS AND METHODS

Animals, diets and management

Experimental procedures including animal cares, cannulation and blood sampling were reviewed and approved by the Animal Care Committee of Iwate University. Four sound healthy crossbred (Corriedale× Suffolk) shorn sheep (*Ovis aries*) weighing 46.6±2.2 kg of body weight (BW) were used. Two dietary treatments were tested; one was mixed hay (MH-diet) of orchardgrass (*Dactylis glomerata*) and reed canarygrass (*Phalaris arundinacea*), other was rice straw (*Oryza sativa*) supplemented with urea and dried molasses. Feed allowance was computed on the basis of maintenance level metabolizable energy (ME). Chemical composition of experimental feed is shown in Table 1. The ME was assumed 1.73 kcal/g for mixed hay (NRC, 1985); 1.30 kcal/g for rice straw and 2.62 kcal/g for molasses (NARO, 2006). Feed allowance was mixed hay 57.8 g/kg$^{0.75}$/d for the MH-diet and rice straw 59.7 g/kg$^{0.75}$/d supplemented with urea 0.84 g/kg$^{0.75}$/d and molasses 7.6 g/kg$^{0.75}$/d for the RSUM-diet. Crude protein supply was given for both diets on dry matter (DM) basis. Urea and molasses were mixed and given on chopped (3 to 4 cm) rice straw immediately before feeding. Feed was given twice a day at 08:00 h and 20:00 h and fresh drinking water was available *ad libitum*. The experiment was performed using crossover design with two 21 days period. Two sheep were fed the MH-diet during the first period and then the RSUM-diet during the second period, and the other two sheep were fed in the reverse order. The sheep were housed in individual pens in an animal barn during the adjustment period (first two weeks), and on day 15, the sheep were moved to a controlled house at an air temperature of 23°C±1°C with lighting from 8:00 h to 22:00 h and maintained in wooden metabolism stalls designed for total collection of feces and urine. The sheep were weighed on day of starting the experiment and every 7 days intervals of each dietary period. All experimental procedures were carried out without noticeable stress to the animals.

Table 1. Chemical composition of experimental feed and supplements on air dry matter (ADM) basis

Item (%)	Mixed hay	Rice straw	Molasses	Urea
Dry matter	93.3	94.7	84.6	ND
Crude protein	12.0	4.6	12.4	288
Crude ash	10.8	14.4	12.5	ND
Crude fiber	28.6	31.9	8.6	ND
NDF	68.8	73.1	30.0	ND
ADF	32.7	41.7	20.6	ND
ADL	2.0	2.4	5.4	ND

ND, not determined; NDF, neutral detergent fiber; ADF, acid detergent fiber; ADL, acid detergent lignin.

Nitrogen balance

Nitrogen balance trial was conducted for 5 days (from day 16 to day 20) of each dietary period as described previously (Alam et al., 2010). Total fecal output was collected daily from each sheep before night feeding. Feces were dried at 60°C in a forced air oven for 48 h and placed at room temperature for 5 days. Then the air dried samples were weighed for measuring the moisture contents and sub-samples were ground to pass through a 1 mm screen, kept them into plastic container and stored at room temperature until analysis. Urine was collected from each sheep every 24 h in a plastic bucket containing 50 mL of 6 N H_2SO_4 solution to prevent the escape of NH_3. Total volume of daily urine output was recorded by measuring cylinder, then the urine was shaken properly and sub-samples (50 mL) were stored at –30°C until analysis.

Collection of rumen fluid

Rumen fluid was collected from each sheep at 2 h after feeding with orally inserted stomach tube on day 20 of each dietary period. The pH value was measured by a pH-meter (HM-10P, Toa Electronics Ltd., Tokyo, Japan) immediately after collection of rumen fluid. A sub-sample was centrifuged at 8,000×g for 10 min at 2°C (RS-18 IV, Tomy, Tokyo, Japan) and then an aliquot of 1 mL supernatant was acidified by 1 mL of 0.1 N HCl for measuring the rumen NH_3 concentration. Finally the prepared samples and residual of rumen fluid were kept frozen at –30°C for later analysis.

Isotope dilution method

Isotope dilution methods using [U-^{13}C]glucose and [1-^{13}C]Leu were conducted to determine the TR of plasma glucose and Leu on day 21 of each dietary period. Two catheters, one for isotope infusion and another for blood sampling were inserted into the left and right jugular veins on the morning of each isotope dilution method. The catheters were filled with sterile solution of tri-sodium citrate (0.13 mol Nacl/L). At 12:00 h, 3.2 μmol/kg$^{0.75}$ of [U-^{13}C]glucose (D-glucose-$^{13}C_6$, 99 atom% excess ^{13}C; Cambridge Isotope Laboratories, Tewksbury, MA, USA) and 7.2 μmol/kg$^{0.75}$ of [1-^{13}C]Leu (L-leucine-1-^{13}C, 99 atom% excess, ^{13}C; Cambridge Isotope Laboratories, USA) dissolved in saline solution (9 g/L) were injected as priming dose injection through the jugular infusion catheter. Immediately after the priming dose injection, [U-^{13}C]glucose and [1-^{13}C]Leu were continuously infused at rates of 3.2 and 7.2 μmol/kg$^{0.75}$/h, respectively, for 4 h through a multichannel peristaltic pump (AC-2120, Atto Co.

Ltd., Tokyo, Japan). Blood samples were collected through the sampling catheter immediately before the priming dose injection (10 mL) and every 30 min intervals (5 mL) over the last 2 h of the primed-continuous infusion of [U-^{13}C]glucose and [1-^{13}C]Leu. The collected blood samples were transferred to the heparinized tubes and stored in crushed ice until centrifugation. Blood samples were centrifuged at 10,000×g for 10 min at 2°C and the plasma samples were then stored at –30°C for further analysis.

Chemical analysis

Dry matter, CP and crude ash contents of the experimental diets were measured according to AOAC (1995). Nitrogen contents in diets, feces, urine and feed refusals were analyzed by Kjeldahl method with the Foss Keltech System (Tecator Digester System and Kjeltec 2300, Foss Tecator, Hoganas, Sweden). Crude fiber, neutral detergent fiber, acid detergent fiber, and acid detergent lignin in diets were determined according to van Soest et al. (1991) using Foss Analytical FiberCap System (Foss Tecator, Sweden). Concentrations of rumen total volatile fatty acid (VFA) were determined by titrating the steam distillate of rumen fluid with 0.1 N NaOH. The titrated distillate was dried and then individual VFA concentrations were determined using gas chromatography (5890A, Hewlett Packard, Avondale, PA, USA). Concentration of rumen NH_3 was determined by colorimetric method (Weatherburn, 1967).

In pre-infusion period of isotope dilution method, plasma free amino acids, NH_3 and urea were determined using an automatic amino acid analyzer (JLC-500/V, JEOL, Akishima, Japan). Plasma concentrations of non-esterified fatty acid (NEFA) were determined enzymatically using a diagnostic kit (NEFA C, Wako Pure Chemicals, Osaka, Japan).

To determine the concentrations and enrichments of plasma glucose and Leu, plasma [U-^{13}C]glucose and isotope enrichments were measured by the procedure of Tserng and Kalhan (1983) with slight modifications as described previously by Sano et al. (1996). The enrichment of plasma [U-^{13}C]glucose was determined using the selected ion monitoring with gas chromatography mass spectrometry system (GC/MS) (QP-2010, Shimadzu, Kyoto, Japan). Concentrations of plasma glucose were determined enzymatically using the method described by Huggett and Nixon (1957). Plasma amino acids were separated and converted to N-methyl-N-t-butyl-dimethylsilyltrifluroacetamide (MTBSTFA; Funakoshi, PCC48920, Tokyo, Japan) derivatives according to the procedures of Calder and Smith (1988) as described previously (Sano et al., 2004). Isotopic enrichments of plasma [1-^{13}C]Leu and concentration of plasma Leu were measured by the selected ion monitoring using the GC/MS.

Calculation

Results were presented as mean values with standard error of the mean. For the isotope dilution methods, the TR of plasma glucose and Leu was calculated using the equation described by Tserng and Kalhan (1983) as follows:

$$TR = I\times(1/E-1)$$

Where, I is the infusion rate of [U-^{13}C]glucose and [1-^{13}C]Leu isotopes and E is the plasma isotopic enrichments of [U-^{13}C]glucose and [1-^{13}C]Leu during the steady state, respectively.

Statistical analysis

All data were statistically analyzed using analysis of variance with the MIXED procedure of SAS (1996). The least square means statement was used to test the effects of period and diet. Results were considered significant at the $p<0.05$ level, and a tendency was defined as $0.05\leq p<0.10$. The repeated measures statement and the Tukey adjustment were used for the time course of changes and the significance level was $p<0.05$.

RESULTS AND DISCUSSION

Daily profile and nitrogen balance

The sheep consumed the RSUM-diet more slowly than the MH-diet. The daily BW gain did not differ ($p = 0.26$) between the diets (Table 2). It was meant that no adverse effect was found on BW gain in sheep throughout the experiment for RSUM-diet. A similar trend was found in lambs fed urea treated rice straw supplemented with molasses (Hue et al., 2008). Dry matter intake was greater ($p = 0.02$) for the RSUM-diet compared to the MH-diet due to difference in feed allowance. The present result was supported by the findings of Singh et al. (1995). Estimated

Table 2. Dietary effects on body weight gain, dry matter intake, estimated metabolizable energy intake, nitrogen intake and nitrogen digestibility in sheep

Item	MH-diet	RSUM-diet	SEM	p-value
No. of sheep	4	4		
BW gain (kg/d)	0.09	0.03	0.05	0.26
DM intake (g/kg$^{0.75}$/d)	54	59	2	0.02
ME intake (kcal/kg$^{0.75}$/d)	99	92	2	0.21
N intake (g/kg$^{0.75}$/d)	1.10	0.94	0.05	0.01
N in feces (g/kg$^{0.75}$/d)	0.37	0.30	0.02	0.01
N in urine (g/kg$^{0.75}$/d)	0.45	0.46	0.01	0.90
N retention (g/kg$^{0.75}$/d)	0.28	0.18	0.04	0.03
N digestibility (%)	67	67	1	0.57

MH, Mixed hay of orchardgrass and reed canarygrass; RSUM, Rice straw supplemented with urea and molasses; BW, body weight; DM, dry matter; ME, metabolizable energy; N, nitrogen.
Estimated according to NRC (1985) and NARO (2006).

ME intake did not differ ($p = 0.21$) between diets. Nitrogen intake and N excretion through feces were lower ($p = 0.01$) for the RSUM-diet than the MH-diet. Lower N intake for the RSUM-diet might be due to loss of some N through residue of rice straw. Nitrogen excretion through urine did not differ between diets and N retention was lower ($p = 0.03$) for the RSUM-diet than the MH-diet. No significant difference ($p = 0.57$) occurred in N digestibility between diets in the present study. Can et al. (2004) reported the lower N digestibility in lambs fed only wheat straw than wheat straw supplemented with urea and molasses. Similar results were obtained in our previous study (Alam et al., 2010). It can be said that N intake as well as N digestibility were improved in sheep fed RSUM-diet than rice straw only. This is probably be due to addition of urea and molasses to rice straw which supplies required NH_3 and energy for microbial activities in the rumen of sheep.

Rumen fermentation characteristics

Rumen pH determined at 2 h after feeding did not differ ($p = 0.81$) between dietary treatments (Table 3). The pH values were within the normal range for both the diets. Similar rumen pH between diets was an indication of balance between the concentrations of VFA and NH_3 in the rumen as described previously (Alam et al., 2010). The numerical values of rumen pH of the present findings were comparable with the data reported in sheep fed urea supplemented diet (Sano et al., 2009). Leng (1990) reported that the critical level of NH_3 is between 2.9 and 14.7 mmol/L of rumen liquor for promoting the rumen fermentation. In the present study rumen NH_3 concentration was within the normal range for promoting the rumen fermentation. Concentration of rumen NH_3 was higher ($p = 0.03$) for the RSUM-diet than the MH-diet in the present study. The higher rumen NH_3 concentration for the RSUM-diet was likely due to the presence of urea and molasses, because the urea is rapidly hydrolyzed and provides NH_3 and molasses provides the required energy substrate for microbial activities in the rumen. This is in accordance with the results of Srinivas and Gupta (1997) and Jain et al. (2005), who reported that in ruminants, supplementation of urea and molasses to low quality roughage diets made better rumen environment for dietary carbohydrate fermentation through supplying adequate NH_3 and energy for rumen microbial growth. Concentrations of rumen total VFA tended to be higher ($p = 0.09$) for the RSUM-diet than the MH-diet. Acetate concentration in the rumen did not differ ($p = 0.38$) between diets, whereas propionate concentration was higher ($p = 0.01$) for the RSUM-diet than the MH-diet. A tendency of higher rumen total VFA for the RSUM-diet has indicated well fermentation of dietary carbohydrate in the rumen. Similarly Jain et al. (2005) observed that rumen VFA concentrations were affected by urea, molasses and mineral granules supplementation with rice straw in goat kids. Propionate concentration in the rumen was affected by the readily fermentable carbohydrate in the diets (van Houtert, 1993). Higher concentration of ruminal propionate for the RSUM-diet was due to presence of molasses as a source of water soluble carbohydrate. Supplementation of molasses to rice straw might activate the microbes which produce propionate in the rumen. The present results were supported by Broderick and Radloff (2004), who mentioned that molasses supplementation to diets influenced the propionate concentration in the rumen.

Table 3. Dietary effects on rumen pH, concentrations of rumen ammonia (NH_3) and volatile fatty acids (VFA) in sheep

Item	MH-diet	RSUM-diet	SEM	p-value
No. of sheep	4	4		
Rumen pH	6.82	6.86	0.10	0.81
NH_3 (mmol/L)	4.56	7.06	1.06	0.03
VFA (mmol/L)				
Total	79.5	88.4	6.4	0.09
Acetic acid	59.5	61.3	4.4	0.38
Propionic acid	13.7	20.9	2.6	0.01
iso-Butyric acid	0.7	0.3	0.1	0.01
Butyric acid	4.6	5.3	0.5	0.17
iso-Valeric acid	0.7	0.3	0.2	0.05
Valeric acid	0.5	0.3	0.1	0.34

MH, mixed hay of orchardgrass and reed canarygrass; RSUM, rice straw supplemented with urea and molasses; SEM, standard error of the mean.

Blood metabolites

Plasma free amino acid concentration is influenced by the several factors such as dietary types and frequency of feeding, protein degradation, microbial protein synthesis and amino acid absorption (Leng and Nolan, 1984; Alam et al., 2013). Plasma free amino acids determined at pre-infusion of isotope dilution did not differ ($p>0.10$) between diets, except that lysine, glutamic acid and glutamine were higher ($p<0.05$) for the RSUM-diet compared to the MH-diet (Table 4). The observation might be due to adequate supply of NH_3 and easily fermentable energy substrates for microbial protein synthesis for the RSUM-diet. Concentrations of plasma NH_3 tended to be higher ($p = 0.07$) for the RSUM-diet compared to the MH-diet, and plasma urea concentration did not differ ($p = 0.18$) between diets. Concentration of NH_3 in plasma is positively associated with the production of NH_3 in the rumen (Nolan and Leng, 1972; Milano and Lobley, 2001). A tendency of higher plasma NH_3 concentration for the RSUM-diet might be reflected by rapid absorption of NH_3 from the rumen. The present result is in accordance with the results by Sano et al. (2009), who suggested that when urea was supplemented to the basal diet, the postprandial plasma NH_3 increased temporally because a large part of the NH_3 produced from the supplemental urea in the rumen and

Table 4. Dietary effects on plasma free amino acids, ammonia (NH_3), urea and non-esterified fatty acid (NEFA) concentrations in sheep

Items	MH-diet	RSUM-diet	SEM	p-value
No. of sheep	4	4		
Amino acids (μmol/L)				
Threonine	194	163	41	0.07
Valine	228	185	34	0.13
Methionine	37	29	14	0.49
Iso-leucine	92	74	16	0.26
Leucine	100	85	20	0.38
Phenylalanine	45	41	6	0.48
Histidine	24	23	2	0.43
Lysine	32	43	5	0.04
Aspartic acid	14	12	2	0.11
Serine	205	176	14	0.33
Asparagine	66	70	11	0.93
Glutamic acid	253	332	24	0.02
Glutamine	84	121	21	0.01
Glycine	610	583	44	0.11
Alanine	183	197	18	0.09
Tyrosine	58	63	11	0.39
Tryptophan	149	136	25	0.41
Arginine	131	101	15	0.06
Proline	67	68	12	0.78
NH_3 (μmol/L)	383	419	16	0.07
Urea (mmol/L)	7.83	8.37	0.89	0.18
NEFA (μEq/L)	284	148	62	0.03

MH, mixed hay of orchardgrass and reed canarygrass; RSUM, rice straw supplemented with urea and molasses; SEM, standard error of the mean.

directly absorbed into portal blood. Concentration of plasma NEFA was lower (p = 0.03) for the RSUM-diet than the MH-diet. Plasma NEFA concentration is the indicator of energy status in ruminants (Fox et al., 1991), because plasma NEFA is mobilized to supply the metabolic needs of animal, primarily the need of energy. In the current study, lower plasma NEFA concentration for the RSUM-diet indicated its improved nutritional status due to nitrogenous substrate and soluble carbohydrate supplementation.

Plasma glucose and leucine kinetics

Plasma glucose concentration and [U-^{13}C]glucose enrichment remained constant during the latter half of the isotope infusion (Figure 1), which indicated the steady state condition. Concentration of plasma glucose determined during the last 2 h continuous infusion of isotope dilution did not differ (p = 0.51) between the RSUM-diet and the MH-diet (Table 5). The numerical values of plasma glucose concentration were similar with the data previously reported in sheep fed rice straw supplemented with corn starch as energy source (Zhang et al., 2009). In ruminants gluconeogenesis takes place mainly in liver and rates of plasma glucose turnover (TR) were influenced with several

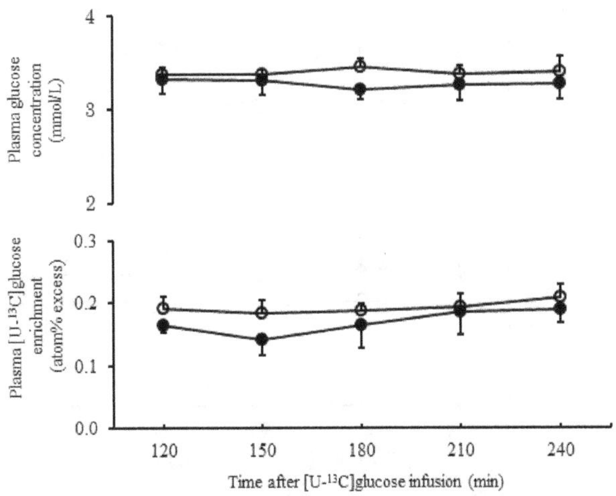

Figure 1. Time course changes in plasma glucose concentration and enrichment of [U-^{13}C]glucose during 120 to 240 min of the primed-continuous infusion of [U-^{13}C]glucose in sheep (n = 4) fed the RSUM-diet (●) and the MH-diet (○) (Means±standard error of the mean).

factors such as the type of diet, energy intake and supply of gluconeogenic substrate to the liver (Ortigues-Marty et al., 2003; Sano and Fujita, 2006). In previous studies it was suggested that the precursor availability is an important factor in regulating gluconeogenesis (Schmidt and Keith, 1983; Oba and Allen, 2003). Plasma glucose TR in the present study did not differ (p = 0.31) between the RSUM-diet and the MH-diet, although the rumen propionate concentration, a major glucose precursor, was higher for the RSUM-diet. This is in accordance with Seal and Parker (1994), who reported that increasing supply of glucogenic substrates did not influence the plasma glucose TR in steers. Although the isotope dilution method was different, the numerical values of plasma glucose TR of the present findings were comparable to the data reported in sheep fed plantain herb (Al-Mamun et al., 2007).

Plasma Leu concentration and [1-^{13}C]Leu enrichment were stable during the latter half of the isotope infusion (Figure 2) which indicated the steady state condition. Concentration of plasma Leu determined during the last 2 h

Table 5. Dietary effects on kinetics of plasma glucose and leucine (Leu) metabolism in sheep

Items	MH-diet	RSUM-diet	SEM	p-value
No. of sheep	4	4		
Glucose				
Concentration (mmol/L)	3.38	3.28	0.11	0.51
TR (mmol/kg$^{0.75}$/h)	1.43	1.52	0.18	0.31
Leu				
Concentration (μmol/L)	93.4	77.3	6.6	0.11
TR (μmol/kg$^{0.75}$/h)	285	272	49	0.76

MH, mixed hay of orchardgrass and reed canarygrass; RSUM, rice straw supplemented with urea and molasses; SEM, standard error of the mean; TR, turnover rate.

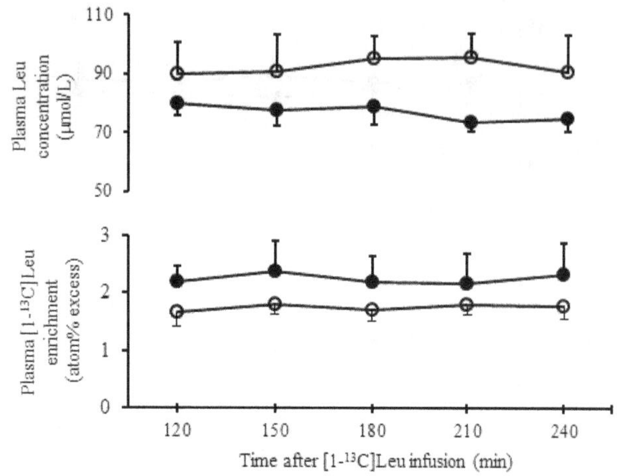

Figure 2. Time course changes of plasma leucine concentration and enrichment of [1-^{13}C]Leu during 120 to 240 min of the primed-continuous infusion of [1-^{13}C]Leu in sheep (n = 4) fed the RSUM-diet (•) and the MH-diet (○) (Means±standard error of the mean). RSUM, rice straw supplemented with urea and molasses; MH, mixed hay.

of the primed-continuous infusion did not differ ($p = 0.11$) between diets. In the present study, although N intake was lower for the RSUM-diet than the MH-diet, plasma LeuTR did not differ ($p = 0.76$) between diets. The present results were supported by Sano et al. (2009), who reported that supplementation of urea and soybean meal to roughage-based diets did not influence the plasma LeuTR in sheep. Moreover, numerical values of plasma LeuTR of the present findings were greater than the data previously found in sheep fed rice straw only (Alam et al., 2010), because the N and ME intake were also greater in sheep fed the RSUM-diet.

In conclusion, RSUM-diet showed improved performance than mixed hay on rumen fermentation characteristics and comparable performance to mixed hay on TR of plasma glucose and Leu in sheep. It can be suggested that rice straw supplemented with nitrogenous substrates in combination with soluble carbohydrate can be used for raising the livestock production as like as mixed hay.

REFERENCES

Alam, M. K., Y. Ogata, Y. Sako, M. Al-Mamun, and H. Sano. 2010. Intermediary metabolism of plasma acetic acid, glucose and protein in sheep fed a rice straw-based diet. Asian Australas. J. Anim. Sci. 23:1333-1339.

Alam, M. K., M. Sasaki, M. Al-Mamun, and H. Sano. 2013. Plasma acetate turnover rate and rumen fermentation characteristics in sheep fed rice straw supplemented with soybean meal. J. Anim. Sci. Adv. 3:65-73.

Al-Mamun, M., C. Tanaka, Y. Hanai, Y. Tamura, and H. Sano. 2007. Effects of plantain (*Plantago lanceolata* L.) herb and heat exposure on plasma glucose metabolism in sheep. Asian Australas. J. Anim. Sci. 20:894-899.

AOAC. 1995. Official Methods of Analysis, 16th edn. Association of Official Analytical Chemists, Arlington, VA, USA.

Broderick, G. A. and W. J. Radloff. 2004. Effect of molasses supplementation on the production of lactating dairy cows fed diets based on alfalfa and corn silage. J. Dairy Sci. 87:2997-3009.

Calder, A. G. and A. Smith. 1988. Stable isotope ratio analysis of leucine and ketoisocaproic acid in blood plasma by gas chromatography/mass spectrometry. Use of tertiary butyldimethylsilyl derivatives. Rapid Commun. Mass Spectrum. 2:14-16.

Can, A., N. Denek, and K. Yazgan. 2004. Effect of urea and molasses supplementation on nutrient intake and digestibility of sheep fed with straw. J. Anim. Vet. Adv. 3:466-469.

Fox, M. T., D. Gerrelli, S. R. Pitt, and D. E. Jacobs. 1991. The relationship between appetite and plasma non-esterified fatty acid levels in housed calves. Vet. Res. Commun. 15:127-133.

Hue, K. T., D. T. T. Van, and I. Ledin. 2008. Effect of supplementing urea treated rice straw and molasses with different forage species on the performance of lambs. Small Rumin. Res. 78:134-143.

Huggett, A. G. and D. A. Nixon. 1957. Enzymatic determination of blood glucose. Biochem. J. 66:12.

Jain, N., S. P. Tiwari, and P. Singh. 2005. Effect of urea molasses mineral granules on rumen fermentation pattern and blood biochemical constituents in goat kids fed sola (*Aeschonomene indica* L.) grass-based diet. J. Vet. Arhiv. 75:521-530.

Leng, R. A. 1990. Factors affecting the utilization of 'poor-quality' forages by ruminants particularly under tropical conditions. Nutr. Res. Rev. 3:277-303.

Leng, R. A. and J. V. Nolan. 1984. Protein nutrition of the lactating dairy cow. J. Dairy Sci. 67:1072-1089.

Milano, G. D. and G. E. Lobley. 2001. Liver nitrogen movements during short-term infusion of high levels of ammonia into the mesenteric vein of sheep. Br. J. Nutr. 86:507-513.

National Agriculture and Food Research Organization. 2006. Japanese feeding standard for dairy cattle. Japan Livestock Industry Association, Tokyo, Japan.

National Research Council. 1985. Nutrient Requirements of Sheep 6th Ed. National Academy Press, Washington, DC, USA.

Nolan, J. V. and R. A. Leng. 1972. Dynamic aspects of ammonia and urea metabolism in sheep. Br. J. Nutr. 27:177-194.

Oba, M. and M. S. Allen. 2003. Extent of hypophagia caused by propionate infusion is related to plasma glucose concentration in lactating dairy cows. J. Nutr. 133:1105-1112.

Ortigues-Marty, I., J. Vernet, and L. Majdoub. 2003. Whole body glucose turnover in growing and non-productive adult ruminants: meta-analysis and review. J. Reprod. Nutr. Dev. 43:371-383.

Rooke, J. A. and D. G. Armstrong. 1989. The importance of the form of nitrogen on microbial protein synthesis in the rumen of cattle receiving grass silage and continuous intrarumen infusions of sucrose. Br. J. Nutr. 61:113-121.

Sano, H. and T. Fujita. 2006. Effect of supplemental calcium propionate on insulin action to blood glucose metabolism in adult sheep. Reprod. Nutr. Dev. 46:9-18.

Sano, H., T. Fujita, M. Murakami, and A. Shiga. 1996. Stimulative effect of epinephrine on glucose production and utilization rates in sheep using a stable isotope. Domest. Anim. Endocrinol. 13:445-451.

Sano, H., M. Kajita, and T. Fujita. 2004. Effect of dietary protein intake on plasma leucine flux, protein synthesis, and degradation in sheep. Comp. Biochem. Physiol. B. 139:163-168.

Sano, H., S. Shibasaki, and H. Sawada. 2009. The effect of the source of nitrogen supplementation on nitrogen balance, rates of plasma leucine turnover, protein synthesis and degradation in sheep. Arch. Anim. Nutr. 63:401-412.

Sarnklong, C., J. W. Cone, W. Pellikaan, and W. H. Hendriks. 2010. Utilization of rice straw and different treatments to improve its feed value for ruminants: A Review. Asian Australas. J. Anim. Sci. 23:680-692.

Schmidt, S. P. and R. K. Keith. 1983. Effects of diet and energy intake on kinetics of glucose metabolism in steers. J. Nutr. 113:2155-2163.

Seal, C. J. and D. S. Parker. 1994. Effect of intraruminal propionic acid infusion on metabolism of mesenteric- and portal-drained viscera in growing steers fed a forage diet: I. Volatile fatty acids, glucose, and lactate. J. Anim. Sci. 72:1325-1334.

Singh, G. P., B. N. Gupta, and M. Madhu. 1995. Effect of supplementation urea molasses mineral licks to straw diet on dry matter intake, volatile fatty acids and methane production. Indian J. Dairy Sci. 48:290-294.

Srinivas, B. and B. N. Gupta. 1997. Rumen fermentation, bacterial and total volatile fatty acid (TVFA) production rates in cattle fed on urea-molasses-mineral block licks supplement. Anim. Feed Sci. Technol. 65:275-286.

SAS. 1996. SAS/STAT Software: Changes and Enhancements through Release 6.11. SAS Inst. Inc. Cary, NC, USA.

Tedeschi, L. O, M. J. Baker, D. J. Ketchen, and D. G. Fox. 2002. Performance of growing and finishing cattle supplemented with a slow-release urea product and urea. Can. J. Anim. Sci. 82:567-573.

Toppo, S., A. K. Verma, R. S. Dass, and U. R. Mehra. 1997. Nutrient utilization and rumen fermentation pattern in crossbred cattle fed different planes of nutrition supplemented with urea molasses mineral block. Anim. Feed Sci. Technol. 64:101-112.

Tserng, K. Y. and S. C. Kalhan. 1983. Calculation of substrate turnover rate in stable isotope tracer studies. Am. J. Physiol. 245:E308-E311.

Van Houtert, M. F. J. 1993. The production and metabolism of volatile fatty acids by ruminants fed roughages: A review. Anim. Feed Sci. Technol. 43:189-255.

van Soest, P. J., J. B. Robertson, and B. A. Lewis. 1991. Methods for dietary fiber, neutral detergent fiber, and nonstarch polysaccharides in relation to animal nutrition. J. Dairy Sci. 74:3583-3597.

Wanapat, M., S. Polyrach, K. Boonnop, C. Mapato, and A. Cherdthong. 2009. Effect of treating rice straw with urea and calcium hydroxide upon intake, digestibility, rumen fermentation and milk yield of dairy cows. Livest. Sci. 125:238-243.

Weatherburn, M. W. 1967. Phenol-hypochlorite reaction for determination of ammonia. Anal. Chem. 39:971-974.

Wu, Y., W. Hu, and J. Liu. 2005. Effects of supplementary urea-minerals lick block on the kinetics of fibre digestion, nutrient digestibility and nitrogen utilization of low quality roughages. J. Zhejiang Univ. Sci. 8:793-797.

Zhang, X. D., W. J. Chen, C. Y. Li, and J. X. Liu. 2009. Effects of protein-free energy supplementation on blood metabolites, insulin and hepatic PEPCK gene expression in growing lambs offered rice straw-based diet. Czech J. Anim. Sci. 54:481-489.

Zinn, R. A., R. Barrajas, M. Montano, and R. A. Ware. 2003. Influence of dietary urea level on digestive function and growth performance of cattle fed steam-flaked barley-based finishing diets. J. Anim. Sci. 81:2383-2389.

8

Influence of Palm Kernel Meal Inclusion and Exogenous Enzyme Supplementation on Growth Performance, Energy Utilization, and Nutrient Digestibility in Young Broilers

M. R. Abdollahi*, B. J. Hosking[1], D. Ning[1], and V. Ravindran

Institute of Veterinary, Animal and Biomedical Sciences, Massey University, Palmerston North 4442, New Zealand

ABSTRACT: The objective of the present study was to investigate the influence of palm kernel meal (PKM) inclusion and exogenous enzyme supplementation on growth performance, nitrogen-corrected apparent metabolizable energy (AMEn), coefficient of apparent ileal digestibility (CAID) and total tract retention of nutrients in young broilers fed corn-based diets. Four inclusion levels of PKM (no PKM [PKM0], 8% [PKM8], 16% [PKM16], and 24% [PKM24]) and two enzyme additions were evaluated in a 4×2 factorial arrangement of treatments. A total of 384, one-d-old male broilers (Ross 308) were individually weighed and allocated to 48 cages (eight broilers/cage), and cages were randomly assigned to eight dietary treatments. Results indicated that the inclusion of 8% and 16% PKM increased ($p<0.05$) the weight gain compared to the PKM0 diet. Birds fed the PKM8 diets had the highest ($p<0.05$) feed intake. Weight gain and feed intake were severely reduced ($p<0.05$) by feeding the PKM24 diet. Enzyme supplementation increased weight gain ($p<0.05$), independent of PKM inclusion level. In PKM0 and PKM8 diets, enzyme addition significantly ($p<0.05$) lowered feed conversion ratio (FCR); whereas enzyme addition had no effect on FCR of birds fed PKM16 and PKM24 diets. In PKM0 and PKM16 diets, enzyme addition significantly ($p<0.05$) increased CAID of nitrogen and energy but had no effect in the PKM8 and PKM24 diets. Inclusion of PKM into the basal diet, irrespective of inclusion level, enhanced ($p<0.05$) starch and fat digestibility. Inclusion of PKM at 16% and 24% resulted in similar CAID of neutral detergent fiber (NDF) but higher ($p<0.05$) than that of the PKM0 and PKM8 diets. Enzyme addition, regardless of the level of PKM inclusion, significantly ($p<0.05$) increased CAID of NDF. There was a significant ($p<0.05$) decrease in AMEn with PKM inclusion of 24%. The present data suggest that inclusion of PKM in broiler diets could be optimized if PKM-containing diets are formulated based on digestible amino acid contents and supplemented with exogenous enzymes. If amino acid digestibility and AME of PKM considered in the formulation, it can be included in broiler diets up to 16% with no deleterious effects on growth performance. (**Key Words:** Palm Kernel Meal, Exogenous Enzyme, Broiler, Performance, Nutrient Utilization)

INTRODUCTION

High demand for conventional feed ingredients due to competition with humans and other livestock species for feed resources and ever-increasing cost of these ingredients have motivated poultry nutritionists to maximize the use of locally available by-products. Palm kernel meal (PKM), an agro-industrial by-product, is a locally available and relatively inexpensive feedstuff in many tropical countries (Perez et al., 2000). Palm kernel meal is produced by extracting the oil from palm kernels using solvent extraction. Palm kernel oil might also be mechanically expelled and the coproduct is referred to as palm kernel cake (PKC; Sundu et al., 2006). Incorporation of PKM in broiler diets is limited and reported to be associated with impaired performance parameters (Mardhati et al., 2011). The deteriorated growth responses has been attributed to high fiber content, grittiness (Sundu et al., 2005; 2006) and low concentration of indispensable amino acids (AA) (Abdollahi et al., 2015).

Published data on the effect of PKM on growth

* Corresponding Author: M. R. Abdollahi.
E-mail: M.Abdollahi@massey.ac.nz
[1] AsiaPac Biotechnology Co. Ltd., Dongguan City, Guangdong 523808, China.

performance of broiler chickens have been contradictory. Panigrahi and Powell (1991) reported that if supplemented with methionine and lysine, PKM can be included in broiler diets up to 40%. Ezieshi and Olomu (2008) replaced 50% of corn in the broiler diet with PKM (30% and 32.5% PKM in starter and finisher diets, respectively) and reported lower body weight and a deteriorated feed conversion ratio (FCR). Shakila et al. (2012) reported that PKM up to 10%, with or without enzyme, might be used in broiler diets without negatively affecting the performance.

Palm kernel meal contains high level of fiber with non-starch polysaccharides (NSP) and β-mannan as the main fiber components of PKM. Almaguer et al. (2014) reported concentrations of 494 and 779 g/kg of acid detergent fiber and neutral detergent fiber (NDF) in PKM. Knudsen (1997) reported values of 466 and 136 g/kg for NSP and lignin in PKM, respectively. Abdollahi et al. (2015) reported that almost 96% of total NSP in PKM being in the form of insoluble NSP. Beta-mannans, also known as β-galactomannans, are linear polysaccharides with repeating units of β-1-4 mannose and side attachments of α-1-6 glucose and/or galactose and commonly found in many feed ingredients (Knudsen, 1997; Jackson et al., 2004). Hsiao et al. (2006) reported β-mannan concentration of 12.6 and 16.1 g/kg in dehulled and non-dehulled soybean meal samples, respectively. Mannans in PKM are hard, highly-insoluble crystalline polymers of mannose and constitute the storage polysaccharides of the endosperm of palms (Daud and Jarvis, 1992). According to Dusterhoft et al. (1991), 78% of the total NSP in PKM is linear mannan. Jackson (2010) reported that PKM has a β-mannan content of 300 to 350 g/kg. Mannans are not depolymerised in the digestive tract of the birds due to the lack of mannan-degrading enzymes.

Poultry are not able to degrade NSP due to lack of endogenous fiber-degrading enzymes and therefore feeding high fiber ingredients such as PKM might impair growth responses. Fiber in PKM has been found to have a complex structure consisting mannans (glucomannan and galactomannan) as the main component, cellulose, glucoronoxylans and arabinoxylans (Dusterhoft et al., 1991; Knudsen, 1997). Due to its complexity, fiber of PKM may benefit more from a combination of enzymes including β-mannanase, xylanase, cellulose, glucosidase and galactosidase. However, studies investigating the effect of enzyme on PKM-containing broiler diets are limited (Soltan, 2009; Shakila et al., 2012). This study sought to investigate the efficacy of an enzyme cocktail in corn-soybean meal-based broiler diets with graded levels of PKM. The objective of the current study was to examine the effects of four dietary inclusion levels of PKM, each without and with enzyme supplementation, on growth performance, nutrient digestibility and energy utilization of young broilers.

MATERIALS AND METHODS

Enzyme

The enzyme was supplied by AsiaPac (Dongguan) Biotechnology Co. Ltd, Guangdong, China, and was a cocktail of β-mannanase and NSP-degrading enzymes (Pokazyme PK516; including: β-mannanase, xylanase, amylase, protease, cellulase and β-glucanase). The enzyme was in powder form and added at the level recommended by the manufacturer (200 g/t of feed).

Diets

The experimental design was a 4×2 factorial arrangement of treatments, which included four different levels of PKM inclusion (0%, 8%, 16%, and 24%) and two enzyme supplementations (without/with enzyme). Whole corn was obtained from a commercial supplier, and ground in a hammer mill (Bisley's Farm Machinery, Auckland, New Zealand) to pass through a screen size of 4.0 mm. Broiler starter diets, based on corn, soybean meal and PKM, were formulated to meet the Ross 308 strain recommendations for major nutrients (Ross, 2007) and to be equivalent in respect of energy density, and digestible protein and AA, calcium and available phosphorus concentrations. Palm kernel meal used in the current study had previously been analysed for nutrients, apparent metabolizable energy (AME) and ileal AA digestibility (Abdollahi et al., 2015) and the data were used to formulate the experimental diets. Four diets were formulated to contain no PKM (PKM0), 8% (PKM8), 16% (PKM16), and 24% (PKM24) PKM (Table 1) and used to develop eight dietary treatments without and with the addition of enzyme. The diets contained 0.3% of titanium dioxide (TiO_2, Merck KGaA, Darmstadt, Germany) as an indigestible marker for the determination of ileal nutrient digestibility. Diets were mixed in a single-screw paddle mixer (Bonser Engineering Co. Pty. Ltd., Merrylands, Australia). Following mixing, all diets were steam-conditioned at 70°C and pelleted using a pellet mill (Model Orbit 15; Richard Sizer Ltd., Kingston-upon-Hull, UK) capable of manufacturing 180 kg of feed/h and equipped with a die ring with 3-mm holes and 35-mm thickness. Conditioning time of the mash was 30 s and the conditioning temperature was measured at the outlet of the conditioner. The diets were run in sequence with no change in the feeder rate, rotation speed or number of knives. Representative samples of all diets were collected after pelleting for chemical analysis.

Birds and housing

Experimental procedures were approved by the Massey University Animal Ethics Committee and complied with the

Table 1. Composition and calculated analysis (%, as fed) of the experimental diets

Item	Palm kernel meal (PKM) inclusion			
	No PKM	8%	16%	24%
Corn	55.10	51.39	42.55	33.68
Soybean meal, 48%	31.06	26.05	20.89	15.60
Wheat bran	4.00	0.00	0.00	0.00
Corn gluten meal	2.30	4.60	6.90	9.30
Meat and bone meal	1.41	2.82	4.23	5.65
Fish meal	1.00	2.00	3.00	4.00
Palm kernel meal	0.00	8.00	16.0	24.0
Soybean oil	1.36	2.04	4.03	6.02
Dicalcium phosphate	1.48	0.98	0.35	0.00
Limestone	0.61	0.42	0.32	0.00
DL-methionine	0.31	0.29	0.27	0.25
L-lysine HCl	0.28	0.34	0.41	0.47
L-threonine	0.09	0.10	0.12	0.13
Sodium chloride	0.21	0.18	0.14	0.10
Sodium bicarbonate	0.26	0.26	0.26	0.27
Trace mineral-vitamin premix[1]	0.23	0.23	0.23	0.23
Titanium dioxide	0.30	0.30	0.30	0.30
Calculated analysis				
Apparent metabolizable energy (MJ/kg)	12.34	12.34	12.34	12.34
Apparent digestible protein	18.80	18.80	18.80	18.80
Apparent digestible methionine	0.62	0.62	0.63	0.63
Apparent digestible methionine+cysteine	0.89	0.89	0.89	0.89
Apparent digestible lysine	1.19	1.19	1.19	1.19
Apparent digestible threonine	0.78	0.78	0.78	0.78
Apparent digestible arginine	1.32	1.33	1.34	1.35
Apparent digestible valine	0.91	0.93	0.95	0.97
Apparent digestible isoleucine	0.80	0.80	0.80	0.80
Crude fat	4.06	5.28	7.69	10.1
Crude fiber	3.01	3.94	5.22	6.51
Calcium	1.00	1.00	1.00	1.00
Available phosphorus	0.50	0.50	0.50	0.50
β-mannan[2]	0.45	2.99	5.52	8.05

[1] Supplied per kilogram of diet: vitamin A (vitamin A acetate), 12,000 IU; vitamin D_3 (cholcalciferol), 4,000 IU; vitamin E (DL-α-tocopherol), 80 IU; biotin, 0.25 mg; pantothenic acid (calcium-D-pantothenate), 15 mg; cyanocobalamin, 0.02 mg; folic acid, 3.0 mg; vitamin K_3 (menadione nicotinamide bisulphite), 4.0 mg; niacin (nicotinic acid), 60 mg; pyridoxine (pyridoxine hydrochloride), 10 mg; riboflavin, 9.0 mg; thiamine (thiamine mononitrate), 3.0 mg; antioxidant (ethoxyquin), 100 mg; choline (choline chloride 60%), 360 mg; Co (cobalt sulfate), 0.15 mg; Cu (copper sulfate), 6.0 mg; organic Cu (B-TRAXIM 2C G/Cu), 3.0 mg; Fe (iron sulfate), 36 mg; I (calcium iodate), 0.93 mg; Mn (manganese oxide), 60 mg; Mo (sodium molybdate), 0.15 mg; Se (sodium selenite), 0.26 mg; organic Se (enriched yeast), 0.14 mg; Zn (zinc sulfate), 48 mg; organic Zn (B-TRAXIM 2C G/Zn), 24 mg.

[2] β-mannan content of feed ingredients from Jackson (2010) was used for calculation.

New Zealand Code of Practice for the Care and Use of Animals for Scientific Purposes. One-d-old male broilers (Ross 308), obtained from a commercial hatchery, were individually weighed and allocated to 48 cages (eight birds per cage) in electrically heated battery brooders so that the average bird weight per cage was similar. Six cages were then randomly assigned to each of the eight dietary treatments. The birds were transferred to grower cages on d 12 and fed the same diets until d 21. The space allocation per bird in brooder and grower cages was 530 and 640 cm^2, respectively. The battery brooders and grower cages were housed in an environmentally controlled room with 20 h of fluorescent illumination per day. The temperature was maintained at 31°C on d 1, and was gradually reduced to 22°C by 21 d of age. Feed was offered *ad libitum* and water was freely available.

Pellet durability

Pellet durability was determined in a Holmen Pellet Tester (New Holmen NHP100 Portable Pellet Durability Tester, TekPro Ltd., Willow Park, North Walsham, Norfolk, UK) using the method described by Abdollahi et al. (2010). Clean pellet samples (100 g; four replicates per diet), with no fines, were rapidly circulated in an air stream around a perforated test chamber for 30 s. Fines were removed continuously through the perforations during the test cycle. After the test cycle, the subject pellets were ejected and weighed manually. The pellet durability index (PDI) was calculated as the percentage of the pellets not passing through the perforations at the end of the test.

Performance data

Body weight and feed intake were recorded on a cage basis at weekly intervals. Mortality was recorded daily. Feed conversion ratio values were corrected for the body weight of any bird that died during the course of the experiment.

Apparent metabolizable energy determination and excreta scoring

Feed intake and total excreta output of each cage were quantitatively measured from d 18 to 21 post-hatch. Daily collections from each cage were pooled, mixed in a blender and sub-sampled. Sub-samples were lyophilized (Model 0610, Cuddon Engineering, Blenheim, New Zealand) ground to pass through a 0.5 mm sieve and stored in airtight plastic containers at 4°C pending analysis. The diets and excreta samples were analysed for dry matter (DM), gross energy (GE), nitrogen (N), fat and NDF. On d 21, excreta were also scored for stickiness on a scale of 1 to 5, with 1 representing normally formed excreta and 5 representing watery and very sticky excreta.

Coefficient of ileal apparent digestibility determination

On d 21, six birds per cage were euthanized by intravenous injection (1 mL per 2 kg live weight) of sodium pentobarbitone solution (Provet NZ Pty Ltd., Auckland, New Zealand) and ileal digesta were collected as described by Ravindran et al. (2005). The ileum was defined as that portion of the small intestine extending from the Meckel's diverticulum to a point ~ 40 mm proximal to the ileo-caecal junction. The ileum was then divided into two halves and the digesta was collected from the lower half towards the ileo-caecal junction. Digesta from birds within a cage were pooled, lyophilized (Model 0610, Cuddon Engineering, Blenheim, New Zealand), ground to pass through a 0.5 mm sieve and stored at 4°C until laboratory analysis. The diets and digesta samples were analysed for GE, titanium dioxide (Ti), N, starch, fat, and NDF.

Chemical analysis

Dry matter was determined using standard procedures (method 930.15; AOAC, 2005). Gross energy was determined by adiabatic bomb calorimetry (Gallenkamp Autobomb, London, UK) standardized with benzoic acid. Samples were assayed for Ti on a UV spectrophotometer following the method of Short et al. (1996). The NDF was determined using standard procedures (AOAC, 2005).

Nitrogen was determined by combustion (method 968.06; AOAC, 2005) using a CNS-200 carbon, N and sulphur auto analyser (LECO Corporation, St. Joseph, MI, USA). Starch was determined using the Megazyme Total Starch Assay kit (Megazyme International Ireland Ltd., Wicklow, Ireland) based on thermostable α-amylase and amyloglucosidase. Fat was determined using Soxhlet extraction procedure (method 991.36; AOAC, 2005).

Calculations

All data were expressed on a DM basis, and the AME was calculated using the following formula:

$$\text{AME (MJ/kg diet)} = [(\text{Feed intake}\times \text{GE}_{\text{diet}}) - (\text{Excreta output}\times \text{GE}_{\text{excreta}})]/\text{Feed intake}$$

Nitrogen-corrected AME (AMEn) was determined by correction for zero N retention by simple multiplication with 36.54 kJ per gram N retained in the body as described by Hill and Anderson (1958).

Coefficient of apparent ileal digestibility (CAID) of nutrients was calculated using the following formula:

$$\text{CAID of diet component} = [(\text{Diet component/Ti})_{\text{diet}} - (\text{Diet component/Ti})_{\text{ileal}}]/(\text{Diet component/Ti})_{\text{diet}}$$

Where $(\text{Diet component/Ti})_{\text{diet}}$ = ratio of diet component to Ti in the diet, and

$(\text{Diet component/Ti})_{\text{ileal}}$ = ratio of diet component to Ti in the ileal digesta

Total tract retention (TTR) of nutrients, as a percentage of intake, was determined as follows:

$$\text{TTR (\%)} = 100\times\{[(\text{Feed intake}\times\text{Nutrient}_{\text{diet}}) - (\text{Excreta output}\times\text{Nutrient}_{\text{excreta}})]/(\text{Feed intake}\times\text{Nutrient}_{\text{diet}})\}$$

Statistical analysis

The data were analysed by two-way analysis of variance to determine the main effects (PKM inclusion and enzyme addition) and their interaction using the General Linear Models procedure of SAS (2004). Cage means served as the experimental unit for all data. Differences were considered to be significant at $p<0.05$ and significant differences between means were separated by the Least Significant Difference test.

RESULTS

Pellet durability index

Increasing the dietary PKM inclusion resulted in lower PDI compared to the diet with no PKM. The PDI values of 85.3, 56.2, 34.4 and 15.4, respectively, were obtained with PKM inclusion of 0%, 8%, 16%, and 24%.

Performance and, excreta score and dry matter

Mortality during the performance experiment was negligible. Only six out of the 384 birds died and the deaths were not related to any specific treatment.

The performance data of broiler starters from d 1 to 21 post-hatch is shown in Table 2. Weight gain of the birds was significantly ($p<0.001$) influenced by PKM inclusion and enzyme addition. Inclusion of 8.0% and 16% PKM increased ($p<0.05$) the weight gain compared to the PKM0 diet. Weight gain was severely reduced by feeding the PKM24 diets. Enzyme supplementation increased weight gain ($p<0.05$), irrespective of PKM inclusion level.

There was a significant ($p<0.001$) effect of PKM inclusion on feed intake with birds fed the PKM8 diets had the highest feed intake ($p<0.05$). Birds fed the PKM0 and PKM16 diets showed similar feed intake. Similar to weight gain, inclusion of 24% PKM resulted in the lowest ($p<0.05$) feed intake. Neither the main effect of enzyme addition nor the PKM inclusion×enzyme addition interaction was significant for feed intake.

A significant ($p<0.01$) interaction between PKM inclusion and enzyme addition was detected for FCR of birds. In PKM0 and PKM8 diets, enzyme addition significantly ($p<0.05$) improved feed efficiency; whereas, enzyme addition had no effect on feed per gain of birds fed the diets with PKM inclusion above 8.0%.

There was a significant ($p<0.001$) effect of PKM inclusion on excreta score and excreta DM (Table 2). As the PKM level was increased, the excreta DM increased and excreta score decreased. Neither the main effect of enzyme addition nor the interaction between PKM inclusion and enzyme addition was significant for excreta score and DM.

Nutrient digestibility

A significant ($p<0.05$) PKM inclusion×enzyme addition interaction was observed for the CAID of N and GE (Table 3). In PKM0 and PKM16 diets, enzyme addition significantly ($p<0.05$) increased N and GE digestibility but had no effect in the PKM8 and PKM24 diets.

The main effect of PKM inclusion was significant

Table 2. Influence of inclusion level of palm kernel meal (PKM) and enzyme addition on the weight gain (g/bird), feed intake (g/bird), feed conversion ratio (FCR; g feed/g gain), excreta score and excreta dry matter (DM; %) of broiler starters[1] (d 1 to 21 post-hatch)

PKM inclusion	Enzyme addition	Weight gain	Feed intake	FCR	Excreta score[2]	Excreta DM[2]
PKM0	–	965	1,138	1.197^a	4.00	24.4
	+	978	1,122	1.146^b	4.00	25.0
PKM8	–	988	1,178	1.189^a	2.75	29.2
	+	1,003	1,164	1.160^b	2.58	29.2
PKM16	–	973	1,110	1.141^b	2.00	34.6
	+	1,003	1,134	1.139^b	2.00	34.0
PKM24	–	914	1,041	1.144^b	1.25	36.4
	+	936	1,067	1.145^b	1.25	37.0
SEM[3]		7.7	11.8	0.0077	0.087	0.71
Main effects						
PKM inclusion						
PKM0		972^b	1,130^b	1.171	4.00^a	24.7^d
PKM8		996^a	1,171^a	1.174	2.67^b	29.2^c
PKM16		988^a	1,122^b	1.140	2.00^c	34.3^b
PKM24		925^c	1,054^c	1.145	1.25^d	36.7^a
Enzyme addition						
–		960^b	1,117	1.168	2.50	31.2
+		980^a	1,122	1.147	2.46	31.3
Probabilities, p≤						
PKM inclusion		0.001	0.001	0.001	0.001	0.001
Enzyme addition		0.001	0.557	0.001	0.504	0.768
PKM inclusion×enzyme addition		0.692	0.137	0.004	0.716	0.806

[1] Each value represents the mean of six replicates (eight birds per replicate).

[2] On d 21 post hatch, excreta were scored for stickiness on a scale of 1 to 5, with 1 representing normally formed excreta and 5 representing watery and very sticky excreta.

[3] Pooled standard error of mean.

Means in a column not sharing a common letter (^{a,b,c,d}) are significantly different ($p<0.05$).

Table 3. Influence of inclusion level of palm kernel meal (PKM) and enzyme addition on coefficient of apparent ileal digestibility of nitrogen (N), starch, fat, gross energy (GE), and neutral detergent fiber (NDF) in broiler starters[1]

PKM inclusion	Enzyme addition	N	Starch	Fat	GE	NDF
PKM0	–	0.745[c]	0.944	0.886	0.693[b]	0.108
	+	0.784[ab]	0.949	0.905	0.718[a]	0.158
PKM8	–	0.787[ab]	0.990	0.921	0.727[a]	0.164
	+	0.779[ab]	0.989	0.917	0.721[a]	0.172
PKM16	–	0.765[bc]	0.986	0.919	0.688[bc]	0.203
	+	0.795[a]	0.984	0.938	0.722[a]	0.292
PKM24	–	0.752[c]	0.990	0.949	0.675[bc]	0.236
	+	0.746[c]	0.984	0.938	0.669[c]	0.259
SEM[2]		0.0087	0.0025	0.0100	0.0078	0.0184
Main effects						
PKM inclusion						
PKM0		0.765	0.947[b]	0.896[c]	0.706	0.133[b]
PKM8		0.783	0.989[a]	0.919[b]	0.724	0.168[b]
PKM16		0.780	0.985[a]	0.928[ab]	0.705	0.248[a]
PKM24		0.749	0.987[a]	0.943[a]	0.672	0.248[a]
Enzyme addition						
–		0.762	0.977	0.919	0.696	0.178[b]
+		0.776	0.977	0.925	0.707	0.220[a]
Probabilities, p≤						
PKM inclusion		0.001	0.001	0.001	0.001	0.001
Enzyme addition		0.032	0.678	0.440	0.045	0.002
PKM inclusion×enzyme addition		0.016	0.170	0.316	0.022	0.152

[1] Each value represents the mean of six replicates (six birds per replicate) measured on d 21 post-hatch.
[2] Pooled standard error of mean.
Means in a column not sharing a common letter ([a,b,c]) are significantly different (p<0.05).

(p<0.001) for the CAID of starch and fat. Inclusion of PKM into the basal diet, irrespective of inclusion level, enhanced (p<0.05) starch and fat digestibility. Whereas the PKM-containing diets resulted in similar starch digestibility, the PKM24 diet had the highest CAID of fat. Neither the main effect of enzyme addition nor the interaction between PKM inclusion and enzyme addition was significant for the CAID of starch and fat.

Inclusion of PKM at 16% and 24% resulted in similar CAID of NDF but higher (p<0.05) than the PKM0 and PKM8 diets. Enzyme addition, regardless of level of PKM inclusion, significantly (p<0.05) increased CAID of NDF.

Nutrient retention and energy utilization

Significant PKM inclusion×enzyme addition interaction was observed for TTR of DM (p<0.05) and NDF (p<0.001; Table 4). Enzyme inclusion had no effect on TTR of DM and NDF in the PKM8 and PKM24 diets, but improved DM and NDF retention in the PKM0 and PKM16 diets.

The main effect of PKM inclusion was significant for the TTR of N (p<0.001) and fat (p<0.01). Diets with no PKM and 8.0% PKM resulted in higher N retention compared to those with 16% and 24% PKM. Inclusion of 24% PKM resulted in the lowest N retention. Birds fed the PKM8 diets had higher (p<0.05) fat retention compared to other PKM inclusion levels.

Increasing the inclusion level of PKM significantly (p<0.001) decreased the TTR of GE. Enzyme addition enhanced (p<0.05) GE retention.

Pam kernel meal inclusion influenced AMEn (p<0.05). There was a significant decrease in AMEn with PKM inclusion of 24%. There was no effect of enzyme addition on energy utilization.

DISCUSSION

Unexpectedly, feeding the PKM8 and PKM16 diets increased weight gain of broilers. Ezieshi and Olomu (2008) reported impaired body weight and feed efficiency in a broiler study with 30% and 32.5% PKM inclusion in starter and finisher diets, respectively. Mardhati et al. (2011) also reported that weight gain and feed efficiency of broilers were poorer in diet with 20% PKM compared to a corn-based control diet. Reduced body weight and feed efficiency have also been reported with PKM inclusion levels above 10% (Soltan, 2009). In most of the studies investigating the effect of PKM inclusion in broiler diets, PKM has been included into a control diet either at the

Table 4. Influence of inclusion level of palm kernel meal (PKM) and enzyme addition on total tract retention of dry matter (DM), nitrogen (N), fat, neutral detergent fiber (NDF), gross energy (GE), and nitrogen-corrected apparent metabolizable energy (AMEn) in broiler starters[1]

PKM inclusion	Enzyme addition	DM	N	Fat	NDF	GE	AMEn
PKM0	–	72.2[b]	67.6	90.0	22.4[d]	76.2	13.35
	+	73.5[a]	69.2	90.7	27.4[b]	77.1	13.52
PKM8	–	71.4[c]	68.4	91.6	23.2[cd]	75.3	13.45
	+	71.5[bc]	68.0	92.1	24.4[c]	75.6	13.50
PKM16	–	67.0[e]	64.7	90.8	23.7[cd]	71.6	13.33
	+	67.7[d]	65.1	90.6	28.7[b]	72.3	13.45
PKM24	–	63.1[f]	60.3	91.0	30.7[a]	68.7	13.36
	+	62.8[f]	59.7	90.0	31.2[a]	68.5	13.27
SEM[2]		0.26	0.49	0.43	0.66	0.25	0.055
Main effects							
PKM inclusion							
PKM0		72.9	68.4[a]	90.4[b]	24.9	76.7[a]	13.43[a]
PKM8		71.4	68.2[a]	91.9[a]	23.8	75.4[b]	13.47[a]
PKM16		67.3	64.9[b]	90.7[b]	26.2	71.9[c]	13.39[ab]
PKM24		63.0	60.0[c]	90.5[b]	30.9	68.6[d]	13.31[b]
Enzyme addition							
–		68.4	65.2	90.9	25.0	72.9[b]	13.37
+		68.9	65.5	90.8	27.9	73.4[a]	13.43
Probabilities, p≤							
PKM inclusion		0.001	0.001	0.005	0.001	0.001	0.036
Enzyme addition		0.013	0.439	0.849	0.001	0.022	0.112
PKM inclusion×enzyme addition		0.016	0.098	0.203	0.001	0.137	0.105

Means in a column not sharing a common letter ([a,b,c,d,e,f]) are significantly different ($p<0.05$).
[1] Each value represents the mean of six replicates (eight birds per replicate) measured from d 18 to 21 post-hatch.
[2] Pooled standard error of mean.

expense of an ingredient (usually the major cereal) in the diet with the consequence of decreased nutrient density or without considering the AA digestibility of PKM. In all the above-mentioned studies, the digestibility of the protein and AA in PKM was not considered and the diets were formulated based on the total protein and AA contents and not the digestible contents. Palm kernel meal has been reported to have a poor protein digestibility (Bryden et al., 2009, CAID of 0.54; Abdollahi et al., 2015, standardized ileal digestibility coefficient of 0.42 to 0.46); which implies that formulating PKM-containing diets based on total crude protein content will fail to meet the birds' requirement for a balanced diet and impair the growth performance of broilers. Use of digestible AA contents of PKM when formulating broiler diets will eliminate this error and hold a promise to increase the inclusion levels of PKM in practical poultry diets. The PKM sample used in the current study had been previously assessed for the AME and digestible AA contents (Abdollahi et al., 2015), and the data was used to formulate the experimental diets based on apparent digestible AA content. It must be noted that both digestible AA systems, apparent or standardized, are superior to the total AA system. The choice of the appropriate system of digestible AA depends on the method of formulating diets, with CAID being suggested as the most appropriate system to use when the diets are formulated to least-cost (Bryden et al., 2009).

Birds fed the PKM8 diets consumed more feed than other dietary treatments. In agreement to the current study, Adrizal et al. (2011) reported an increased feed intake with increasing PKM level from 0% to 15% and 30% in laying hens. The higher feed intake in birds fed PKM-containing diets has also been reported by other researchers (Onifade and Babatunde, 1998; Sundu et al., 2005; Ezieshi and Olomu, 2008). This observation was attributed to its faster passage rate in the digestive tract (Onifade and Babatunde, 1998), high bulk density and low water holding capacity (Sundu et al., 2006). Onifade and Babatunde (1998) speculated that the higher feed intake in birds fed high levels of PKM in diets was an attempt by the bird to compensate the dilution of the diets and to consume enough nutrients to achieve comparable growth.

Inclusion of 24% PKM in the current study, however, was associated with a marked depression in weight gain, a finding which does not support the implication from several researchers (Panigrahi and Powell, 1991; Perez et al., 2000; Sundu et al., 2006; Adrizal et al., 2011) reporting that PKM can be included in poultry diets up to 40% without impairing performance parameters. Weight gain deterioration of birds fed PKM24 diets although corresponded with the lowest TTR of DM, N and GE, and AMEn, might be explained largely by the feed intake depression at this PKM level. Pellet durability was found to deteriorate with increasing dietary PKM inclusions. This is mostly evidenced by the lowest PDI value determined in the PKM24 diet. Deteriorated pellet quality is also likely to be responsible, in part, for the lowest feed intake and weight gain observed with feeding PKM24 diets.

The present data demonstrated a weight gain response to dietary enzyme addition, even in the PKM0 diet. Enzyme supplementation improved feed efficiency at inclusion levels of 0% and 8% PKM. Although viscosity is not a major nutritional obstacle in corn-based diets, the advantages of exogenous NSP-degrading enzymes on nutrient and energy utilization and broiler performance in corn-based diets have been previously reported (Zanella et al., 1999; Kiarie et al., 2014). In a recent study, Kiarie et al. (2014) reported that xylanase supplementation improved growth performance, nutrient and energy utilization in both wheat- and corn-based diets. The release of previously entrapped nutrients, and not the viscosity reduction, due to hydrolysis of structural arabinoxylans accounted for the positive response to xylanase addition in corn-based diets. These researchers suggested that young birds are sensitive to both soluble and insoluble NSP and could benefit from NSP-degrading enzymes independent of the source of cereal in the diets. In agreement with the current study, Williams et al. (2014) found that supplementation of a corn-soy diet with β-mannanase and a carbohydrase cocktail, separately and intermittently, improved body weight and feed efficiency in broilers. These researchers speculated that the application of a cocktail carbohydrase in corn-soy diets might be the most effective strategy to eliminate the negative effects of various NSP components and to improve the nutrient availability in these conventional feedstuffs.

The beneficial effects of supplementing the poultry diets with β-mannanase have been shown in several studies (Lee et al., 2003; Daskiran et al., 2004; Jackson et al., 2004; Zou et al., 2006; Williams et al., 2014). Lee et al. (2003) reported that supplementation of β-mannanase to corn-based broilers diets containing guar meal reduced intestinal viscosity and enhanced body weight and feed efficiency. Improved feed efficiency as a result of degrading β-mannans through β-mannanase supplementation in a corn-soy-based broiler diet has also been reported by Daskiran et al. (2004). Jackson et al. (2004) reported an improved weight gain and feed efficiency in broilers fed corn-soybean meal based diets supplemented with β-mannanase. The positive growth responses might be, partly, explained by a reduction in energy-draining immune stimulation associated with a decrease in the concentration of β-mannan in the digestive tract of bird. In the current study, calculated β-mannan contents of the PKM0, PKM8, PKM16, and PKM24 diets were 4.50, 29.9, 55.2, and 80.5 g/kg, respectively. It has been shown that β-mannans are components of the surface of many pathogens and perceived by animals as a pathogen-associated antigens stimulating the innate immune system of animals (Jackson et al., 2004; Hsiao et al., 2006; Williams et al., 2014). This immune response, termed as feed-induced immune response, deteriorates animal performance by diverting the energy from growth toward an energy-consuming immune response (Hsiao et al., 2006; Williams et al., 2014). It has been suggested that the inclusion of exogenous β-mannanase hydrolyses the high molecular weight β-mannans into mannose-oligosaccharides which are not recognisable by immune systems. By reducing the immune responses induced by feed, β-mannanase has been shown to spare the energy for production (Williams et al., 2014).

The fact that enzyme supplementation had no effect on the FCR of birds fed PKM16 and PKM24 diets is probably due, in large part, to the fact that the inclusion of PKM at these levels significantly improved feed efficiency compared to the PKM0 and PKM8 diets with no enzyme supplementation. Moreover, there is possibility that β-mannanase inclusion in diets with PKM levels above 8% was not sufficient to hydrolyse β-1,4-glucosidic linkages in β-mannan and to generate performance responses. Daskiran et al. (2004), in a broiler study with diets varying in β-mannan contents, found that diets with higher β-mannan content require higher levels of endo-β-D-mannanase supplementation. These researchers suggested that when ingredients high in β-mannan, such as PKM, copra meal or guar gum, are included in broiler diets, both the dietary enzyme and β-mannan levels should be monitored.

Enzyme supplementation increased the CAID of NDF and TTR of GE. Surprisingly, enzyme addition enhanced CAID of N and GE, and TTR of DM and NDF only in PKM0 and PKM16 diets and not in PKM8 and PKM24; a finding which is not readily explainable. Degradation of NSP by the enzyme cocktail, containing β-mannanase, xylanase, amylase, protease, cellulase and glucanase, might partially explain the positive effect of enzyme addition on nutrient utilization. Corn and soybean meal, though the most common feedstuffs in poultry diets, contain considerable amount of NSP. While corn contains a negligible amount of soluble NSP, it has been reported to have a concentration of insoluble NSP of 80 g/kg which consists mainly of arabinoxylans and β-glucans (Choct, 2006). Ravindran et al. (2014) reported average insoluble and soluble NSP contents of 171 and 16 g/kg, respectively, in soybean meal. A typical corn-soy diet has been reported to contain 123 g NSP/kg, with 111 and 12 g/kg of the NSP being in the form of insoluble and soluble NSP, respectively

(Abdollahi et al., 2010). Palm kernel meal sample used in the current study contained 686 and 610 g/kg of NDF and total NSP, respectively (Abdollahi et al., 2015). Saenphoom et al. (2011; 2013) reported that pre-treating palm kernel expeller with exogenous enzyme reduced hemicellulose and cellulose contents by 26% and 33%, respectively. Olaniyi (2014) reported significant reductions in fiber content of PKM as a result of hydrolysing PKM with β-mannanase at 30°C for 60 h. It is also possible that NSP-degrading enzymes by releasing the β-mannans encapsulated in intact cell walls, which can be then depolymerised by β-mannanase, might synergize with β-mannanase as shown by Williams et al. (2014). Aya et al. (2013), using a combination of enzymes containing β-mannanase, α-galactosidase, xylanase, cellulase and glucanase activity, reported higher nutrient (including fiber) retention in enzyme-supplemented diets independent of PKM inclusion level (10%, 20%, 30%, and 40%); an improvement which contributed to the higher weight gain and feed efficiency.

CONCLUSIONS

The present data are suggestive of the potential for PKM to be included in broiler diets. If PKM is to be used in broiler diets, formulating diets based on digestible AA contents and enzyme supplementation are strategies to support optimum growth performance of broilers. When PKM-containing diets are balanced on a digestible AA basis, PKM can be included in broiler diets up to 16% with no deleterious effects on growth performance. Further studies are warranted to elucidate the effects of higher β-mannanase activity, individually or in combination, with NSP-degrading enzymes in broiler diets containing PKM.

ACKNOWLEDGMENTS

We would like to gratefully acknowledge the financial support of the AsiaPac (Dongguan) Biotechnology Co. Ltd. (Guangdong, China) under project APD13001 "Use of β-mannanase in enzyme complexes to enhance the utilization of PKM in poultry diets".

REFERENCES

Abdollahi, M. R., B. Hosking, and V. Ravindran. 2015. Nutrient analysis, metabolizable energy and ileal amino acid digestibility of palm kernel meal for broilers. Anim. Feed Sci. Technol. 206:119-125.

Abdollahi, M. R., V. Ravindran, T. J. Wester, G. Ravindran, and D. V. Thomas. 2010. Influence of conditioning temperature on performance, apparent metabolisable energy, ileal digestibility of starch and nitrogen and the quality of pellets, in broiler starters fed maize- and sorghum-based diets. Anim. Feed Sci. Technol. 162:106-115.

Adrizal, A., Y. Yusrizal, S. Fakhri, W. Haris, E. Ali, and C. R. Angel. 2011. Feeding native laying hens diets containing palm kernel meal with or without enzyme supplementations: 1. Feed conversion ratio and egg production. J. Appl. Poult. Res. 20:40-49.

Almaguer, B. L., R. C. Sulabo, Y. Liu, and H. H. Stein. 2014. Standardized total tract digestibility of phosphorus in copra meal, palm kernel expellers, palm kernel meal, and soybean meal fed to growing pigs. J. Anim. Sci. 92:2473-2480.

AOAC. 2005. Official Methods of Analysis. 18th edn. Association of Official Analytical Chemists, Gaithersburg, MD, USA.

Aya, V. E., B. A. Ayanwale, A. T. Ijaiya, and A. Aremu. 2013. Performance and nutrient digestibility in broiler chicks as influenced by multi enzyme addition to starter diets containing palm kernel meal. Biotechnol. Anim. Husb. 29:93-104.

Bryden, W. L., X. Li, G. Ravindran, L. I. Hew, and V. Ravindran. 2009. Ileal digestible amino acid values in feedstuffs for poultry. RIRDC Publication No 09/071. Rural Industries Research and Development Corporation, Canberra, Australia.

Choct, M. 2006. Enzymes for the feed industry: Past, present and future. World's Poult. Sci. J. 62:5-16.

Daskiran, M., R. G. Teeter, D. Fodge, and H. Y. Hsiao. 2004. An evaluation of endo-β-D-mannanase (Hemicell) effects on broiler performance and energy use in diets varying in β-mannan content. Poult. Sci. 83:662-668.

Daud, M. J. and M. C. Jarvis. 1992. Mannan of oil palm kernel. Phytochemistry 31:463-464.

Dusterhoft, E. M., A. G. J. Voragen, and F. M. Engles. 1991. Non-starch polysaccharides from sunflower (*Helianthus annuus*) meal and palm kernel (*Elaeis guineensis*) meal- preparation of cell wall material and extraction of polysaccharide fractions. J. Sci. Food Agric. 55:411-422.

Ezieshi, E. V. and J. M. Olomu. 2008. Nutritional evaluation of palm kernel meal types: 2. Effects on live performance and nutrient retention in broiler chicken diets. Afr. J. Biotechnol. 7:1171-1175.

Hill, F. W. and D. L. Anderson. 1958. Comparison of metabolizable energy and productive energy determinations with growing chicks. J. Nutr. 64:587-603.

Hsiao, H. Y., D. M. Anderson, and N. M. Dale. 2006. Levels of β-mannan in soybean meal. Poult. Sci. 85:1430-1432.

Jackson, M. E. 2010. Mannanase, alpha-galactosidase and pectinase. In: Enzymes in Farm Animal Nutrition, 2nd Ed. (Eds. M. R. Bedford and G. G. Partridge). CABI, Wallingford, Oxfordshire, UK. pp. 54-84.

Jackson, M. E., K. Geronian, A. Knox, J. McNab, and E. McCartney. 2004. A dose-response study with the feed enzyme β-mannanase in broilers provided with corn-soybean meal based diets in the absence of antibiotic growth promoters. Poult. Sci. 83:1992-1996.

Kiarie, E., L. F. Romero, and V. Ravindran. 2014. Growth performance, nutrient utilization, and digesta characteristics in broiler chickens fed corn or wheat diets without or with supplemental xylanase. Poult. Sci. 93:1186-1196.

Knudsen, K. E. B. 1997. Carbohydrate and lignin contents of plant materials used in animal feeding. Anim. Feed Sci. Technol. 67:319-338.

Lee, J. T., C. A. Bailey, and A. L. Cartwright. 2003. β-mannanase ameliorates viscosity-associated depression of growth in broiler chickens fed guar germ and hull fractions. Poult. Sci. 82:1925-1931.

Mardhati, M., H. K. Wong, and S. Noraini. 2011. Growth performance and carcass quality of broilers fed with palm kernel meal-based rations. J. Trop. Agric. Food Sci. 39:157-166.

Olaniyi, O. O. 2014. Effect of beta-mannanse treatment on nutritive quality of palm kernel meal. Afr. J. Microbiol. Res. 8:2405-2410.

Onifade, A. A. and G. M. Babatunde. 1998. Comparison of the utilization of palm kernel meal, brewers' dried grains and corn offal by broiler chicks. Br. Poult. Sci. 39:245-250.

Panigrahi, S. and C. J. Powell. 1991. Effects of high rates of inclusion of palm kernel meal in broiler chick diets. Anim. Feed Sci. Technol. 34:37-47.

Perez, J. F., A. G. Gernat, and J. G. Murillo. 2000. The effect of different levels of palm kernel meal in layer diets. Poult. Sci. 79:77-79.

Ravindran, V., L. I. Hew, G. Ravindran, and W. L. Bryden. 2005. Apparent ileal digestibility of amino acids in feed ingredients for broiler chickens. Anim. Sci. 81:85-97.

Ravindran, V., M. R. Abdollahi, and S. M. Bootwalla. 2014. Nutrient analysis, metabolizable energy, and digestible amino acids of soybean meals of different origins for broilers. Poult. Sci. 93:2567-2577.

Ross, 2007. Ross 308 Broiler: Nutrition Specification, June 2007. Ross Breeders Limited, Newbridge, Midlothian, Scotland, UK.

Saenphoom, P., J. B. Liang, Y. W. Ho, T. C. Loh, and M. Rosfarizan. 2013. Effects of enzyme treated palm kernel expeller on metabolizable energy, growth performance, villus height and digesta viscosity in broiler chickens. Asian Australas. J. Anim. Sci. 26:537-544.

Saenphoom, P., J. B. Liang, Y. W. Ho, T. C. Loh, and M. Rosfarizan. 2011. Effect of enzyme treatment on chemical composition and production of reducing sugars in palm (Elaeis guineenis) kernel expeller. Afr. J. Biotechnol. 10:15372-15377.

SAS Institute Inc. 2004. SAS/STAT User's Guide. Version 9.1.2. SAS Institute Inc., Cary, NC, USA.

Shakila, S., P. Sudhakara Reddy, P. V. V. S. Reddy, J. V. Ramana, and A. Ravi. 2012. Effect of palm kernel meal on the performance of broilers. Tamilnadu J. Vet. Anim. Sci. 8:227-234.

Short, F. J., P. Gorton, J. Wiseman, and K. N. Boorman. 1996. Determination of titanium dioxide added as an inert marker in chicken digestibility studies. Anim. Feed Sci. Technol. 59:215-221.

Soltan, M. A. 2009. Growth performance, immune response and carcass traits of broiler chicks fed on graded levels of palm kernel cake without or with enzyme supplementation. Livest. Res. Rural Dev. 21, Article #37.

Sundu, B., A. Kumar, and J. Dingle. 2006. Palm kernel meal in broiler diets: effect on chicken performance and health. World's Poult. Sci. J. 62:316-325.

Sundu, B., A. Kumar, and J. Dingle. 2005. Response of birds fed increasing levels of palm kernel meal supplemented with enzymes. Proceedings of the 17th Australian Poultry Science Symposium, February 7-9, 2005; Sydney, New South Wales, Australia. 17:227-228.

Williams, M. P., B. Brown, S. Rao, and J. T. Lee. 2014. Evaluation of beta-mannanase and nonstarch polysaccharide-degrading enzyme inclusion separately or intermittently in reduced energy diets fed to male broilers on performance parameters and carcass yield. J. Appl. Poult. Res. 23:715-723.

Zanella, I., N. K. Sakomura, F. G. Silversides, A. Fiqueirdo, and M. Pack. 1999. Effect of enzyme supplementation of broiler diets based on corn and soybeans. Poult. Sci. 78:561-568.

Zou, X. T., X. J. Qiao, and Z. R. Xu. 2006. Effect of β-mannanase (Hemicell) on growth performance and immunity of broilers. Poult. Sci. 85:2176-2179.

9

Toxicity of Mycotoxins from Contaminated Corn with or without Yeast Cell Wall Adsorbent on Broiler Chickens

Q. H. Shang, Z. B. Yang, W. R. Yang, Z. Li, G. G. Zhang, and S. Z. Jiang*

Department of Animal Sciences and Technology, Shandong Agricultural University, Tai'an, Shandong 271018, China

ABSTRACT: This study investigated the effects of feeds naturally contaminated with mycotoxins on growth performance, serum biochemical parameters, carcass traits, and splenic heat shock protein 70 (Hsp70) mRNA expression levels in broiler chickens. The efficacy of yeast cell wall (YCW) adsorbent in preventing mycotoxicosis was also evaluated. Three hundred 1-d-old Arbor Acres broiler chicks were randomly allotted to 3 treatments in completely randomized design for 42 d. Each treatment group had 5 replicate pens with 20 birds. The treatments were as follows: i) basal diet (control), ii) naturally contaminated diet (NCD), and iii) NCD+0.2% YCW adsorbent (NCDD). The NCD decreased average daily gain (ADG) ($p<0.01$) of 0 to 21 d, 22 to 42 d, and 0 to 42 d, and increased feed conversion ratio ($p<0.01$) of 22 to 42 d and 0 to 42 d. Both the breast meat percentage and thigh meat percentage of the NCD group were significantly higher ($p<0.01$) than that of the control group on d 21. The NCD group showed significantly increased levels of triglycerides ($p<0.05$) and cholesterol ($p<0.05$) on both d 21 and d 42 compared to the control group. However, the NCD significantly reduced ($p<0.01$) the high-density lipoprotein (HDL) on d 42 compared to controls. Compared with the NCD, supplementation with YCW significantly improved ($p<0.01$) the ADG of 0 to 21 d and 0 to 42 d, and increased ($p<0.01$) concentrations of HDL on d 42, and on d 21, and triglycerides ($p<0.05$) on d 21 and d 42. Supplementation with YCW reduced ($p<0.01$) the breast meat percentage, the thigh meat percentage, the concentrations of cholesterol ($p<0.01$) and the low-density lipoprotein ($p<0.05$) on d 21, and improved ($p<0.01$) the splenic Hsp70 mRNA expression levels compared with the NCD group. The results of this study indicated that feeding NCD for 42 d had adverse effects on broiler chickens, and that YCW might be beneficial in counteracting the effects of mycotoxins. (**Key Words:** Broiler, Mycotoxin, Toxicity, Yeast Cell Wall Adsorbent)

INTRODUCTION

Mycotoxins are toxic secondary metabolites of various fungi that ubiquitously exist in cereal crops (Bryden, 2012). Globally, food and feeds have been seriously contaminated with mycotoxins among which aflatoxin (AFL), zearalenone (ZEN), fumonisin (FUM), deoxynivalenol (DON), and ochratoxin A are the most commonly found (Schatzmayr and Streit, 2013). Furthermore, food and feeds are frequently co-contaminated with two or more mycotoxins, and their synergistic interaction may exert additive effects (Che et al., 2011; Schatzmayr and Streit, 2013). The consumption of mycotoxin-contaminated food and feeds leads to reduced nutrient absorption (Maresca et al., 2002), poor growth performance (Chowdhury and Smith, 2004), immunosuppression (Swamy et al., 2003), disturbed reproductive performance (Young et al., 1990), and residues in animal products (Akande et al., 2006), resulting in a massive economic impact worldwide in human health, animal health, and agricultural trade. However, information on the effects of feeds naturally contaminated with mycotoxins on carcass traits and the splenic heat shock protein 70 (Hsp70) mRNA expression levels in broiler chickens is limited.

So far, one of the most practical and effective methods to detoxify mycotoxin-contaminated feed is the use of adsorbents. A polymeric glucomannan mycotoxin adsorbent derived from the cell wall of yeast has been shown to counteract some of the deleterious effects of mycotoxins in swine (Swamy et al., 2002a) and chickens (Aravind et al.,

* Corresponding Author: S. Z. Jiang.
E-mail: shuzhen305@163.com

2003). Because of the extensive use of this adsorbent as a feed additive in China, we carried out this *in vivo* experiment to evaluate the efficiency of yeast cell wall (YCW) in sequestering mycotoxins and its safety towards broiler chickens.

Therefore, this study was conducted to evaluate the effects of naturally contaminated diets (NCD) on growth performance, serum biochemical parameters, carcass traits, and the splenic Hsp70 mRNA expression levels in broiler chickens, and to determine the efficacy of YCW in alleviating mycotoxin-induced effects.

MATERIALS AND METHODS

Adsorbents

Detoxza, a commercial YCW adsorbent, was obtained from AB Co Products Asia (Harbin, China). The main component of YCW adsorbent used in this study is a kind of highly refined polysaccharide complex with a unique adsorptive capacity.

Experimental animals, diets, and management

Three hundred 1-d-old Arbor Acres broiler chicks (mixed sexes) were obtained from a local commercial hatchery (Shandong, China). All broiler chicks were individually weighed, wing-banded, and randomly allotted to 3 treatments in completely randomized design. Each treatment group has 5 replicate pens ($1.22\times1.08\times0.45$ m^3) with 20 birds. The treatments were as follows: i) basal diet (control), ii) NCD with mycotoxins, and iii) NCD+0.2% YCW adsorbent (NCDD). The NCD treatment were formulated by replacing non-contaminated corn that used in the control diet with naturally mycotoxin-contaminated corn, and the NCDD treatment diet was completed by adding adsorbent into the contaminated diets. All diets were prepared in a single batch per treatment group and then stored in covered containers prior to feeding (Table 1). Nutrient concentration of all diets met or exceeded the minimum requirements according to Feeding Standards for Chicken in the People's Republic of China (NY/T 33-2004).

Throughout the experiment, all broiler chickens were kept in an environmentally controlled room with *ad libitum* feeding and watering. The room temperature was maintained at 32°C for the first week, and then reduced 3°C weekly until the temperature reached 26°C. All animals used in this study were cared for strictly following the animal care and use protocol approved by the Animal Nutrition Research Institute of Shandong Agricultural University.

Mycotoxin quantification

Representative control and the contaminated diets were taken before and at the end of the starter and grower diet periods and analyzed for mycotoxin concentrations by the Asia Mycotoxin Analysis Center (Chaoyang University of Technology, Taichung, Taiwan). DON was analyzed using high-performance liquid chromatography. Fluorometry and enzyme-linked immune-assay were used to measure ZEN, FUM, and AFL concentrations. Mycotoxin concentrations are presented in Table 1. The detection limits for these mycotoxins were 1 μg/kg for AFL, 0.1 mg/kg for ZEN, 0.1 mg/kg for DON (including 3-acetyl deoxynivalenol, 15-acetyl deoxynivalenol, and nivalenol), and 0.25 mg/kg for FUM.

Growth parameters

Daily mortality was recorded. All broiler chickens were weighed individually each week to determine the average daily gain (ADG). Orts and spillages were collected and weighed daily to determine the average daily feed intake (ADFI). The feed conversion ratio (FCR, g of feed/g of gain) was calculated based on these data at the end of each diet period.

Sample collection

After fasting for 12 h, ten broilers from each treatment (2 per pen) were randomly selected in the morning of d 21 and d 42. Approximately 10 mL of blood were collected from the jugular vein of each broiler into non-heparinized tubes. After incubated at 37°C for 2 h, the serum was separated by centrifugation at 1,500×*g* for 10 min and stored in 1.5-mL centrifuge tubes at –20°C until biochemical analysis. Broiler chickens were then humanely euthanized by cervical dislocation. Spleen tissues were collected from each broiler chicken and another part was stored at –80°C for Hsp70 analysis.

Carcass traits

After blood sampling, the birds were weighed, slaughtered, de-feathered, processed (removal of head and feet), and eviscerated. Then eviscerated yield, breast meat and thigh meat of each broiler chicken were weighed. Eviscerated yield percentage was calculated as a percentage of live body weight. Breast meat percentage and thigh meat percentage were calculated as a percentage of eviscerated yield.

Serum biochemical parameters

Cholesterol, triglyceride, high-density lipoprotein (HDL), low-density lipoprotein (LDL), and uric acid levels in the serum were determined by a diagnostic kit (Nanjing Jiancheng Bioengineering Institute, Nanjing, China) using an automatic clinical chemistry analyzer (Roche, Cobus-Mira-Plus, Roche Diagnostic System Inc., Los Angeles, CA, USA).

Table 1. Composition (%) and dietary mycotoxin concentrations of the experimental diets

Item	Starter (0 to 21 d)			Grower (22 to 42 d)		
	Control[1]	NCD[1]	NCDD[1]	Control	NCD	NCDD
Ingredients						
Moldy corn	-	56.12	56.12	-	56.25	56.25
Normal corn	56.12	-	-	61.25	5.00	5.00
Soybean meal	32.50	32.50	32.50	26.00	26.00	26.00
Corn gluten meal	4.50	4.50	4.50	5.00	5.00	5.00
Soybean oil	2.00	2.00	2.00	3.00	3.00	3.00
Calcium hydrogen phosphate	1.60	1.60	1.60	1.50	1.50	1.50
Limestone	1.30	1.30	1.30	1.40	1.40	1.40
Sodium chloride	0.30	0.30	0.30	0.30	0.30	0.30
Lysine	0.24	0.24	0.24	0.15	0.15	0.15
Methionine	0.24	0.24	0.24	0.20	0.20	0.20
Vitamin-mineral premix[2]	1.00	1.00	1.00	1.00	1.00	1.00
Sand	0.20	0.20	-	0.20	0.20	-
Yeast cell wall adsorbent	-	-	0.20	-	-	0.20
Calculated composition						
Metabolizable energy (MJ/kg)	12.64	12.64	12.64	13.01	13.01	13.01
Crude protein (%)	21.36	21.36	21.36	19.35	19.35	19.35
Calcium (%)	0.96	0.96	0.96	0.95	0.95	0.95
Total phosphorus (%)	0.66	0.66	0.66	0.63	0.63	0.63
Sodium chloride (%)	0.30	0.30	0.30	0.30	0.30	0.30
Lysine (%)	1.20	1.20	1.20	1.00	1.00	1.00
Methionine (%)	0.54	0.54	0.54	0.48	0.48	0.48
Total sulfur amino acids (%)	0.88	0.88	0.88	0.79	0.79	0.79
Threonine (%)	0.81	0.81	0.81	0.73	0.73	0.73
Tryptophan (%)	0.26	0.26	0.26	0.23	0.23	0.23
Analyzed mycotoxins (μg/kg)						
Aflatoxin B_1	6.68	102.08	102.08	8.48	101.18	101.18
Zearalenone	18.32	281.92	281.92	18.81	285.20	285.20
Fumonisin	952.59	5,874.38	5,874.38	1,276.88	5,977.36	5,977.36
Deoxynivalenol	642.16	2,038.96	2,038.96	545.73	2,051.08	2,051.08

[1] Control, basal diet; NCD, naturally contaminated diet; NCDD, NCD+0.2% yeast cell wall adsorbent.

[2] Supplied per kilogram of diet: vitamin A, 8,050 IU; cholecalciferol, 1,800 IU; vitamin E, 20 IU; vitamin K_3, 5.1 mg; thiamin, 2.4 mg; riboflavin, 8.2 mg; pantothenic acid, 15.3 mg; pyridoxine, 3.1 mg; cobalamin, 0.02 mg; niacin, 32 mg; choline chloride, 1,000 mg; biotin, 0.20 mg; folic acid, 1.2 mg; Mn, 68 mg; Fe, 85 mg; Zn, 58 mg; Cu, 8.6 mg; I, 0.27 mg; Se, 0.20 mg.

Determination of spleen Hsp70 mRNA expression levels by real-time polymerase chain reaction

Total RNA was extracted from splenocytes using an ultra pure RNA extraction kit (Cat#CW0581, CWbio. Co. Ltd., Beijing, China). The RNA concentrations and integrity were determined by agarose gel electrophoresis using 5 μL samples. The DNAse treated RNA was used for cDNA synthesis using a HiFi-MMLV cDNA First Strand Synthesis kit (Cat#CW0744, CWbio. Co. Ltd., China) with random primers and stored at –80°C. The real-time polymerase chain reaction (RT-PCR) reactions were performed by an RT-PCR system (Roche 480, Roche Diagnostic System Inc., Los Angeles, CA, USA) using UltraSYBR Mixture, ROX (Cat#CW0956, CWbio. Co. Ltd., China). After denaturing at 95°C for 10 min, the PCR amplification was performed for 40 cycles (95°C for 15 s and 60°C for 1 min), followed by a final extension step (72°C for 10 min). The splenic Hsp70 mRNA expression levels were normalized with the relative intensity of the *β-actin* gene. The sequences of oligonucleotide primers used in RT-PCR analysis are presented in Table 2.

Statistical analysis

Data were analyzed as a completely randomized design using the GLM procedure of SAS (Version 9.0; SAS Inst. Inc., Cary NC, USA). The pen was used as the experimental unit. The data were analyzed as a completely randomized design to examine the overall effect of treatments. Differences among treatments were separated using Duncan's multiple range test and accepted as significant if $p<0.05$.

Table 2. Sequences of oligonucleotide primers used in real-time PCR analysis

Gene		Oligonucleotide sequences (5′ to 3′)	Accession number
Hsp70	F	GACAAGTCAAAGCCGCACAT	NM_001006685
	R	AAGTCGTTCATCGGGAGC	
β-actin	F	CAACACAGTGCTGTCTGGTGG	NM_205518
	R	ATCGTACTCCTGCTTGCTGAT	

PCR, polymerase chain reaction; Hsp70, Heat shock protein 70; F, forward primer; R, reverse primer.

RESULTS

Growth performance

The ADFI did not differ significantly among treatments during the starter or grower diet period (Table 3). However, compared with the control, the NCD decreased ADG ($p<0.01$) of 0 to 21 d, 22 to 42 d, and 0 to 42 d, and increased FCR ($p<0.01$) of 22 to 42 d and 0 to 42 d. Supplementation with YCW significantly improved ($p<0.01$) the ADG of 0 to 21 d and 0 to 42 d compared with the NCD. The mortalities of chickens fed the NCD were higher ($p<0.05$) compared with the control. But mortality was reduced ($p<0.05$) significantly by adding YCW in basal diets.

Carcass traits

Both the breast meat percentage and thigh meat percentage of the NCD group were significantly higher ($p<0.01$) than that of the control group on d 21, but not on d 42 (Table 4). All broiler chickens had similar eviscerated yield percentage during the entire experimental period. Supplementation with YCW reduced ($p<0.01$) the effect of mycotoxins on the breast meat percentage and thigh meat percentage on d 21 compared to the NCD.

Serum biochemical parameters

The NCD group showed significantly increased levels of triglycerides ($p<0.05$) and cholesterol ($p<0.05$) on both d 21 and d 42 compared to the control group (Table 5). However, the NCD significantly reduced ($p<0.01$) the HDL on d 42 compared to controls. No differences were noted in uric acid among treatments. Supplementation of YCW at the dosage of 0.2% to the NCD increased ($p<0.01$) concentrations of HDL on d 42, and decreased concentrations of cholesterol ($p<0.01$) and the LDL ($p<0.05$) on d 21, and triglycerides ($p<0.05$) on d 21 and d 42.

Splenic Hsp70 mRNA expression levels

The splenic Hsp70 mRNA expression levels of control, NCD, and NCDD groups on d 42 are presented in Figure 1. The NCD group showed significantly higher ($p<0.01$)

Table 3. The effects of basal diet and naturally mycotoxin-contaminated diets with or without yeast cell wall adsorbent, on growth performance of broiler chickens (n = 5)

Item	Control[1]	NCD[1]	NCDD[1]	SEM	p-value
BW (g)					
0 d	50.34	52.03	52.55	2.06	0.735
21 d	798.99[a]	740.83[c]	772.85[b]	7.345	<0.001
42 d	2,296.50[a]	2,117.59[b]	2,184.47[ab]	42.868	0.036
ADG (g/d)					
0 to 21 d	35.65[a]	32.80[c]	34.30[b]	0.433	0.002
22 to 42 d	71.30[a]	65.56[b]	67.23[b]	1.046	0.006
0 to 42 d	53.48[a]	49.18[c]	50.76[b]	0.413	0.001
ADFI (g/d)					
0 to 21 d	47.72	46.6	47.23	0.502	0.324
22 to 42 d	133.6	133.52	134.58	0.787	0.588
0 to 42 d	90.66	90.06	90.90	0.540	0.543
FCR					
0 to 21 d	1.34[b]	1.42[a]	1.38[ab]	0.023	0.075
22 to 42 d	1.88[b]	2.04[a]	2.00[a]	0.026	0.003
0 to 42 d	1.70[b]	1.83[a]	1.79[a]	0.015	<0.001
Mortality (%)					
0 to 42 d	1.00[b]	6.00[a]	1.00[b]	1.350	0.034

SEM, standard error of the mean; BW, body weight; ADG, average daily gain; ADFI, average daily feed intake; FCR, g of feed/g of gain.

[1] Control, basal diet; NCD, naturally contaminated diet; NCDD, NCD+0.2% yeast cell wall adsorbent.

[a,b] Means within a row with different letters differ significantly at $p<0.05$.

Table 4. The effects of basal diet and naturally mycotoxin-contaminated diets with or without yeast cell wall adsorbent, on carcass rates of broiler chickens (n = 5)

Item	Control[1]	NCD[1]	NCDD[1]	SEM	p-value
Breast meat percentage (%)					
21 d	61.56[b]	64.42[a]	58.48[b]	0.315	0.004
42 d	83.47	85.97	83.45	0.290	0.219
Thigh meat percentage (%)					
21 d	62.65[b]	67.89[a]	63.38[b]	0.233	0.001
42 d	71.11	71.95	70.46	0.348	0.568
Eviscerated yield percentage (%)					
21 d	708.58	706.12	707.76	1.996	0.315
42 d	764.30	755.32	760.74	1.356	0.616

SEM, Standard error of the mean.

[1] Control, basal diet; NCD, naturally contaminated diet; NCDD, NCD+0.2% yeast cell wall adsorbent.

[a,b] Means within a row with different letters differ significantly at $p<0.05$.

Hsp70 mRNA expression levels than that of the control group. Supplementation with YCW improved ($p<0.01$) the splenic Hsp70 mRNA expression levels compared with the NCD group.

DISCUSSION

Although each ingredient had been carefully selected, the AFL, ZEN, FUM, and DON were detected in the basal diet, which may suggest that chicken feeds are extensively contaminated with mycotoxins in China. However, the levels of AFL, ZEN, FUM, and DON were considered acceptable, since they were in compliance with the regulations of the Food and Agriculture Organization of the United Nations (FAO, 2006; AFL, 20 μg/kg; ZEN, 1 mg/kg; FUM, 3 mg/kg; and DON, 2 mg/kg).

The common symptom of mycotoxins in poultry is poor growth performance (Chowdhury and Smith, 2004), which results in massive economic losses. The effect of mycotoxins on growth performance of broiler chickens was contradictory. In the current study, ADG of broilers in NCD was significantly reduced and the mortalities of chickens fed the NCD were higher compared to those provided the basal diet, but without any changes in ADFI. These results indicated that the adverse effect of mycotoxins on ADG was probably due to impairment of nutrient absorption and inhibition of protein synthesis (Klasing and Barnes, 1988), other than the attenuation of feed intake. Xu et al. (2011) also observed that body weight gain reduced with

Table 5. The effects of basal diet and naturally mycotoxin-contaminated diets with or without yeast cell wall adsorbent, on serum biochemical parameters of broiler chickens (n = 5)

Item	Control[1]	NCD[1]	NCDD[1]	SEM	p-value
Cholesterol (mmol/L)					
21 d	3.06[b]	3.22[a]	3.09[b]	0.79	0.001
42 d	3.05[b]	3.13[a]	3.10[a]	1.061	0.019
Triglycerides (mmol/L)					
21 d	0.42[b]	0.87[a]	0.54[b]	0.248	0.017
42 d	0.88[b]	1.03[a]	0.68[c]	1.19	0.038
HDL (mmol/L)					
21 d	1.68	1.82	1.47	0.356	0.098
42 d	1.60[a]	1.43[b]	1.59[a]	0.018	0.001
LDL (mmol/L)					
21 d	1.18[a]	1.21[a]	1.04[b]	0.564	0.036
42 d	0.87	0.95	0.86	2.01	0.235
Uric acid (mmol/L)					
21 d	1.03	1.05	1.01	0.674	0.063
42 d	0.50	0.42	0.46	1.039	0.217

SEM, standard error of the mean; HDL, high-density lipoprotein; LDL, low-density lipoprotein.

[1] Control, basal diet; NCD, naturally contaminated diet; NCDD, NCD+0.2% yeast cell wall adsorbent.

[a,b] Means within a row with different letters differ significantly at $p<0.05$.

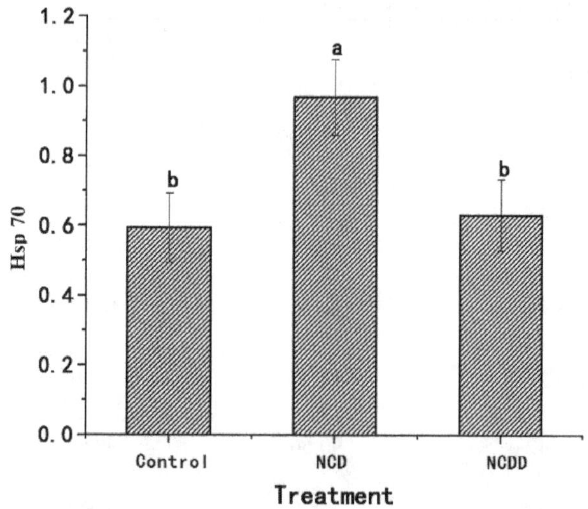

Figure 1. Splenic heat shock protein 70 (Hsp70) mRNA expression levels of 42-d broiler chickens that provided basal diet and naturally mycotoxin-contaminated diets with or without yeast cell wall adsorbent ($p<0.01$, SEM = 0.054). Values are the mean±SEM (n = 5). NCD, naturally contaminated diet; NCDD, NCD+0.2% yeast cell wall adsorbent. SEM, standard error of the mean. Bars with different letters differ significantly at $p<0.05$.

increasing concentrations of dietary DON without affecting the overall feed consumption of broilers as the mucosal lining of the gastrointestinal tract was altered. In contrast, Awad et al. (2006) indicated that body weight, body weight gain, and feed conversion were not affected by the inclusion of 5 mg/kg of DON in the feed. It has been reported that synergistic effects among mycotoxins can multiply their individual adverse effects (Conkova et al., 2003). Therefore, this inconsistency of results may be due to the presence of other mycotoxins and their synergistic effect in the liver, which therefore inhibited protein and DNA syntheses (Huff et al., 1988).

Very limited information is available on the effect of feeding mycotoxin-contaminated grains on carcass traits of broilers. In the present study, the beast meat percentage and thigh meat percentage of the NCD group were significantly higher than that of the control group on d 21, without significantly changes in eviscerated yield percentage, breast meat percentage and thigh meat percentage on d 42. This may be because Zearalanol promotes protein synthesis and increases the lean meat ratio in a manner similar to estrogen (Wang, 2013). However, its mechanisms still need further research.

Changes in serum biochemical parameters can be used to predict toxic effects of aflatoxicosis before the appearance of major symptoms (Kececi et al., 1998). In this study, the NCD resulted in an increase in triglycerides and cholesterol, and a decrease in the concentration of HDL as compared to the control diet. Hedayati et al. (2014) have reported that cholesterol level has been increased significantly by addition of AFL into the diet. Chowdhury and Smith (2007) also observed plasma cholesterol concentration increased compared with controls when birds were fed contaminated grains. Increased cholesterol levels may be attributed to the hepatotoxic effects of AFL, ZEN, FUM, and DON or their synergetic effect that characterized by impairment of transport and lipid metabolism of liver (Rosa et al., 2001). The HDL can remove cholesterol from the blood vessels and carries it back to the liver, where it can be processed and sent out of the body. The liver is an important place for synthesis of HDL. The reduced HDL may suggest that mycotoxins exert a toxic effect on liver of the chickens.

One of the key cellular responses to toxicant exposure, which could potentially be used as early marker of toxicity, is the heat shock response (Carnevali and Maradonna, 2003). The Hsp70 is closely related to cytoprotection and induced in response to toxic stress (Hassen et al., 2007). In this study, the results clearly indicated that mycotoxins induced an increase in the splenic Hsp70 mRNA expression levels. Previous reported studies also showed significant changes in the expression levels of Hsp 70 mRNA in different cell lines (El Golli et al., 2006) that may be due to systemic inflammation, oxidative stress (Hassen et al., 2007), or tissue damage caused by mycotoxins.

A variety of adsorbents that prevent mycotoxicosis have been extensively studied in livestock (Huwig et al., 2001). It has been shown that glucomannan polymer was efficacious in preventing some adverse effects of *Fusarium* mycotoxins in broiler chickens (Swamy et al., 2002b), swine (Swamy et al., 2003), and laying hens (Chowdhury and Smith, 2004). Yeast β-D-glucan has been also shown to be a suppressor of mycotoxin effects (Yiannikouris et al., 2004) because it has a large surface area and high adsorption capacity. In the current study, the improved results of ADG on 0 to 42 d, cholesterol on d 21, triglycerides in serum on both d 21 and d 42, HDL in serum on d 42 and the splenic Hsp70 mRNA expression levels on d 42 in NCDD group compared to NCD group may due to YCW partially trapping the mycotoxin molecule in its glucomannan matrix and preventing toxin absorption from the gastrointestinal tract (Che et al., 2011). Li et al. (2012) reported that supplementation of YCW to the NCD showed a positive effect against oxidative stress (total superoxide dismutase, malondialdehyde in serum) and on immunological parameters (splenic mRNA expression of interleukin1-1β and interleukin-6). Therefore, the YCW could potentially be an effective method to adsorb and sequester mycotoxins, leading to the reduction of toxin bioavailability (Kogan and Kocher, 2007).

In summary, this study showed that diets naturally contaminated with mycotoxins negatively affected growth performance, carcass traits, and some serum biochemical parameters. The YCW that added to NCD prevented some adverse effects of mycotoxins on broiler chickens. These results suggested that YCW might be beneficial in counteracting the effects of mycotoxins. However, further research is required to evaluate the addition of YCW in feeds at different concentrations, in order to define the optimum application rate for the reduction of mycotoxin effects.

REFERENCES

Akande, K. E., M. M. Abubakar, T. A. Adegbola, and S. E. Bogoro. 2006. Nutritional and health implications of mycotoxins in animal feeds: A review. Pakistan J. Nutr. 5:398-403.

Aravind, K. L., V. S. Patil, G. Devegowda, B. Umakantha, and S. P. Ganpule. 2003. Efficacy of esterified glucomannan to counteract mycotoxicosis in naturally contaminated feed on performance and serum biochemical and hematological parameters in broilers. Poult. Sci. 82:571-576.

Awad, W. A., J. Bohm, E. Razzazi-Fazeli, and J. Zentek. 2006. Effects of feeding deoxynivalenol contaminated wheat on growth performance, organ weights and histological parameters of the intestine of broiler chickens. J. Anim. Physiol. Anim. Nutr. 90:32-37.

Bryden, W. L. 2012. Mycotoxin contamination of the feed supply chain: Implications for animal productivity and feed security. Anim. Feed. Sci. Technol. 173:134-158.

Carnevali, O. and F. Maradonna. 2003. Exposure to xenobiotic compounds: Looking for new biomarkers. Gen. Comp. Endocrinol. 131:203-208.

Che, Z., Y. Liu, H. Wang, H. Zhu, Y. Hou, and B. Ding. 2011. The protective effects of different mycotoxin adsorbents against blood and liver pathological changes induced by mold-contaminated feed in broilers. Asian Australas. J. Anim. Sci. 24:250-257.

Chowdhury, S. R. and T. K. Smith. 2004. Effects of feeding blends of grains naturally contaminated with Fusarium mycotoxins on performance and metabolism of laying hens. Poult. Sci. 83:1849-1856.

Chowdhury, S. R. and T. K. Smith. 2007. Effects of feed-borne *Fusarium* mycotoxins on performance, plasma chemistry and hepatic fractional protein synthesis rates of turkeys. Can. J. Anim. Sci. 87:543-551.

Conkova, E., A. Laciakova, G. Kovac, and H. Seidel. 2003. Fusarial toxins and their role in animal diseases. Vet. J. 165: 214-220.

El Golli, E., W. Hassen, A. Bouslimi, C. Bouaziz, M. M. Ladjimi, and H. Bacha. 2006. Induction of Hsp 70 in Vero cells in response to mycotoxins: Cytoprotection by sub-lethal heat shock and by Vitamin E. Toxicol. Lett. 166:122-130.

Food and Agriculture Organization of the United Nations (FAO). 2006. Worldwide regulations for mycotoxins in food and feed in 2003. FAO Food and Nutrition Paper No. 81. Rome, Italy.

Hassen, W., I. Ayed-Boussema, A. A. Oscoz, A. De Cerain Lopez, and H. Bacha. 2007. The role of oxidative stress in zearalenone-mediated toxicity in Hep G2 cells: Oxidative DNA damage, gluthatione depletion and stress proteins induction. Toxicology 232:294-302.

Hedayati, M., M. Manafi, M. Yari, and S. V. Mousavipour. 2014. Commercial broilers exposed to aflatoxin b_1: Efficacy of a commercial mycotoxin binder on internal organ weights, biochemical traits and mortality. Int. J. Agric. For. 4:351-358.

Huff, W. E., R. B. Harvey, L. F. Kubena, and G. E. Rottinghaus. 1988. Toxic synergism between aflatoxin and t-2 toxin in broiler chickens. Poult. Sci. 67:1418-1423.

Huwig, A., S. Freimund, O. Kappeli, and H. Dutler. 2001. Mycotoxin detoxication of animal feed by different adsorbents. Toxicol. Lett. 122:179-188.

Kececi, T., H. Oguz, V. Kurtoglu, and O. Demet. 1998. Effects of polyvinylpolypyrrolidone, synthetic zeolite and bentonite on serum biochemical and haematological characters of broiler chickens during aflatoxicosis. Br. Poult. Sci. 39:452-458.

Klasing, K. C. and D. M. Barnes. 1988. Decreased amino acid requirements of growing chicks due to immunologic stress. J. Nutr. 118:1158-1164.

Kogan, G. and A. Kocher. 2007. Role of yeast cell wall polysaccharides in pig nutrition and health protection. Livest. Sci. 109:161-165.

Li, Z., Z. B. Yang, W. R. Yang, S. J. Wang, S. Z. Jiang, and Y. B. Wu. 2012. Effects of feed-borne Fusarium mycotoxins with or without yeast cell wall adsorbent on organ weight, serum biochemistry, and immunological parameters of broiler chickens. Poult. Sci. 91:2487-2495.

Maresca, M., R. Mahfoud, N. Garmy, and J. Fantini. 2002. The mycotoxin deoxynivalenol affects nutrient absorption in human intestinal epithelial cells. J. Nutr. 132:2723-2731.

Ministry of Agriculture of China. 2004. Feeding standard of Chicken of the People's Republic of China. NY/T 33-2004. Ministry of Agriculture, Beijing, China.

Rosa, C. A. R., R. Miazzo, C. Magnoli, M. Salvano, S. M. Chiacchiera, S. Ferrero, M. Saenz, E. C. Q. Carvalho, and A. Dalcero. 2001. Evaluation of the efficacy of bentonite from the south of Argentina to ameliorate the toxic effects of aflatoxin in broilers. Poult. Sci. 80:139-144.

Schatzmayr, G. and E. Streit. 2013. Global occurrence of mycotoxins in the food and feed chain: Facts and figures. World Mycotoxin J. 6:213-222.

Swamy, H. V. L. N., T. K. Smith, E. J. MacDonald, N. A. Karrow, B. Woodward, and H. J. Boermans. 2003. Effects of feeding a blend of grains naturally contaminated with Fusarium mycotoxins on growth and immunological measurements of starter pigs, and the efficacy of a polymeric glucomannan mycotoxin adsorbent. J. Anim. Sci. 81:2792-2803.

Swamy, H. V., T. K. Smith, E. J. MacDonald, H. J. Boermans, and E. J. Squires. 2002a. Effects of feeding a blend of grains naturally contaminated with Fusarium mycotoxins on swine performance, brain regional neurochemistry, and serum chemistry and the efficacy of a polymeric glucomannan mycotoxin adsorbent. J. Anim. Sci. 80:3257-3267.

Swamy, H. V., T. K. Smith, P. F. Cotter, H. J. Boermans, and A. E. Sefton. 2002b. Effects of feeding blends of grains naturally contaminated with Fusarium mycotoxins on production and metabolism in broilers. Poult. Sci. 81:966-975.

Wang, W. 2013. Phytoestrogen α-Zearalanol in an Animal Model of Menopause. In: Nutrition and Diet in Menopause (Eds. C. H. Martin, R. R. Watson, and V. R. Preedy). Humana Press, New York, NY, USA. pp. 407-417.

Xu, L., S. D. Eicher, and T. J. Applegate. 2011. Effects of increasing dietary concentrations of corn naturally contaminated with deoxynivalenol on broiler and turkey poult performance and response to lipopolysaccharide. Poult. Sci. 90:2766-2774.

Yiannikouris, A., J. Francois, L. Poughon, C. G. Dussap, G. Bertin, G. Jeminet, and J. P. Jouany. 2004. Alkali extraction of β-D-glucans from Saccharomyces cerevisiae cell wall and study of their adsorptive properties toward zearalenone. J. Agric. Food Chem. 52:3666-3673.

Young, L. G., H. Ping, and G. J. King. 1990. Effects of feeding zearalenone to sows on rebreeding and pregnancy. J. Anim. Sci. 68:15-20.

10

Effect of L- or DL-methionine Supplementation on Nitrogen Retention, Serum Amino Acid Concentrations and Blood Metabolites Profile in Starter Pigs

Q. Y. Tian, Z. K. Zeng, Y. X. Zhang, S. F. Long, and X. S. Piao*

State Key Laboratory of Animal Nutrition, Ministry of Agriculture Feed Industry Centre, China Agricultural University, Beijing 100193, China

ABSTRACT: The objective of the current study was to evaluate the effect of supplementation of either L-methionine (L-Met) or DL-methionine (DL-Met) to diets of starter pigs on nitrogen (N) balance, metabolism, and serum amino acid profile. Eighteen crossbred (Duroc×Landrace×Yorkshire) barrows weighing 15.45±0.88 kg were randomly allotted to 1 of 3 diets with 6 pigs per treatment. The diets included a basal diet (Met-deficient diet) containing 0.24% standardized ileal digestibility Met with all other essential nutrients meeting the pig's requirements. The other two diets were produced by supplementing the basal diet with 0.12% DL-Met or L-Met. The experiment lasted for 18 days, consisting of a 13-day adaptation period to the diets followed by a 5-day experimental period. Pigs were fed *ad libitum* and free access to water throughout the experiment. Results showed that the supplementation of either L-Met or DL-Met improved N retention, and serum methionine concentration, and decreased N excretion compared with basal diet ($p<0.01$). The N retention of pigs fed diets supplemented with the same inclusion levels of DL-Met or L-Met were not different ($p>0.05$). In conclusion, on equimolar basis DL-Met and L-Met are equally bioavailable as Met sources for starter pigs. (**Key Words:** Amino Acids Profile, Methionine, Pigs, Nitrogen Balance)

INTRODUCTION

Methionine (Met) is one of the most important essential amino acids (AA) in livestock nutrition. It is a limiting amino acid in complex pig diets containing spray dried blood products or dried whey (Cromwell, 2004). Apart from its nutritional function, Met is also important for the metabolism and gut health of animals. Through transmethylation Met can be converted to S-adenosylmethionine, a primary methyl donor that methylates compounds to form such products as creatine and phosphatidylcholine (Finkelstein, 1990; Martín-Venegas et al., 2006). Methionine can be converted to cysteine, a precursor of glutathione, and taurine. These molecules may regulate intestinal epithelial oxidative status, and may contribute to intestinal mucosal integrity and gut function (Shoveller et al., 2005; Riedijk, et al., 2007; Chen et al., 2014; Shen et al., 2014).

Methionine is primarily produced by either chemical synthesis or hydrolyzing proteins. The product by chemical synthesis is DL-mixture of the amino acid (Mannsfeld et al., 1978; Gomes and Kumar, 2005) whereas hydrolysis of proteins leads to a complex mixture from which Met must be separated (Kumar and Gomes, 2005). It is commonly supplemented in diets as dry DL-methionine (DL-Met) or as liquid DL-Met hydroxy analog-free acid (MHA-FA). However, both the DL-Met and MHA-FA have to be converted to L-Met before it can be used by animals, and L-Met is the only form used for protein synthesis and metabolism (Dibner and Ivey, 1992; Stoll et al., 1998; Martín-Venegas et al., 2006). The conversion of D-Met to L-Met is not a limiting factor due to the existence of substantial D-amino acid oxidase activity in different tissues, such as kidney, liver, stomach, duodenum, jejunum and ileum, of pigs (Fang et al., 2010). Recently a

* Corresponding Author: X. S. Piao.
E-mail: piaoxsh@mafic.ac.cn

fermentation approach has been used to produce L-Met (Odunfa et al., 2001; Ikeda, 2003; Kumar and Gomes, 2005). During the fermentation process, L-Met precursor was obtained from microorganism strain, and mixed with converting enzyme and methyl mercaptan which converted the L-Met precursor to L-Met (Kim et al., 2015). Studies have been conducted on comparison on L-Met and DL-Met for pigs in the past (Cho et al., 1980; Kim and Bayley, 1983; Chung and Baker, 1992; Shen et al., 2014). Some of them were conducted more than 30 years ago. Shen et al. (2014) conducted two experiments to test the effect of dietary L-Met supplementation on growth performance and gut health in nursery pigs compared with DL-Met containing a basal diet (0.18%, standardized ileal digestibility [SID] Met) supplemented with L-Met or DL-Met 0.048%, 0.096%, and 0.144%, respectively. However, little information is available on the effect of L-Met and DL-Met on nitrogen (N) balance, and serum amino acid profile in starter pigs.

The objective of the present study was to determine the effect of L-Met as a Met source compared with DL-Met on N balance, serum protein, and amino acid profile in starter pigs.

MATERIALS AND METHODS

General

The experiment was conducted in the Metabolism Laboratory of Ministry of Agriculture Feed Industry Center (Beijing, China). The Institutional Animal Care and Use Committee at China Agricultural University (Beijing, China) reviewed and approved the protocols used in this study. L- and DL-methionine were obtained from CJ (CheilJedang Corporation, Seoul, Korea). L-methionine was produced by fermentation processes. The analyzed purity of L-Met and DL-Met was 99.3% and 99.2%, respectively.

Animals, diets, and experimental design

Eighteen crossbred (Duroc×Landrace×Yorkshire) barrows with initial body weight of 15.45±0.88 kg were randomly allotted to 3 diets. The basal (BD) diet was formulated according to the recommendations of NRC (1998) with the exception of Met. The Met concentration in the basal diet was 0.24%, which is below the Met requirement for piglets between 10 to 20 kg body weight according to NRC (1998). All other nutrients and energy were adequate for the basal diet. The other two experimental diets were formulated based on the basal diet, 0.12% either DL-Met or L-Met was supplemented. The SID value of Met in the experimental diets was 0.36%, which met the recommendation of NRC (1998). The composition of all diets is presented in Table 1. Before the beginning of experiment, the concentrations of AA in test diets were analyzed for confirmation. The pigs were individually housed in stainless-steel metabolism crates (1.4×0.7×0.6 m) and maintained in an environmentally controlled room with ambient temperature of 24±2°C. All pigs were fed *ad libitum* and free access to water throughout the experiment.

Table 1. Ingredients and nutrient contents of experimental diets (%, as-fed basis)

Items	Basal diet	L-Met	DL-Met
Ingredients			
Corn	31.26	31.14	31.14
Wheat	40.00	40.00	40.00
Soybean meal	23.30	23.30	23.30
Soybean oil	2.00	2.00	2.00
Dicalcium phosphate	1.20	1.20	1.20
Limestone	0.80	0.80	0.80
Sodium chloride	0.30	0.30	0.30
Lysine	0.48	0.48	0.48
Methionine	-	0.12	0.12
Threonine	0.16	0.16	0.16
Vit. Min. Mix[1]	0.50	0.50	0.50
Total	100.00	100.00	100.00
Analyzed nutritional content			
Crude protein	18.94	18.55	18.52
Calcium	0.70	0.71	0.70
Total phosphorous	0.62	0.60	0.61
Lysine	1.24	1.26	1.25
Methionine	0.27	0.38	0.38
Methionine+cysteine	0.60	0.72	0.74

L-Met, L-methionine; DL-Met, DL-methionine.

[1] Provided per kilogram complete feed: vitamin A, 12,000 IU; vitamin D_5, 2,500 IU; vitamin E 30 IU; vitamin K_3, 3 mg; vitamin B_{12}, 0.012 mg; vitamin K_3, 3 mg; riboflavin, 4 mg; niacin, 40 mg; pantothenic acid, 15 mg; Choline chloride, 400 mg; folic acid, 0.7 mg; vitamin B_1, 1.5 mg; vitamin B_6, 1.5 mg; Biotin, 0.1 mg; Zn, 105 mg; Mn, 22 mg; Fe, 84 mg; Cu, 225 mg; I, 0.50 mg; Se, 0.35 mg.

Management procedures

The experiment lasted for 18 days, consisting of a 13-day (d) adaptation period to the diets followed by a 5-day (d) experimental period (total collection of feces and urine). Feed refusals and spillage were collected, dried, and weighed to correct feed intake. During the experimental period feed intake was recorded on daily basis.

Sample collection

Total feces and total urine were collected during the last 5 d of the study. During the 5-d collection period, all feces were collected into plastic bags and stored at –20°C immediately. At the end of the collection period, fecal samples from each pig were pooled and weighed and a 300-g sample was taken and dried in a forced-draft oven at 65°C for 72 h. After drying and grinding, samples were stored at

Table 2. Effect of different type of Met supplementation on N retention and excretion in pigs[1]

Item	Basal diet	L-Met	DL-Met	SEM	p-value
Initial BW (kg)	15.43	15.47	15.45	0.38	0.99
Finial BW (kg)	25.67	27.63	26.95	0.73	0.20
Nitrogen utilization					
N intake (g/d)	26.48[b]	28.66[a]	28.47[a]	0.55	0.04
Fecal N excreted (FN, g/d)	5.76	4.90	5.04	0.49	0.47
Urine N excreted (UN, g/d)	8.34[a]	5.33[b]	5.37[b]	0.50	<0.01
Retained N (RN, g/d)	12.38[b]	18.43[a]	18.06[a]	0.77	<0.01
N digestibility (%)	78.30	82.91	82.58	1.66	0.16
N retention rate (%)	46.94[b]	64.32[a]	63.06[a]	1.84	<0.01

L-Met, L-methionine; DL-Met, DL-methionine; SEM, standard error of the mean; BW, body weight.

[1] Data are means of 6 observations per treatment.

[ab] Means in a row followed by different letters are significantly different (p<0.05).

–20°C for further chemical analysis. Total urine samples were collected into plastic buckets attached to funnels located under the metabolism cages at the same time as the fecal collection. Approximately 50 mL of 6 *N* HCl were added to each bucket to limit microbial growth and reduce loss of ammonia. Urine volume was recorded daily and a subsample of 10% of the urine excreted from each pig was collected and stored at –20°C. At the end of the collection period, urine samples were pooled for each pig and a subsample was prepared for further analysis.

On the second day after the collection period, blood samples were collected from the precava of each pig into 10 mL heparin-free tubes (Greiner Bio-One Company, Monroe, NC, USA) at 0800 h. Blood samples were then centrifuged (3,000×*g* at 4°C for 15 min) within 1 h of collection, and serum samples were stored at –20°C until ready for serum urea nitrogen (SUN) and AA analysis.

Chemical analysis

Dry matter (method 934.01), crude protein (method 990.03), ash (method 942.05), and Ca and P (method 985.01) content of the diets were analyzed according to the procedures of the AOAC International (2007).

Amino acids in diets were analyzed according to the AOAC International (2007; method 151 982.30). Samples were hydrolyzed with 6 *N* HCl at 110°C for 24 h and analyzed for lysine using an Amino Acid Analyzer (Hitachi L-8900, Tokyo, Japan). Methionine and cysteine were determined as methionine sulfone and cysteic acid after oxidized with cold performic acid overnight and hydrolyzed with 7.5 *N* HCl at 110°C for 24 h. An Amino Acid Analyzer (Hitachi L-8900, Tokyo, Japan) was used for the measurements.

Blood urea nitrogen, total protein (TP), albumin (ALB), and glucose concentrations were measured using an Automatic Biochemistry Analyzer (Hitachi 7020, Japan). Serum AA concentrations were determined by ion-exchange chromatography with physiological fluid analysis conditions (S-433D AA Analyzer, Sykam, Germany) according to Zhang et al. (2013). Frozen serum samples were first thawed at 4°C and then deproteinized with 120 mg of salicylic acid/mL of serum. After samples were placed an ice bath for 20 min, the reaction system was adjusted for pH by adding lithium hydroxide solution (2 mol/L) and then centrifuged at 12,000 *g* (L-80 XP, Beckman, Brea, CA, USA) for 30 min. The supernatant fluid was collected and then passed through a filter (0.1 μm) before use for AA analysis.

Statistical analysis

Data were analyzed using the general linear model procedure of SAS (SAS Inst. Inc., Cary, NC, USA) followed by Student-Newman-Keuls multiple tests. In all analyses, the differences were considered to be significant if p<0.05.

RESULTS

Nitrogen-balance

Supplementation of either L-Met or DL-Met improved N retention (p<0.01) and N retention rate (p<0.01) when compared with the BD (Table 2). No difference was found in N retention and N retention rate between the DL-Met and L-Met treatments.

Supplementation of either L-Met or DL-Met significantly decreased (p<0.01) urine nitrogen when compared with the BD group. No difference in total N excretion was observed between the DL-Met treatment and the L-Met treatment (p>0.05).

Serum amino acid profile and metabolites

The serum amino acid profile of pigs fed the different diets is presented in Table 3. Compared with the BD group, supplementation of either L-Met or DL-Met increased (p<0.01) the serum Met concentration. Supplementation of L-Met significantly decreased (p<0.05) serum glycine

Table 3. Effect of different type of Met supplementation on serum amino acids concentrations in pigs[1]

Serum amino acids (nmol/mL)	Basal diet	L-Met	DL-Met	SEM	p-value
Indispensable amino acid					
Lysine	297.72	260.35	272.11	5.89	0.14
Threonine	446.01	442.19	443.30	9.26	0.96
Methionine	33.04[b]	50.70[a]	53.18[a]	0.83	<0.01
Arginine	145.65	139.91	157.05	5.77	0.14
Histidine	86.75	85.27	80.51	3.00	0.33
Leucine	243.45	214.21	229.07	9.26	0.17
Isoleucine	137.01	123.62	116.74	6.80	0.13
Phenylalanine	97.18	110.68	113.71	5.51	0.11
Tryptophan	66.91	70.06	70.59	3.10	0.67
Valine	205.47	179.45	186.84	11.23	0.27
Dispensable amino acid					
Cysteine	17.54	18.36	19.52	0.72	0.18
Aspartic acid	153.50	140.82	146.19	8.54	0.58
Glutamic acid	228.89	201.58	193.79	11.97	0.13
Alanine	656.74	580.52	607.85	26.25	0.15
Glycine	1,362.67[a]	1,186.71[b]	1,277.71[ab]	37.63	0.02
Serine	195.68	167.36	180.45	10.07	0.17
Tyrosine	115.76	131.24	125.92	5.72	0.18

L-Met, L-methionine; DL-Met, DL-methionine; SEM, standard error of the mean.

[1] Data are means of 6 observations per treatment.

[ab] Means in a row followed by different letters are significantly different (p<0.05).

concentration when compared with the BD control. There were no differences in most serum AA concentrations and in treatments.

The results of SUN, serum ALB, and TP concentration are shown in Table 4. Supplementation of either L-Met or DL-Met increased TP concentration and decreased SUN concentration when compared with the BD group ($p<0.01$) without differences between the DL-Met treatment and the L-Met treatment.

DISCUSSION

Studies have been conducted to evaluate the effects of supplementing of different sources of Met in monogastric feed since 1980. The results of those studies were not consistent. Kim and Bayley (1983) has used the oxidation of phenylalanine as an indicator to determine the requirement of Met and the efficacy of L-Met compared with D-Met. The results indicated that D-Met was 50% efficacious relative to L-Met in young pigs. In contrast, it was also reported that L-Met and DL-Met has the same efficacy on growth performance in pigs (Chung and Baker, 1992) and in chicks (Garlich, 1985; Dilger and Baker, 2007), or even L-Met was found to have less efficacy than D-Met and DL-Met in chicks (Tipton et al., 1966). Some studies found that efficacy of Met utilization is different with supplementation of dietary Met levels and whether Met was added as the sole source sulfur amino acid. For example, Katz and Baker (1975) conducted four experiments to evaluate the relative efficacy of different sources of Met and to estimate the requirement for D- and L-Met of broiler chickens when was Met serving as the sole source of sulfur AA or when added to a diet containing 0.27% L-cystine. Their results showed that at lower levels of supplementation, L-Met is a better source of sulfur AA than D-Met on growth performance of broiler chickens. L- and D-Met have equal efficacy when incorporated into diets that are only marginally deficient in sulfur-containing AA.

Table 4. Effect of different type of Met supplementation on serum urea nitrogen (SUN), albumin, and total protein concentration in pigs[1]

Item	Basal diet	L-Met	DL-Met	SEM	p-value
Albumin (g/L)	37.58	39.27	39.04	0.71	0.22
Total protein (g/L)	58.22[b]	70.30[a]	69.83[a]	1.62	<0.01
SUN (mmoL/L)	5.10[a]	3.90[b]	3.97[b]	0.18	<0.01

L-Met, L-methionine; DL-Met, DL-methionine; SEM, standard error of the mean.

[1] Data are means of 6 observations per treatment.

[ab] Means in a row followed by different letters are significantly different (p<0.05).

Our data provided evidence to support earlier results indicating that the efficacy of L-Met and DL-Met is indistinguishable (Cho et al., 1980; Chung and Baker, 1992). However, Shen et al. (2014) reported that the relative bioavailability (RBA) of L-Met to DL-Met in nursery pigs for average daily gain (ADG) and gain:feed ratio was 143.8% and 122.7%, respectively. The inconsistency in different reports is likely due to many factors. Some studies pointed out that the body weight of animals could lead to the difference of results (Chung and Baker, 1992; Shen et al., 2014). Moreover, at lower levels of Met supplementation below the requirement supplementation as the sole source of sulfur AA or the difference of relative contribution of Met and cysteine to treatment diet may also cause discrepancy (Katz and Baker, 1975; Christensen et al., 1980). Ball et al. (2006) reported that more than 40% of the sulfur amino acid requirement can be met by dietary cysteine. Cysteine has the sparing effect on Met requirement which may alleviate the deficiency of Met (Shoveller et al., 2003).

Nitrogen retention or N balance reflects the utilization of proteins, the balance between the body protein synthesis and body protein degradation (Metayer et al., 2008). In the current study, when compared with BD group, both L-Met and DL-Met diets improved retained N ($p<0.01$) and N retention rate ($p<0.01$) demonstrating a better utilization of nitrogen in L-Met and DL-Met treatments than the control. There was no difference in N retention and N retention rate between the DL-Met and L-Met treatments. In contrast, Shen et al. (2014) reported that the RBA of L-Met to DL-Met estimated from concentrations of plasma urea nitrogen on d 10 was 160.2%. However, on d 20 no difference in ADG was observed between the L-Met and DL-Met treatments which indicated that utilization of Met isomers may be a function of age. The initial body weight of pigs in their study was 7.15±0.97 kg (weanling pigs). In our study it was 15.45±0.88 kg. This may partly explain the inconsistency between the results presented by Shen et al. (2014), and those showed in this study.

The estimation of plasma urea nitrogen is a rapid method to estimate amino acid requirement in pigs (Coma et al., 1995). A rapid response in nitrogen metabolism was found when there was a change in the concentration of dietary AA. The reduction of plasma urea nitrogen reflects more efficient nitrogen utilization and, consequently, decreased the urea synthesis (Brown and Cline, 1974). In present study, the low level of SUN in Met supplemented group reflects a low level of urea synthesis and a higher efficiency of AA or nitrogen utilization. To our knowledge, D-Met must be converted to L-Met before it can be used by the gastrointestinal tract (Dibner and Ivey, 1992). The rate-limiting enzyme for conversion of D-Met to L-Met is D-AA oxidase (Fang et al., 2010). The existence of substantial D-AA oxidase activity in different tissues, not only liver and kidney (major sites conversion) but also the gastrointestinal tract (stomach, duodenum, jejunum and ileum) can efficiently convert D-Met to L-Met in pigs (Fang et al., 2010). Thus, the effective conversion of D-Met to L-Met might possibly explain the result that the supplementation of L-Met did not affect SUN and serum amino acid profile compared with DL-Met treatment.

Although this absorption and metabolism of Met from different sources were not evaluated in this study, it is necessary to conduct more experiments to investigate the metabolism of amino acid utilization.

CONCLUSION

In present study, the results indicated that supplementation of either L-Met or DL-Met improved N retention, and decreased N excretion. The N retention of pigs fed diets supplemented with the same inclusion levels of DL-Met or L-Met were not different indicating that DL-Met and L-Met are equally bioavailable as Met sources for starter pigs.

ACKNOWLEDGMENTS

The study was financially supported by the National Natural Science Foundation of China (No. 31372316) and CJ Corporation, Seoul, Korea.

REFERENCES

AOAC. 2007. Official Methods of Analysis. 18th ed. Association of Official Analytical Chemists, Arlington VA, USA.

Ball, R. O., G. Courtney-Martin, and P. B. Pencharz. 2006. The *in vivo* sparing of methionine by cysteine in sulfur amino acid requirements in animal models and adult humans. J. Nutr. 136:1682S-1693S.

Brown, J. A. and T. R. Cline. 1974. Urea excretion in the pig: an indicator of protein quality and amino acid requirements. J. Nutr. 104:542-545.

Chen, Y., D. Li, Z. Dai, X. Piao, Z. Wu, B. Wang, Y. Zhu, and Z. Zeng. 2014. L-Methionine supplementation maintains the integrity and barrier function of the small-intestinal mucosa in post-weaning piglets. Amino Acids 46:1131-1142.

Cho, E. S., D. W. Andersen, L. J. Filer, and L. D. Stegink. 1980. D-methionine utilization in young miniature pigs, adult rabbits, and adult dogs. J. Parenter. Enteral. Nutr. 4:544-547.

Christensen, A. C., J. O. Anderson, and D. C. Dobson, 1980. Factors affecting efficacy of methionine hydroxy analogue for chicks fed amino acid diets. Poult. Sci. 59:2480-2484.

Chung, T. K. and D. H. Baker. 1992. Utilization of methionine isomers and analogs by the pig. Can. J. Anim. Sci. 72:185-188.

Coma, J., D. Carrion, and D. R. Zimmerman. 1995. Use of plasma urea nitrogen as a rapid response criterion to determine the lysine requirement of pigs. J. Anim. Sci. 73:472-481.

Cromwell, G. L. 2004. Identifying the limiting amino acids in complex and cereal grain-based diets to minimize nitrogen excretion. In Midwest Swine Nutr. Conf. Proc. Indianapolis, IN, USA. The Ohio Univ. Press, Columbus, OH, USA. pp. 69-83.

Dibner, J. J. and F. J. Ivey. 1992. Capacity in the liver of the broiler chick for conversion of supplemental methionine activity to L-methionine. Poult. Sci. 71:700-708.

Dilger, R. N. and D. H. Baker. 2007. DL-Methionine is as efficacious as L-methionine, but modest L-cystine excesses are anorexigenic in sulfur amino acid-deficient purified and practical-type diets fed to chicks. Poult. Sci. 86:2367-2374.

Fang, Z., H. Luo, H. Wei, F. Huang, Z. Qi, S. Jiang, and J. Peng. 2010. Methionine metabolism in piglets fed DL-methionine or its hydroxy analogue was affected by distribution of enzymes oxidizing these sources to keto-methionine. J. Agric. Food Chem. 58:2008-2014.

Finkelstein, J. D. 1990. Methionine metabolism in mammals. J. Nutr. Biochem. 1:228-237.

Gomes, J. and D. Kumar. 2005. Production of L-methionine by submerged fermentation: A review. Enzyme Microb. Tech. 37:3-18.

Garlich, J. D. 1985. Response of broilers to DL-methionine hydroxy analog free acid, DL-methionine, and L-methionine. Poult. Sci. 64:1541-1548.

Ikeda, M. 2003. Amino acid production processes. In: Microbial Production of L-amino Acids. Springer Berlin Heidelberg, Germany. pp. 1-35.

Katz, R. S. and D. H. Baker. 1975. Efficacy of D-, L- and DL-methionine for growth of chicks fed crystalline amino acid diets. Poult. Sci. 54:1667-1674.

Kim, K. I. and H. S. Bayley. 1983. Amino acid oxidation by young pigs receiving diets with varying levels of sulphur amino acids. Br. J. Nutr. 50:383-390.

Kim, S. Y., K. M. Cho, Y. U. Shin, H. W. Um, K. O. Choi, J. S. Chang, Y. W. Cho, and Y. H. Park. 2015. Microorganism producing L-methionine precursor and method of producing L-methionine and organic acid from the L-methionine precursor. US Patent 9029105.

Kumar, D. and J. Gomes. 2005. Methionine production by fermentation. Biotechnol. Adv. 23:41-61.

Mannsfeld, S. P., A. Pfeiffer, H. Tanner, H. Wagner, and E. Liebertanz. 1978. Continuous process for the manufacture of methionine. US Patent 04069251.

Martín-Venegas, R., P. A. Geraert, and R. Ferrer. 2006. Conversion of the methionine hydroxy analogue DL-2-hydroxy-(4-methylthio) butanoic acid to sulfur-containing amino acids in the chicken small intestine. Poult. Sci. 85:1932-1938.

Metayer, S., I. Seiliez, A. Collin, S. Duchene, Y. Mercier, P. A. Geraert, and S. Tesseraud. 2008. Mechanisms through which sulfur amino acids control protein metabolism and oxidative status. J. Nutr. Biochem. 19:207-215.

NRC. 1998. Nutrient Requirements of Swine, 10th edition. National Academic Press, Washington, DC, USA.

Odunfa, S. A., S. A. Adeniran, O. D. Teniola, and J. Nordstrom. 2001. Evaluation of lysine and methionine production in some lactobacilli and yeasts from *Ogi*. Int. J. Food Microbiol. 63:159-163.

Riedijk, M. A., B. Stoll, S. Chacko, H. Schierbeek, A. L. Sunehag, J. B. van Goudoever, and D. G. Burrin. 2007. Methionine transmethylation and transsulfuration in the piglet gastrointestinal tract. Proc. Natl. Acad. Sci. USA. 104:3408-3413.

Shen, Y. B., A. C. Weaver, and S. W. Kim. 2014. Effect of feed grade L-methionine on growth performance and gut health in nursery pigs compared with conventional DL-methionine. J. Anim. Sci. 92:5530-5539.

Shoveller, A. K., J. A. Brunton, J. D. House, P. B. Pencharz, and R. O. Ball. 2003. Dietary cysteine reduces the methionine requirement by an equal proportion in both parenterally and enterally fed piglets. J. Nutr. 133:4215-4224.

Shoveller, A. K., B. Stoll, R. O. Ball, and D. G. Burrin. 2005. Nutritional and functional importance of intestinal sulfur amino acid metabolism. J. Nutr. 135:1609-1612.

Stoll, B., J. Henry, P. J. Reeds, H. Yu, F. Jahoor, and D. G. Burrin. 1998. Catabolism dominates the first-pass intestinal metabolism of dietary essential amino acids in milk protein-fed piglets. J. Nutr. 128:606-614.

Tipton, H. C., B. C. Dilworth, and E. J. Day. 1966. A comparison of D-, L-, DL-methionine and methionine hydroxy analogue calcium in chick diets. Poult. Sci. 45:381-387.

Zhang, S., S. Qiao, M. Ren, X. Zeng, X. Ma, Z. Wu, P. Thacker, and G. Wu. 2013. Supplementation with branched-chain amino acids to a low-protein diet regulates intestinal expression of amino acid and peptide transporters in weanling pigs. Amino Acids 45:1191-1205.

11

Yeast Culture and Vitamin E Supplementation Alleviates Heat Stress in Dairy Goats

Lizhi Wang[1,2], Zhisheng Wang[1,2], Huawei Zou[1,2], and Quanhui Peng[1,2,*]

[1] Animal Nutrition Institute, Sichuan Agricultural University, Ya'an 625014, China

ABSTRACT: This study was conducted to determine and compare the effects of yeast yeast culture (YC) and vitamin E (VE) supplementation on endotoxin absorption and antioxidant status in lactating dairy goats suffering from heat stress (HS). Three first lactation Saanen dairy goats (body weight 30±1.5 kg) were surgically fitted with indwelling catheters in the portal vein, mesenteric vein and carotid artery, and were randomly assigned to a 3×3 Latin square design. Dietary treatments were the basal diet, and the basal diet supplemented with either 100 IU VE or 30 g YC. Goats were kept in temperature and humidity-controlled room at 35°C from 8:00 to 20:00 and at 24°C from 20:00 till the next morning at 8:00. The relative humidity was kept at 55%. HS increased dairy goats' rectum temperature and respiration frequency ($p<0.01$). HS reduced plasma flux rate of milk goats ($p<0.01$), but the plasma flux rate increased when the animal was under the conditions of the thermo-neutral period ($p<0.01$). The VE supplementation lowered dairy goats' rectum temperature during thermo-neutral period ($p<0.01$). Meanwhile, no significant differences were observed between the control and YC treatment in rectum temperature and respiration frequency ($p>0.05$). Dietary supplementation of VE and YC reduced heat stressed dairy goats' endotoxin concentration of the carotid artery and portal vein ($p<0.01$). However, the endotoxin concentration of the YC treatment was higher than that of the VE treatment ($p<0.01$). Both VE and YC supplementation decreased heat stressed dairy goats' absorption of endotoxin in portal vein ($p<0.01$). The endotoxin absorption of YC treatment was higher than the VE treatment ($p<0.01$). The addition of VE and YC decreased dairy goats' superoxide dismutase (SOD) concentration during HS and the whole experiment period ($p<0.01$). The addition of VE lowered SOD concentration during thermo-neutral period ($p<0.01$). Likewise, the addition of VE and YC lowered dairy goats' malonaldehyde (MDA) concentration during HS and the whole experimental period, and the MDA concentration in the VE treatment was lower than the YC treatment ($p<0.05$). The addition of VE decreased MDA concentration during thermo-neutral period. On the contrast, the addition of VE increased dairy goats total antioxidant potential (TAP) concentration during HS, thermo-neutral and the whole experimental period ($p<0.01$). The addition of YC increased TAP concentration only during HS period ($p<0.01$). It is concluded that both VE and YC are useful in alleviating HS of dairy goats by weakening endotoxin absorption and promoting antioxidant capacity. Compared with YC, VE is much more powerful in easing dairy goats HS. (**Key Words:** Dairy Goats, Heat Stress, Vitamin E, Yeast Culture, Endotoxin Absorption)

INTRODUCTION

The gastrointestinal tract of dairy goat presents an extensive surface area providing direct contact between the animal and a large assortment of nutrients, microbes and exogenous toxins. The intestine should permit the exchange of nutrients between the gut lumen and the systemic circulation, while preventing penetration of pathogenic organisms and toxic compounds (Marai et al., 2007). However, it had been documented that heat stress (HS) may cause injury to the intestinal epithelium (Bouchama and Knochel, 2002; Fan et al., 2014) which leads to the entrance of bacterial endotoxin into the systemic circulation (Gathiram et al., 1987; Hall et al., 2001; Wang et al., 2011). The entrance of endotoxin in systemic circulation leads to a wide range of injuries including inflammation, damage of internal organs, disseminated intravascular coagulation, and even death (Gathiram et al., 1987; Mani et al., 2012). Therefore, preventing endotoxin absorption during HS may

* Corresponding Author: Quanhui Peng.

E-mail: pengquanhui@126.com

[2] Key Laboratory of Low Carbon Culture and Safety Production in Cattle in Sichuan, Ya'an 625014, China.

be greatly beneficial to animal health. Huber (1998) reported that the supplementation of yeast culture (YC) increased dry matter intake (DMI) and milk production of Holstein cows during HS. However, how YC achieves this effect is still unknown. It was reported that HS reduced the antioxidant capacity by reducing the activity of antioxidant enzymes (Harmon et al., 1997), which increased the amount of free radicals (Flanagan et al., 1998; Pan et al., 2012), and the reduction of the antioxidant capacity was the main cause of injury to the gastrointestinal tract (Hall et al., 2001). On the other hand, vitamin E (VE) is a proven effective antioxidant. It is well documented VE is active in alleviating poultry HS (Bollengier-Lee et al., 1998; 1999). Although, whether VE alleviates HS in dairy goats is still unknown. Therefore, the objectives of this study were to evaluate and compare the effects of YC and VE supplementation on endotoxin absorption and antioxidant status in lactating dairy goats suffering from HS.

MATERIAL AND METHODS

Animals and surgery

All experimental procedures involving animals were approved by the Animal Care and Use Committee of Animal Nutrition Institute, Sichuan Agricultural University, and were carried out in accordance with the National Research Council's Guide for the Care and Use of Laboratory Animals.

Three lactating Saanen dairy goats (body weight 30±1.5 kg) were surgically fitted with indwelling catheters in the portal vein, the mesenteric vein and carotid artery. Portal vein catheters were silicone tube, and the 2 mesenteric catheters were made of Tygon. Two mesenteric venous catheters were inserted into branches between the major venous arch and the small intestine, one for sampling and the other for para-aminohippurate infusion. The establishment of catheters (portal vein and mesenteric vein) was described by Huntington et al. (1989). Surgically peel off the truncus vagosympathicus which is concomitant to the right carotid, and then the carotid was wrapped and sealed in skin by stitch. Carotid catheter was inserted into the dissociative carotid one day before sampling.

Experimental design and feeding management

The animals were housed in individual metabolic cages in a temperature and humidity-controlled house with *ad libitum* access to water and were hand milked twice daily. The goats were fed a basal diet during the preliminary period. After a 2-wk preliminary period the three goats were randomly assigned to a 3×3 Latin square design. The 3 dietary treatments were basal diet (Control), basal diet supplemented with 100 IU VE, and basal diet supplemented with 30 g Original XPTM (A kind of YC, the product of Diamond V Biological Fermentation Engineering & Technologies Shenzhen Co., Ltd., Shenzhen, Guangdong, China). Each period lasted 14 days and the animals were fed the experimental diets. During the 2-wk of each period, the last 4 days were for sample collection. Basal diet was fed twice daily in equal amounts. VE and YC were supplemented at 8:00. The formulation of the basal diet was summarized in Table 1. The room temperature during experiment period was 35°C from 8:00 to 20:00 and 24°C from 20:00 till the next morning at 8:00. The relative humidity kept stable (55%). Feed intake was recorded daily.

Sampling and analyses

Each sample collection period last 4 days. On the eleventh day of experiment, a continuous mesenteric venous infusion of para-aminohippurate (15 mg/mL formulated according to Huntington [1982]) was initiated via the distal mesenteric vein catheter to determine portal and mesenteric venous blood flows. Continuous infusion was performed by using a HL-2B calibrated syringe pump (Made in Lanngue Instrument Company, Shijiazhuang, China) with a speed of 12 mg para-aminohippurate per min (0.8 mL/min). The infusion commenced at 13:00, 2 h after which a spot sampling procedure was administrated. A total of 12 spot samplings were performed at 1 h intervals. At each

Table 1. Ingredients and nutrient content of the basal diet fed to goats

Items	g/100 g
Ingredients	
Oat hay	50.0
Corn	20.0
Wheat bran	15.0
Soybean meal	8.0
Rapeseed meal	4.5
Calcium carbonate	0.8
Dicalcium phosphate	0.1
Salt	0.5
Sodium bicarbonate	1.0
Premix[1]	0.1
Nutrition level, dry matter basis	
NE_L (MJ/kg)	5.52
CP (%)	12.6
Ca (%)	0.55
Tp (%)	0.32
VE[2] (IU/kg)	28.7
Concentrate/roughage[3]	51:49

NE_L, net energy for lactation; CP, crude protein; Ca, Calcium; Tp, total phosphorus; VE, vitamin E.

[1] Premix contained $FeSO_4{\cdot}7H_2O$ 170 g/kg; $CuSO_4{\cdot}5H_2O$ 70 g/kg; $MnSO_4{\cdot}5H_2O$ 290 g/kg; $ZnSO_4{\cdot}7H_2O$ 240 g/kg; $CoCl_2{\cdot}6H_2O$ 510 mg/kg; KI 220 mg/kg; Na_2SeO_3 130 mg/kg; vitamin A 1, 620,000 IU/kg; vitamin D_3 324,000 IU/kg; vitamin E 540 IU/kg.

[2,3] VE concentration and ratio of concentrate to roughage was determined, and all the other parameters were calculated.

sampling spot, simultaneous arterial, portal and mesenteric venous blood samples were collected (5 mL for each sample) and were slowly dropped into a cuvette containing heparin (200 IU/mL). Samples were centrifuged at 1,500 rpm for 15 min at 4°C. Serum was frozen at –20°C for further analyses.

The para-aminohippurate concentration of plasma was determined according to Huntington (1982). Plasma concentration of superoxide dismutase (SOD), total antioxidant potential (TAP) and malonaldehyde (MDA) were measured using the methods of xanthine oxidation, ferric reducing/antioxidant power, and thiobarbituric acid, respectively. Reagent kits were bought from Jiancheng Bioengineering Institute, Nanjing, China. Plasma endotoxin concentration was determined by kinetic turbidimetric assay (Yokota et al., 1989), using reagent kit bought from Xiamen tachypleus amebocyte lysate Co. LTD, Xiamen, China. The brief description of the procedure is: The standard endotoxin (*E. coli* 0111: B4) was dissolved in pyrogen-free distilled water. A BECKMAN VIS-723 Spectrophotometer (Beckman Coulter, Inc., Brea, CA, USA) was used and the absorbance of different concentrations of endotoxin dilutions at 545 μm were recorded. The absorbance data was used as the ordinate and the endotoxin concentrations as the abscissa to establish standard curve. The plasma samples were pretreated with moderate heating. Then 0.2 mL pyrogen-free water and 0.2 mL Tris-Hcl was added to 0.1 mL plasma and mixed, the mixture was placed into 100°C water bath for 10 min, centrifuged for 10 min (3,000 rpm), then 0.1 mL supernatant was taken for further determination. Next, 0.05 mL Limulus amebocyte lysate was added to the supernatant and mixed, and the tube placed into 37°C water bath for 25 min, and then 0.05 mL shark tripeptide was added and placed into 37°C water bath for 3 min, and 0.05 mL houjiaruya nitrate solution added and mixed, then 0.5 mL amino acid amine added and mixed, and finally added 0.5 mL benzathine. The absorbance of the mixture at 545 μm was recorded. The endotoxin concentration could be obtained using the standard curve.

Rectum temperature and respiration frequency

Rectum temperature and respiration frequency were measured at 16:00, 18:00, 20:00, 22:00, 24:00, 02:00 on the 4th day of each period. Rectum temperature was measured by inserting thermometer into anus for 3 cm, and respiration frequency was calculated by counting the undulant times of flank within certain time slot measured by stopwatch.

Parameters calculation

Portal and mesenteric venous blood flows were calculated from the following equations respectively:

Portal blood flows
= Para-aminohippurate infusion rate
/(Para-aminohippurate concentrations in the portal venous blood – Para-aminohippurate concentrations in the arterial blood)

Mesenteric blood flows
= Para-aminohippurate infusion rate
/(Para-aminohippurate concentrations in the mesenteric venous blood – Para-aminohippurate concentrations in the arterial blood)

Net fluxes of endotoxin across portal-and mesenteric-drained viscera were calculated from the following equations:

Net absorption of endotoxin via portal-drained viscera
= (concentrations of endotoxin in the portal blood – concentrations of endotoxin in the arterial blood) ×portal blood flow×1,000

Net absorption of endotoxin via mesenteric-drained viscera
= (concentrations of endotoxin in the mesenteric blood – concentrations of endotoxin in the arterial blood) ×mesenteric blood flow×1,000

Net fluxes of antioxidants (including TAP, SOD, and MDA across portal-and mesenteric-drained viscera were calculated from the following equations:

Net flux of endotoxin via portal-drained viscera
= plasma concentrations of antioxidants in the portal blood ×portal blood flow

Net flux of endotoxin via mesenteric-drained viscera
= concentrations of antioxidants in the mesenteric blood ×mesenteric blood flow

Statistical analyses

The MIXED model $Y_{ij} = \mu+T_i+e_{ij}$ of SAS (SAS Institute, 2009) was used to analyze all of the data, where Y_{ij} is an observation on the dependent variable ij, μ is the overall population mean, T_i is the fixed effect of treatments (VE vs YC), e_{ij} is the random error associated with the observation ij. The data of plasma flux rate, endotoxin concentration in the portal vein and carotid, endotoxin absorption in the portal vein, plasma concentration of SOD, TAP, and MDA across different time spots were also analyzed using the same model, where the T_i is the fixed effect of different time spots. When significant differences ($p<0.05$) were detected, post-hoc analyses were carried out using least significant difference test to compute pairwise differences in the means. Means with different superscript letter groups

Table 2. Dry mater intake of the goats in different treatments

Item	Treatments		
	Control	VE	YC
DMI (kg/d)	1.52±0.06	1.60±0.09	1.51±0.08

VE, vitamin E; YC, yeast culture; DMI, dry matter intake.

were obtained with PDMIX 800 SAS macro (SAS, Cary, NC, USA). Data were presented as mean±standard deviation.

RESULTS

Dry matter intake

The effect of supplementation YC and VE on the DMI of dairy goats is reported in Table 2. Feed intake of goats did not differ among treatments during the whole experiment period ($p = 0.402$).

Rectum temperature and respiration frequency

Rectum temperature of the goats in the 3 treatments did not differ from each other ($p = 0.125$) during HS period, and decreased significantly ($p<0.01$) when ambient temperature dropped from 35°C to 24°C (Figure 1a). During thermo-neutral period, rectum temperature of the goats in VE and YC treatments kept on falling. Rectum temperature of the goats in YC treatment did not differ ($p = 0.478$) from VE treatment from 20:00 to 24:00, but greater ($p<0.05$) than VE treatment from 24:00 to 02:00 (Figure 1). Average rectum temperature of the goats during thermo-neutral period in the control was similar ($p = 0.096$) as that in YC treatment, and was greater ($p<0.05$) than VE treatment (Table 3). There was no difference ($p = 0.329$) among the 3 treatments on rectum temperature during HS period and during the whole experiment period (Table 3). Respiration frequency of the goats in the 3 treatments kept stable from 16:00 to 20:00, and dropped significantly ($p<0.01$) when ambient temperature decreased from 35°C to 24°C (Figure 1b). Average respiration frequency of the goats in the 3 treatments during thermo-neutral period ($p = 0.205$), HS period ($p = 0.224$), and the whole experiment period ($p = 0.897$), was not different from each other (Table 3).

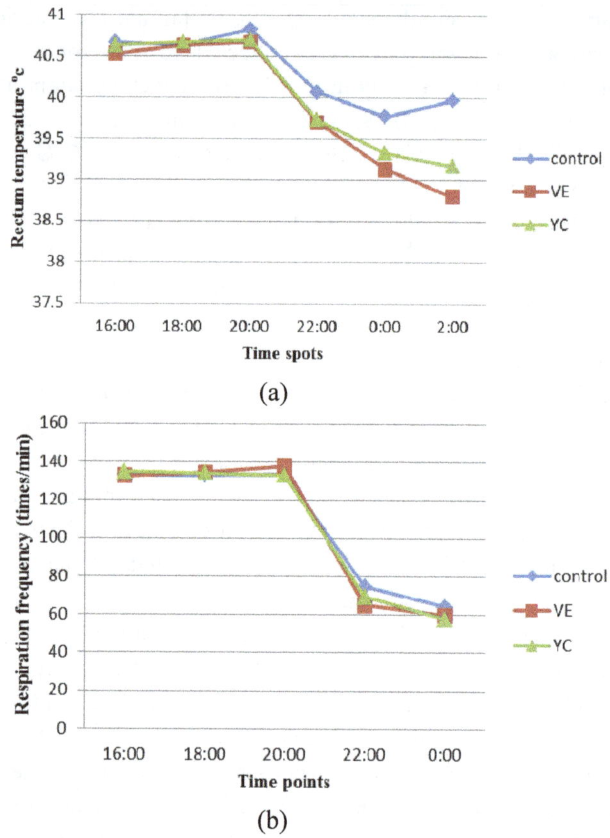

Figure 1. Rectum temperature (a) and respiration frequency (b) of the goats in different treatments at different time spots. VE, vitamin E; YC, yeast culture.

Plasma flux rate

There was no significant difference ($p>0.05$) in the rate of plasma flux of the goats among the 3 treatments during HS period (15:00 to 20:00) (Figure 2). However, with the prolonged HS, plasma flux rate in portal vein kept on decreasing. When the ambient temperature decreased from 35°C to 24°C, plasma flux rate of the goats in 3 treatments

Table 3. Average rectum temperature and respiration frequency of the goats in different treatments and in different periods

Item	Treatments		
	Control	VE	YC
Rectum temperature (°C)			
Average in heat stress period	40.72±0.06	40.61±0.04	40.67±0.02
Average in thermo-neutral period	39.93±0.09 ^{a}	39.21±0.26^{b}	39.41±0.17 ^{ab}
Average of whole period	40.33±0.18	39.91±0.33	40.04±0.29
Respiration frequency (times/min)			
Average in heat stress period	133.11±0.11	134.89±1.47	133.89±0.48
Average in thermo-neutral period	67.67±3.50	60.11±2.70	60.56±4.47
Average of whole period	100.39±14.72	97.50±16.78	97.22±16.52

VE, vitamin E; YC, yeast culture.
^{a,b} Means within a row without a similar superscript are different ($p<0.05$).

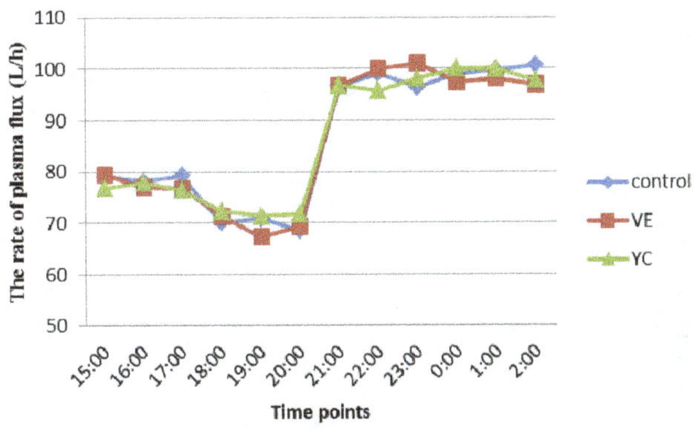

Figure 2. The rate of plasma flux of the goats in different treatments at different time spots. VE, vitamin E; YC, yeast culture.

increased significantly (p<0.01). The average plasma flux rate of the goats did not differ among treatments during thermo-neutral period (p = 0.769), HS period (p = 0.717), and the whole experiment period (p = 0.863) (Table 4).

Table 4. The rate of plasma flux of the goats in different treatments and in different periods

Item	Treatments		
	Control	VE	YC
Heat stress period (L/h)	74.24±2.04^{a}	73.50±1.98^{a}	74.35±1.19^{a}
Thermo-neutral period (L/h)	98.54±0.69^{b}	98.31±0.74^{b}	98.02±0.71^{b}
Whole experimenta period (L/h)	86.39±3.80^{b}	85.90±3.87^{b}	86.18±3.63^{b}

VE, vitamin E; YC, yeast culture.
a,b Means within column without a similar superscript are different (p<0.05).

Endotoxin concentration and absorption

Endotoxin concentration in the portal vein and carotid of the goats kept stable from 15:00 to 20:00 (HS period) within each treatment, and decreased (p<0.01) sharply after 21:00 (Figure 3a,b). The endotoxin concentration in the portal vein and carotid of the goats during HS period was the highest in the control (p<0.01), followed by YC treatment (p<0.05), and the lowest in VE treatment (p<0.01). However it was not different during thermo-neutral period (p = 0.175 for portal vein, and p = 0.185 for carotid) among the 3 treatments (Table 5).

The endotoxin absorption was the highest (p<0.01) in the control treatment and lowest (p<0.01) in VE treatment at any time spot (Figure 3c). The maximum absorption of endotoxin in the three treatments happened at 21.00, followed by 22.00, significantly greater (p<0.01) than any other time spot (Table 5). The average endotoxin absorption during the whole experiment period and during HS period was the highest (p<0.01) in the control and lowest (p<0.01) in VE treatment, but it was not different (p = 0.161) among the 3 treatments during thermo-neutral period (Table 5).

Antioxidants concentrations in portal vein

Plasma concentration of SOD in portal vein in the control kept stable from 15:00 to 02:00 the next day, except for time spot (20:00), while the VE treatment kept stable before 19:00 and increased (p<0.01) significantly from 23:00 (Figure 4a). Plasma concentration of TAP in portal vein kept stable within each treatment, except for few prominent time spots (Figure 4b). Across treatments, plasma concentration of TAP in portal vein was the highest (p<0.01) in VE and the lowest (p<0.01) in the control at any time spots (Figure 4b). Plasma concentration of MDA in the portal vein of the goats kept stable across any time spots within each treatment, except for that at 21:00 (Figure 4c). Plasma MDA concentration was always the highest (p<0.01) in the control and the lowest (p<0.01) in VE treatment (Figure 4c).

Table 5. The average concentration of endotoxin in carotid and portal vein, and the average absorption of endotoxin in portal vein

Item	Treatments		
	Control	VE	YC
Average endotoxin concentration in carotid (EU/mL)			
Heat stress period	0.63±0.04^{c}	0.43±0.02^{a}	0.52±0.01^{b}
Thermo-neutral period	0.55±0.08	0.41±0.07	0.44±0.06
Whole experiment period	0.59±0.04^{c}	0.42±0.03^{a}	0.48±0.03^{b}
Average endotoxin concentration in portal vein (EU/mL)			
Heat stress period	3.98±0.24^{c}	2.74±0.12^{a}	3.33±0.08^{b}
Thermo-neutral period	3.53±0.50	2.60±0.44	2.80±0.37
Whole experiment period	3.76±0.27^{c}	2.67±0.22^{a}	3.06±0.20^{b}
Average absorption of endotoxin in portal vein (10^4 EU/h)			
Heat stress period	24.71±0.84^{c}	16.86±0.39^{a}	20.83±0.26^{b}
Thermo-neutral period	29.24±4.01	21.54±3.56	23.08±2.90
Whole experiment period	26.97±2.07^{c}	19.20±1.85^{a}	21.95±1.43^{b}

VE, vitamin E; YC, yeast culture.
a,b,c Means within row without a similar superscript are different (p<0.05).

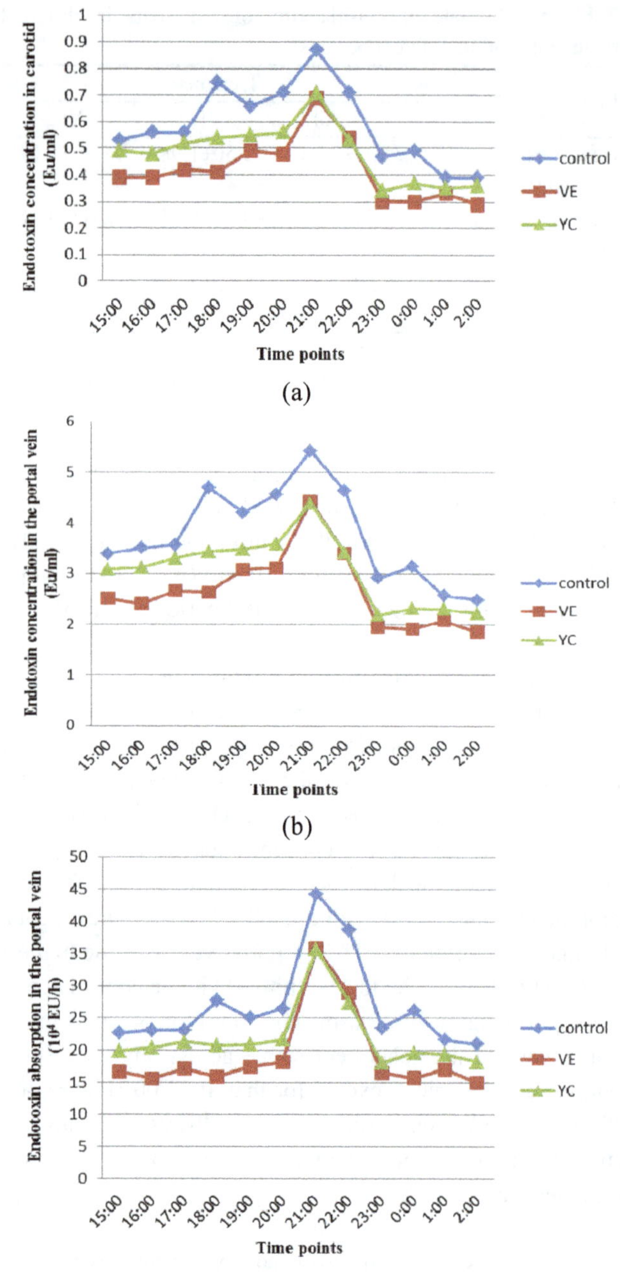

Figure 3. Endotoxin concentration in the portal vein (a) and carotid (b) and the endotoxin net absorption in the portal vein (c). VE, vitamin E; YC, yeast culture.

The average plasma concentration of TAP in portal vein was the highest ($p<0.01$) in VE and the lowest ($p<0.01$) in the control during HS period (Table 6). The average plasma concentration of TAP in portal vein in the control was significantly lower ($p<0.05$) than VE treatment, and the YC treatment was not significantly different from the control ($p = 0.06$) and VE ($p = 0.479$) treatment during thermo-neutral period (Table 6). The average plasma concentration of TAP during the whole experiment period in the control did not differ ($p = 0.089$) from the YC treatment, but was significantly lower than VE treatment ($p<0.01$) (Table 6).

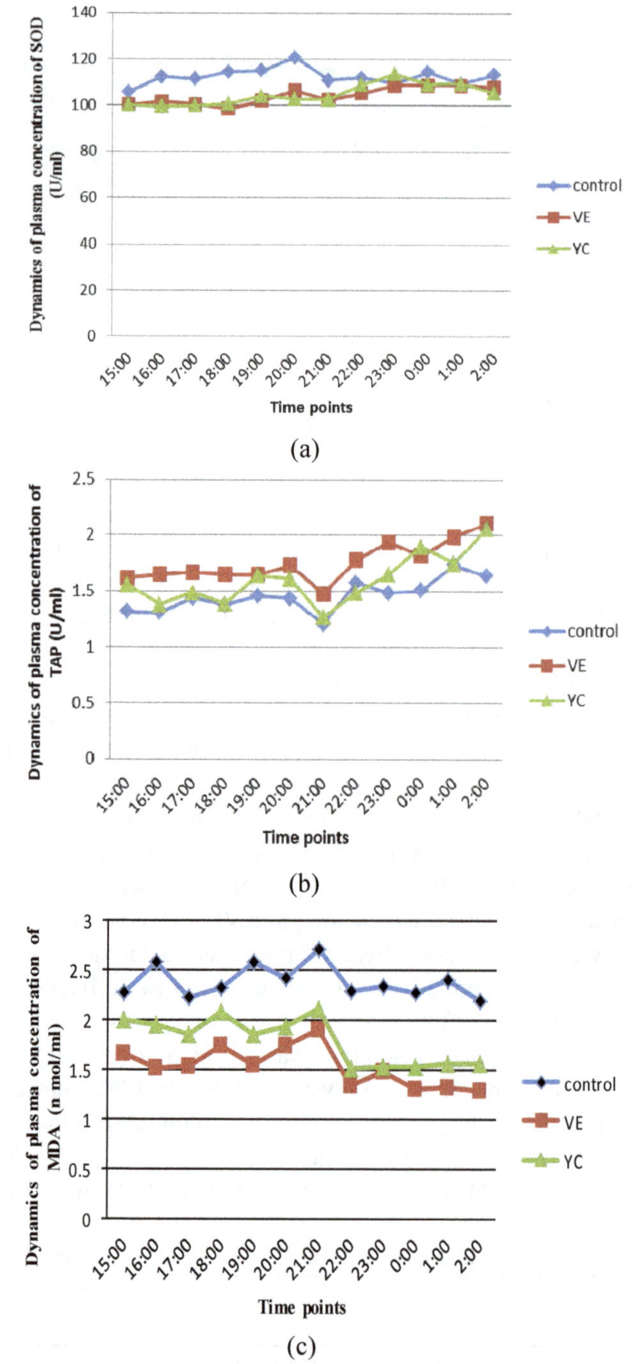

Figure 4. Dynamics of plasma concentration of superoxide dismutase (SOD, a), total antioxidant potential (TAP, b), and malonaldehyde (MDA, c) in different treatments in different time spots. VE, vitamin E; YC, yeast culture.

The average plasma concentration of SOD in VE and YC treatment during HS period was significantly lower ($p<0.01$) than the control during HS period and whole experiment period, while no significant difference was observed between VE and YC treatments (Table 6). The SOD concentration of VE treatment was significantly lower than the control during the thermo-neutral period ($p<0.01$) (Table 6).

Table 6. The average concentration of plasma antioxidants in different treatments

Item	Treatments		
	Control	VE	YC
SOD (U/mL)			
Heat stress period	113.48±1.98[b]	101.67±1.11[a]	101.51±0.80[a]
Thermo-neutral period	111.74±0.82[b]	107.16±1.06[a]	108.36±1.53[ab]
Whole experiment period	112.61±1.05[b]	104.42±1.10[a]	104.93±1.32[a]
TAP (EU/mL)			
Heat stress period	1.39±0.03[a]	1.66±0.01[c]	1.51±0.04[b]
Thermo-neutral period	1.53±0.07[a]	1.85±0.09[b]	1.69±0.12[ab]
Whole experiment period	1.46±0.04[a]	1.76±0.05[b]	1.60±0.07[a]
MDA (n mol/mL)			
Heat stress period	2.40±0.06[c]	1.62±0.04[a]	1.94±0.03[b]
Thermo-neutral period	2.37±0.08[b]	1.44±0.10[a]	1.64±0.09[ab]
Whole experiment period	2.39±0.05[c]	1.53±0.06[a]	1.79±0.07[b]

VE, vitamin E; YC, yeast culture; SOD, superoxide dismutase; TAP, total antioxidant potential; MDA, malonaldehyde.

[a,b,c] Means within row without a similar superscript are different ($p<0.05$).

Plasma MDA concentration during HS period and the whole experiment period was the highest ($p<0.01$) in the control treatment, followed ($p<0.01$) by YC treatment, and the lowest ($p<0.01$) in VE treatment (Table 6). There was no difference ($p = 0.382$) on plasma MDA concentration during thermo-neutral period between VE and YC treatments, but the VE treatment was lower ($p<0.01$) than the control treatment (Table 6).

DISCUSSION

Yeast culture and vitamin E supplementation on feed intake, rectum temperature and respiration frequency

One of the most noticeable consequences of HS is to reduce feed intake (Baumgard and Rhoads, 2012). Many studies with dairy cows (Schingoethe et al., 2004; Bruno et al., 2009; Shwartz et al., 2009) reported that YC supplementation increased feed intake during HS. Furthermore, Sahin et al. (2006) reported that VE supplementation increased feed intake of quail during HS. However, feed intake of the heat stressed dairy goats was not significantly affected by the supplementation of YC and VE in the present study. No study on the effect of YC and VE supplementation on feed intake of dairy goats has been previously reported.

In this study, rectum temperature and respiration frequency of goats increased during HS, which was in accordance with Darcan and Guney (2008). Researches on dairy cows revealed that the supplementation of YC (Bruno et al., 2009) and other fungal cultures (Huber et al, 1985; Gomez-Alarcon et al., 1991; Higginbotham et al., 1993) decreased rectum temperature and respiration frequency during HS. Research on poultry revealed that the high dosage of VE supplementation (30 mg/kg body weight) significantly decreased rectum temperature (Sinkalu et al., 2008). In addition, Zeidan et al. (2006) reported that the injection of VE once per week (100 IU/rabbit) significantly decreased rectum temperature and respiration frequency of New Zealand rabbits. But this effect was not found in the present study on dairy goats. In present study, the VE intake of the control was above 42 IU/d, and the plasma concentration of VE was 2.52 mg/L. This indicated the animals VE nutrition status in the control was adequate. Although VE and YC supplementation did not decrease rectum temperature and respiration frequency of dairy goats during HS, the supplementation may help with the rehabilitation of rectum temperature and respiration frequency when ambient temperature decreased to thermo-neutral zone.

Yeast culture and vitamin E supplementation on plasma flux rate and endotoxin absorption

Plasma flux rate in the 3 treatments decreased with the prolonged HS, whereas it increased significantly when temperature dropped from 35°C to 24°C. These results were in agreement with Kregel et al. (1988). Results from the present study showed that the supplementation of VE or YC did not significantly affect the plasma flux in portal vein of dairy goats during HS. It should be noticed that the endotoxin concentration in portal vein and carotid, and that the endotoxin absorption in portal vein in the 3 treatments increased significantly from 20:00 to 21:00 when the temperature decreased from 35°C to 24°C. This could be due to the phenomenon of ischemia reperfusion injury, which refers to the functional damage of an organ caused by blood reperfusion after ischemia. The plasma flux rate in portal vein in goats during HS was 70 L/h, which was in line with Smuts et al. (1995), who reported 1,047 to 1,290 mL/min in Alpine goats. When the temperature dropped from 35°C to 24°C, blood flux rehabilitated to 100 L/h in present study, an increase of more than 40%. It was recently reported that the mesenteric ischemia–reperfusion induced oxidative injury of the ileum (Şen et al., 2015). The authors suspected that blood reperfusion further led to the injury of intestinal tract, and the damage of intestinal tract led to the increased permeability (Grotz et al., 1999), and this incurred the increased absorption of endotoxin. It was proven that endotoxin absorption increased during HS (Cronjé, 2005), which was not confirmed in the present study. It seemed that the average concentration of endotoxin in the portal vein and carotid of the goats was affected by dietary treatments, rather than HS. The amount of endotoxin absorption in portal vein and endotoxin concentration in portal vein and carotid was always the lowest in the VE treated and greatest in the control both in HS period and in thermo-neutral period, indicating that VE supplementation prevented endotoxin absorption.

Yeast culture and vitamin E supplementation on plasma antioxidants

The lowest TAP concentration and the greatest MDA concentration in the 3 treatments in the present study were attained during a narrow time slot from 20:00 to 21:00 when ambient temperature dropped from 35°C to 24°C, which indicated the increased oxidative stress occurred during this period. This could also be caused by ischemia reperfusion as Horton and Walker (1993) proved that ischemia reperfusion may lead to oxidative injury.

VE supplementation significantly increased the plasma TAP concentration, while significantly decreased plasma MDA concentration of dairy goats during HS and thermo-neutral periods, which meant that VE supplementation effectively strengthened the antioxidant capacity of dairy goats. VE exerts antioxidant function by breaking the free radical reaction chain (Yu, 1994). There was no report on the effect of VE supplementation on antioxidants of goat, however, a report on poultry indicated that plasma concentration of MDA decreased during HS when VE was supplemented (Sahin et al., 2006). YC supplementation only decreased plasma MDA concentration, but did not increase plasma TAP concentration during HS and thermo-neutral periods, which meant that YC was less powerful than VE in alleviating oxidative stress. It was surprising that VE and YC supplementation lead to decreased plasma SOD concentration during HS. This was possibly attributed by the greater amount of endotoxin absorption in the control which depleted antioxidants and induced the compensatory expression of *SOD* gene (Ceriello et al., 1996). After 3 h of ischemia reperfusion, plasma SOD concentration in portal vein in VE and YC treatments, but not in the control, increased significantly. These results reflected the alleviation of oxidative stress when VE or YC was supplemented.

CONCLUSION

The portal vein plasma flux rate of heat stressed dairy goats is decreased, which leads to the increase of endotoxin absorption, and thereby decreasing the antioxidant capacity. The supplementation of VE and YC reduces the absorption of endotoxin and effectively strengthens the antioxidant capacity of dairy goats, and hence alleviates HS. Compared to YC, VE is much more powerful in preventing endotoxin absorption and promoting antioxidant capacity.

ACKNOWLEDGMENTS

We would like to thank the financial support provided by the Department of Sciences and Technology of Sichuan Province (2014SZ0067) and National Natural Science Foundation of China (31402104).

REFERENCES

Baumgard, L. H. and R. P. Rhoads. 2012. Ruminant Nutrition Symposium: Ruminant production and metabolic response to heat stress. J. Anim. Sci. 90:1855-1865.

Bouchama, A. and J. P. Knochel. 2002. Heat stroke. New Engl. J. Med. 346:1978-1988.

Bollengier-Lee, S., M. A. Mitchell, D. B. Utomo, P. E. V. Williams, and C. C. Whitehead. 1998. Influence of high dietary Vitamin E supplementation on egg production and plasma characteristics in hens subjected to heat stress. Br. Poult. Sci. 39:106-112.

Bollengier-Lee, S., P. E. V. Williams, and C. C. Whitehead. 1999. Optimal dietary concentration of vitamin E for alleviating the effect of heat stress on egg production in laying hens. Br. Poult. Sci. 40:102-107.

Bruno, R. G. B., H. M. Rutigliano, R. L. Cerri, P. H. Robinson, and J. E. P. Santos. 2009. Effect of feeding Saccharomyces cerevisiae on performance of dairy cows during summer heat stress. Anim. Feed Sci. Technol. 150:175-186.

Ceriello, A., P. dello Russo, P. Amstad, and P. Cerutti. 1996. High glucose induces antioxidant enzymes in human endothelial cells in culture: Evidence linking hyperglycemia and oxidative stress. Diabetes 45:471-477.

Cronjé, P. B. 2005. Heat stress in livestock–the role of the gut in its aetiology and a potential role for betaine in its alleviation. Recent Adv. Anim. Nutr. Australia 15:107-122.

Darcan, N. and O. Guney. 2008. Alleviation of climatic stress of dairy goats in Mediterranean climate. Small Rumin. Res. 74:212-215.

Fan, L., L. Hu, B. Yang, X. Fang, Z. Gao, W. Li, Y. Sun, Y. Shen, X. Wu, Y. Shu, Y. Gu, X. Wu, and Q. Xu. 2014. Erlotinib promotes endoplasmic reticulum stress-mediated injury in the intestinal epithelium. Toxicol. Appl. Pharmcol. 278:45-52.

Flanagan, S. W., P. L. Moseley, and G. R. Buettner. 1998. Increased flux of free radicals in cells subjected to hyperthermia: Detection by electron paramagnetic resonance spin trapping. FEBS Lett. 431:285-286.

Gathiram, P., M. T. Wells, J. G. Brock-Utne, and S. L. Gaffin. 1987. Antilipopolysaccharide improves survival in primates subjected to heat stroke. Circ. Shock 23:157-164.

Gomez-Alarcon, R. A., J. T. Huber, G. E. Higginbotham, F. Wiersma, D. Ammon, and B. Taylor. 1991. Influence of feeding Aspergillus oryzae fermentation extract on the milk yields, eating patterns, and body temperatures of lactating cows. J. Anim. Sci. 69:1733-1740.

Grotz, M. R., E. A. Deitch, J. Ding, D. Xu, Q. Huang, and G. Regel. 1999. Intestinal cytokine response after gut ischemia: Role of gut barrier failure. Ann. Surg. 229:478-486.

Hall, D. M., G. R. Buettner, L. W. Oberley, L. Xu, R. D. Matthes, and C. V. Gisolfi. 2001. Mechanisms of circulatory and intestinal barrier dysfunction during whole body hyperthermia. Am. J. Physiol. Heart Circ. Physiol. 280:H509-H521.

Harmon, R. J., M. Lu, D. S. Trammel, and B. A. Smith. 1997. Influence of heat stress and calving on antioxidant activity in bovine blood. J. Dairy Sci. 80(Suppl. 1):264.

Higginbotham, G. E., D. L. Bath, and L. J. Butler. 1993. Effect of feeding an Aspergillus oryzae extract on milk production and related responses in a commercial dairy herd. J. Dairy Sci. 76: 1484-1489.

Horton, J. W. and P. B. Walker. 1993. Oxygen radicals, lipid peroxidation, and permeability changes after intestinal ischemia and reperfusion. J. Appl. Physiol. 74:1515-1520.

Huber, J. T. 1998. Yeast products help cattle handle heat. Hoard's Dairyman 143:367.

Huber, J., G. Higginbotham, and D. Ware. 1985. Influence of feeding Vitaferm, containing an enzyme-producing culture from Aspergillus oryzae, on performance of lactating cows. J. Dairy Sci. 68(Suppl):122(Abstr.).

Huntington, G. B., C. K. Reynolds, and B. H. Stroud. 1989. Techniques for measuring blood flow in splanchnic tissues of cattle. J. Dairy Sci. 72:1583-1595.

Huntington, G. B. 1982. Portal blood flow and net absorption of ammonia-nitrogen, urea-nitrogen, and glucose in nonlactating Holstein cows. J. Dairy Sci. 65:1155-1162.

Kregel, K. C., P. T. Wall, and C. V. Gisolfi. 1988. Peripheral vascular responses to hyperthermia in the rat. J. Appl. Physiol. 64:2582-2588.

Mani, V., T. E. Weber, L. H. Baumgard, and N. K. Gabler. 2012. Growth and Development Symposium: Endotoxin, inflammation, and intestinal function in livestock. J. Anim. Sci. 90:1452-1465.

Marai, I. F. M., A. A. El-Darawany, A. Fadiel, and M. A. M. Abdel-Hafez. 2007. Physiological traits as affected by heat stress in sheep–A review. Small Rumin. Res. 71:1-12.

Pan, P. J., C. F. Hsu, J. J. Tsai, and J. H. Chiu. 2012. Musculoskeletal: The role of oxidative stress response revealed in preconditioning heat stimulation in skeletal muscle of rats. J. Surg. Res. 176:108-113.

Sahin, K., M. Onderci, N. Sahin, F. Gulcu, N. Yıldız, M. Avcı, and O. Kucuk. 2006. Responses of quail to dietary vitamin E and zinc picolinate at different environmental temperatures. Anim. Feed Sci. Technol. 129:39-48.

Saxton, A. M. 1998. A macro for converting mean separation output to letter groupings in PROC MIXED. In Proc. 23rd SAS User Group Intl. SAS Institute, Cary, NC, USA. pp. 1243-1246.

SAS Institute. 2009. User's Guide: Statistics. 9th ed. SAS Inst. Inc., Cary, NC, USA.

Schingoethe, D. J., K. N. Linke, K. F. Kalscheur, A. R. Hippen, D. R. Rennich, and I. Yoon. 2004. Feed efficiency of mid-lactation dairy cows fed yeast culture during summer. J. Dairy Sci. 87:4178-4181.

Şen, L. S., B. Karakoyun, C. Yeğen, M. Akkiprik, M. Yüksel, F. Ercan, A. Özer, and B. Ç. Yeğen. 2015. Treatment with either obestatin or ghrelin attenuates mesenteric ischemia–reperfusion-induced oxidative injury of the ileum and the remote organ lung. Peptides 71:8-19.

Shwartz, G., M. L. Rhoads, M. J. Van Baale, R. P. Rhoads, and L. H. Baumgard. 2009. Effects of a supplemental yeast culture on heat-stressed lactating Holstein cows. J. Dairy Sci. 92:935-942.

Sinkalu, V. O., J. O. Ayo, A. B. Adelaiye, and J. O. Hambolu. 2008. Effects of vitamin E on diurnal variation in rectal temperature of Black Harco pullets during the hot-dry season. J. Therm. Biol. 33:32-36.

Smuts, M., S. G. Pierzynowski, R. Puchala, A. Al-Dehneh, T. Sahlu, J. M. Fernandez, and R. N. Heitmann. 1995. Effect of mimosine on portal-drained visceral net flux and concentrations of amino acids and minerals in plasma of Alpine goats. Small Rumin. Res. 18:43-49.

Wang, L., B. Xue, K. Wang, S. Li, and Z. Li. 2011. Effect of heat stress on endotoxin flux across mesenteric-drained and portal-drained viscera of dairy goat. J. Anim. Physiol. Anim. Nutr. 95:468-477.

Yokota, M., J. Kambayashi, T. Tanaka, T. Tsujinaka, M. Sakon, and T. Mori. 1989. A simple turbidimetric time assay of the endotoxin in plasma. J. Biochem. Biophys. Methods 18:97-104.

Yu, B. P. 1994. Celluler defenses against damage from reactive oxygen species. Physiol. Rev. 74:139-162.

Zeidan, A. E. B., G. M. A. Solouma, M. H. El Nenaey, M. S. Shoeib, and M. M. El Taher. 2006. Reproductive activity of heat stressed rabbit bucks and its improvement using vitamin E and selenium under subtropical Egyptian condition. Assiut Vet. Med. J. 52:360-382.

12

The Effects of Freezing and Supplementation of Molasses and Inoculants on Chemical and Nutritional Composition of Sunflower Silage

Y. Konca, S. Buyukkilic Beyzi*, T. Ayaşan[1], M. Kaliber, and A. Bozkurt Kiraz[2]
Agricultural Faculty, Animal Science Department, Erciyes University, Kayseri 38039, Turkey

ABSTRACT: This study was conducted to determine the effects of freezing and supplementation of molasses (M), lactic acid bacteria (LAB) and LAB+enzyme mixture on chemical and nutritional composition of sunflower silage (SF). Sunflower crops were harvested (at about 29.2%±1.2% dry matter) and half of fresh sunflower was ensiled alone and half was frozen (F) at –20°C for 7 days. Silage additives were admixed into frozen SF material. All samples were ensiled in glass jars with six replicates for 90 days. The treatments were as follows: i) positive control (non-frozen and no additives, NF), ii) negative control (frozen, no additives, F), iii) F+5% molasses (FM), iv) F+LAB (1.5 g/tons, *Lactobacillus plantarum* and *Enterococcus faecium*, FLAB); v) F+LAB+enzyme (2 g/tons *Lactobacillus plantarum* and *Enterococcus faecium* and cellulase and amylase enzymes, FLEN). Freezing silage increased dry matter, crude ash, neutral detergent fiber, and acid detergent lignin. The organic matter, total digestible nutrient, non-fiber carbohydrate, metabolizable energy and *in vitro* dry matter digestibility were negatively influenced by freezing treatments ($p<0.05$). In conclusion, freezing sunflower plants prior to ensiling may negatively affect silage quality, while molasses supplementation improved some quality traits of frozen silage. Lactic acid bacteria and LAB+enzyme inoculations did not effectively compensate the negative impacts of freezing on sunflower silage. (**Key Words:** Enzyme, Freezing, Lactobacillus Bacteria, Molasses, Sunflower Silage)

INTRODUCTION

Sunflower is generally planted for seed production; however the green sunflower plant is also used as silage and forage by livestock producers. Sunflower is known to be drought and cold-resistant, therefore it can be planted as both first and second crop (Tan et al., 2015). The high fiber content of sunflower silage decreases digestibility of nutrients and low dry matter content at maturity makes ensilage difficult (Demirel et al., 2006).

Silage plants, cultivated as a second crop, are usually harvested and ensiled in fall. However, in recent years due to global climate changes, silage crops may sometimes freeze either before the harvest or after ensilage. Thus, plants are frozen in some cases before sufficient fermentation was achieved. With freezing, ice crystals may result in physical and microbiological spoilage of the silage. Freezing reduces water activity, increases intracellular solute concentration, promotes protein degradation, decreases pH levels, and initiates a thermal shock on microorganisms (Speck and Ray, 1977). All these events ultimately affect the maturation of ensilaged material and may alter chemical and nutrient composition of the silage. Together with freezing, enormous changes are observed in lipid oxidation, enzymatic esterification, taste, protein degradation, cellulose content, loss of pigments and vitamins (Fennoma and Powrie, 1963).

At slow freezing rates (0.2 to 1 cm/h), ice crystals grow slowly and settle in the vacuoles of feed. This can damage the cell wall and alter consequently the shape is distorted and degradation is accelerated (Martino and Zaritzky, 1988). Freeze and thaw cycles may also breakup the cell walls and release non-nutritious and toxic substances (Gleadow et al., 2012). Such cycles particularly alter cellulose content of the silage. However, such a condition may vary based on

* Corresponding Author: S. Buyukkilic Beyzi.
E-mail: sbuyukkilic@hotmail.com
[1] East Mediterranean Agricultural Research Institute, Adana 01321, Turkey.
[2] Agricultural Faculty, Animal Science Department, Harran University, Şanlıurfa 38039, Turkey.

soluble protein concentrations during thawing. Similarly, severe water losses occur during thawing and dehydration of water-soluble substances resulting in significant losses of carbohydrates. Such conditions ultimately lead to loss of energy and carbohydrates due to reductions in total digestible nutrients (TDN) of feeds (Martino and Zaritzky, 1988).

On the other hand, when the plants freeze, certain bacteria and other microorganisms naturally present in plant may be killed by freezing temperatures. So, following the freezing process, microbial fermentation and thus the silage quality will be altered. Freezing is usually applied under laboratory conditions to preserve the silage samples until analysis.

The present study was conducted to determine the effects of slow freezing (0.2 to 1 cm/h) and different supplements on chemical and nutritional composition of frozen sunflower silage.

MATERIALS AND METHODS

Experimental designs and procedures

Sunflower (*Helianthus annuus L.*) plants were grown over the experimental fields of Erciyes University Agricultural Research and Implementation Center (ERUTAM) between May 21 and September 17, 2013. Sunflowers were harvested and chopped in 1.5 to 3 cm pieces using a conventional corn silage machine (Çelikel Challenger, Turkey). A portion of fresh silage material was ensiled directly in 2-liter glass jars in 6 replications without any treatment to create positive control group, and another portion refrigerated at –20°C for 7 days in a plastic bag (45×45 cm), then ensiled alone or with additives. This freezing process is called a slow freezing. In brief, materials were frozen 0.2 to 1 cm per hour at –10°C to –20°C with still air freezers and cold stores type freezers (George, 1993). After seven days, frozen bags were thawed at +4°C and immediately mixed with additives and filled two lt glass jars with six replicates. Then, sampled materials were kept for 90 days at room temperature (about 20°C to 26°C). Each treatment group had six replicates.

The experimental treatments were as follows: i) positive control (NF, non-frozen and no additive), ii) negative control (F, frozen and no additive), iii) F+5% molasses (FM), iv) F+lactic acid bacteria (FLAB, 1.5 g/ton, consisting of *Lactobacillus plantarum* and *Enterococcus faecium* applied at a rate of 6.00 log10 cfu LAB/g of material, Pioneer 1174, DuPont Pioneer, Johnston, IA, USA) (5) F+LAB and enzyme mixture (FLEN, 2 g/ton, *Lactobacillus plantarum* bacterium (6.00 log10 cfu/g) and cellulose (150,000 carboxymethyl cellulose unit/kg) and amylase (200,000 Sandstedt, Kneen, and Blish (SKB)/kg) enzymes, Silaid WS, Global Nutritech Co., Richmond, VA, USA).

Chemical analyses

At the end of the 90-day ensilage period, samples were taken for chemical and nutritional analyses. For pH measurements, 25 g of silage samples were taken into a beaker and 100 ml distilled water was added. Then the mixture was mixed in a blender for 5 minutes and resultant mixture was filtered through Whatman filter paper and pH measurements were performed on this filtrate (Akyildiz, 1986). The dry matter (DM, AOAC, 2000 Method 934.01), crude ash (CA, AOAC, 2000, Method 942.05). Crude protein (CP, AOAC, 1996, Method 954.01), crude fiber (CF, AOAC, 1996, Method 978.10) and ether extract (EE, AOAC, 2000, Method 920.39) were determined in accordance with the methods specified in AOAC (1996, 2000). Neutral detergent fiber (NDF) was analyzed with heat-stable amylase and without Na-sulfite, acid detergent fiber (ADF) and acid detergent lignin (ADL) were determined according to the sequential method of Van Soest et al. (1991) by an ANKOM fiber analyzer ($ANKOM_{220}$ Technology, Macedon, NY, USA), and expressed inclusive of residual ash. Hemicellulose was defined as NDF – ADF.

The fleig point (FP) was calculated with the equation of FP = 220+(2×DM%–15)–40×pH (Akyildiz, 1986). The TDN were calculated according to the equation proposed by Chandler (1990), where TDN% = 105.2–0.68×NDF%. The non-fiber carbohydrates (NFC) were calculated by the equation proposed by Weiss et al. (1992): NFC% = 100–(NDF%+CP%+EE%+CA%). Total carbohydrates (TC) were determined according to Sniffen et al. (1992) with the equation TC% = 100–(CP%+EE%+CA%). The metabolizable energy (ME) was calculated by the equation proposed by Robinson et al. (2004): ME = 14.03–(0.01386×CF%)–(0.1018×CA%)

To determine water soluble carbohydrate (WSC) content, liquid extractions were prepared with 40 g silage. Samples were placed into a beaker, 360 mL distilled water was added and mixed in a blender. The resultant slurry was filtered through Whatman 54 filter paper and then centrifuged. Samples were stored at –20°C until the analyses. The WSC of samples was determined by phenol sulfuric acid method (Dubois et al., 1956).

The lactic acid (LA) was determined by the colorimetric method of Barker and Summerson (1941) and volatile fatty acid analyses (Fussell, 1987) (acetic [Chem Service O-4], propionic [Chem Service O-25] and butyric acid [Chem Service O-5]) were carried out in a gas chromatograph (Shimadzu GC-2010+, Kyoto, Japan) with a capillary column (30 m×0.25 mm×0.25 μm, Restek) and with flame ionization dedector over a temperature range of 45°C to 230°C.

In vitro dry matter digestibility (IVDMD) was determined in accordance with Tilley and Terry (1963). The rumen fluid was obtained from a steer slaughtered at a local slaughterhouse. All treatment groups were analyzed with the same rumen fluid. After incubation (Daisy incubator, ANKOM Technology, Macedon, NY, USA), sealed bags were extracted sequentially with neutral detergent solution (without either alpha-amylase or sodium sulfite) to determine the amount of undigested NDF remaining in each bag.

Statistical analysis

Data were analyzed using the general linear model procedure of the SPSS (1998). Differences between reported means were determined using Duncan's multiple range tests with a 5% level of probability. The results of statistical analysis were presented as mean values and standard error of the means in tables.

RESULTS AND DISCUSSION

Effects on chemical composition

The effects of freezing and supplementation of molasses, LAB and LAB+enzyme on chemical composition are provided in Table 1. The data revealed the greatest DM content in group F and the least DM content in group NF. The CA, NDF, and ADL values of group F were also higher than group NF ($p<0.05$). Kohn and Allen (1992) investigated the effects of freezing at –25°C for 12 months on quality of lucerne silage and reported significant decreases ($p<0.05$) in dry matter content of silage samples with freezing. The pH has an important effect on silage mass conservation and its decline is due primarily to the production of LA from bacteria action on soluble carbohydrates. The pH value of group F was also higher than group NF ($p<0.05$). Park et al. (2002) reported that there were no significant differences between Near Infrared Spectrophotometer (NIRS) predictions for freezing treatments and the reference predictions (pH, 3.50 to 5.25; mean 4.18). The NDF contents of groups NF and F were respectively found to be 37.11% and 39.62%. In a study conducted to investigate the chemical composition of fresh and frozen lucerne silage, Kohn and Allen (1992) reported significant increases in NDF and ADF contents of the samples with freezing. However, Deinum and Maassen (1994) reported that freezing fodder radish, lucerne, Italian ryegrass and maize silage at –20°C for 244 days, thawing and subsequent freeze-drying cycles had little effect on chemical composition. The ADL contents of groups NF and F were respectively found to be 11.87% and 12.33%. The average of ADL values of group F and NF was 12.10%.

The pH of group NF (3.55) was lower than those of F and F+supplemented groups (varied between 4.21 and 4.26) ($p<0.01$). However, there were no significant differences among the frozen groups. Meeske et al. (1993) reported that when LAB and LAB+enzyme were used in the sorghum silage, pH dropped rapidly compared to the control group. Silage quality is highly related to DM content of silage. In this experiment, DM contents of the groups varied between 29.40% and 35.51% and the effects of supplements were found to be significant ($p<0.01$). The DM content of silages was lower in group NF than in groups F and FM, FLAB and FLEN. The DM content of group FM was higher than those of groups FLAB and FLEN but similar with group F. The DM content of FLAB group was lower than those of F and FLEN groups. On the other hand, Koc et al. (2009) and

Table 1. The effects of freezing, molasses, LAB and LAB+enzyme supplementation on chemical composition of sunflower silage

Parameters	Treatments[1]					SEM	p
	NF	F	FM	FLAB	FLEN		
pH	3.55^{Bb}	4.26^{Aa}	4.23^{a}	4.24^{a}	4.21^{a}	0.07	**
DM (%)	29.40^{Bd}	34.74^{Aab}	35.51^{a}	31.89^{c}	33.65^{b}	0.53	**
CP (% DM)	7.96^{Aa}	7.94^{Aa}	7.96^{a}	6.33^{b}	7.27^{a}	0.19	*
CA (% DM)	13.44^{Ba}	13.63^{Aa}	12.72^{a}	10.05^{b}	13.72^{a}	0.37	**
EE (% DM)	5.23	4.53	4.35	5.36	4.58	0.17	NS
CF (% DM)	20.56^{Ab}	21.94^{Aab}	19.16^{b}	22.90^{a}	21.53^{ab}	0.42	*
NDF (% DM)	37.11^{Bbc}	39.62^{Aab}	35.70^{c}	42.14^{a}	39.82^{a}	0.620	**
ADF (% DM)	32.43^{Abc}	34.27^{Aab}	30.81^{c}	36.41^{a}	34.92^{a}	0.540	**
HEM (% DM)	4.68^{B}	5.35^{A}	4.89	5.73	4.90	0.23	NS
ADL (% DM)	11.87^{B}	12.34^{A}	11.66	13.51	13.39	0.360	NS

LAB, lactic acid bacteria; SEM, pooled standard error of means; p, probability; DM, dry matter; CP, crude protein; CA, crude ash; EE, ether extract; CF, crude fiber; NDF, neutral detergent fiber; ADF, acid detergent fiber; HEM, hemicellulose; ADL, acid detergent lignin.

[1] NF, non-frozen; F, frozen; FM, Frozen+%5 molasses; FLAB, frozen+LAB; FLEN, frozen+LAB+Enzyme.

* $p<0.05$; ** $p<0.01$; NS, not significant.

^{a,b,c,d} Values with different superscript in a line differ significantly between treatment groups.

^{A,B} Values with different superscript in a line differ significantly between freezing treatment.

Kamarloiy and Teimouri Yansari (2008) reported that addition of LAB to sunflower silage did not change the silage dry matter content. Considering the frozen groups, some water leakage and thus increased DM contents were observed in F, FM, FLAB, and FLEN groups. Martino and Zaritzky (1988) reported that water loss through thawing and dehydration with water-soluble substances may alter DM contents of the foods. Especially during freezing, intracellular and extracellular water expands and the cell wall is broken. Then, tissue integrity of frozen plants is destroyed and intracellular and extracellular water will be unbound and leak outside the tissue and drip loss will occur. CP content of FLAB group was lower than those of the other groups ($p<0.05$). On the other hand, CP content of NF, F, FM, and FLEN groups were similar. LAB inoculation might have increased bacterial growth, fermentation, required protein or nitrogen source of silage and consumed by the microorganisms. The LAB generally need various amino acids and vitamins for growth (Pahlow et al., 2003). It is assumed that bacterial growth would increase via LAB inoculation to silage during the ensiling process and therefore it would reach a better fermentation and consequently better silage quality. On the other hand, lipid oxidation, enzymatic esterification, protein degradation and change may occur with the freezing of foods (Fennoma and Powrie, 1963). Some experimental results revealed that sunflower silage protein ratio may vary between 11.60% and 13.45% (Tan et al., 2015). However many factors such as cultivar, harvest time and fertilization may also affect the protein content of sunflower silages (Demirel et al., 2006). The CA of silages was significantly affected by the treatments ($p<0.05$) and CA ratio of the LAB treatment group was lower than those of the other groups. Also LAB treatment lowered, DM and CP compared to the other groups. These results may be due to decreasing protein and ash contents and increasing fermentable carbohydrate ratio of silages (Mehmet, 2006). On the other hand, crude ash is related to mineral content of feeds. Exposure of feeds to freeze-thaw processes may result in water leakage and consequently minerals losses from the silage. There was no significant differences among the treatment groups in terms of ether extract ratio ($p>0.05$). Koc et al. (2009) reported that LAB and enzyme inoculants did not affect silage ether extract in sunflower silages.

Effects on fermentation and nutritive value

The effects of freezing and supplementation of molasses, LAB and LAB+enzyme on IVDMD, fermentation metabolites and nutritive value of sunflower silages is provided in Table 2. The OM was influenced negatively by freezing treatment at –20°C. Also TDN, NFC, and ME were negatively affected by freezing treatment. Such results were probably because of the loss of easily soluble carbohydrates during the thawing of frozen material. There was also a significant decrease in IVDM digestibility of silage in the freezing group which might be due to higher levels of cell wall components (especially ADL content) in the freezing group. Fleig points of the treatment groups were not significantly different from each other ($p>0.05$). The FP of group NF and F was respectively determined as 121.80 and

Table 2. The effects of freezing, molasses, LAB and LAB+enzyme supplementation on IVDMD, fermentation metabolites and nutritive value of sunflower silages

Parameters	Treatments[1]					SEM	p
	NF	F	FM	FLAB	FLEN		
IVDMD (%)	77.85^{Aa}	72.89^{Bb}	77.92^{a}	71.49^{b}	70.56^{b}	0.822	**
TDN (% DM)	79.96^{Aab}	78.26^{Bbc}	80.93^{a}	76.55^{c}	78.13^{c}	0.420	**
OM (% DM)	86.56^{Ab}	86.37^{Bb}	87.28^{b}	89.95^{a}	86.28^{b}	0.370	**
NFC (% DM)	36.27^{Ab}	34.28^{Bb}	39.28^{a}	36.12^{b}	34.61^{b}	0.540	*
TC (% DM)	73.38^{Ab}	73.90^{Ab}	74.97^{b}	78.26^{a}	74.43^{b}	0.520	*
ME (Mcal/kg DM)	12.38^{Ab}	12.34^{Bb}	12.47^{b}	12.69^{a}	12.34^{b}	0.040	**
FP	121.80^{Aa}	104.18^{Bbc}	107.03^{b}	99.28^{c}	103.80^{c}	1.980	**
WSC (% DM)	2.43^{Ab}	2.65^{Bb}	3.20^{a}	2.50^{b}	2.69^{b}	0.06	**
LA (% DM)	3.52^{B}	4.64^{A}	3.77	5.01	3.16	0.37	NS
AA (% DM)	1.34^{B}	1.62^{A}	1.55	2.47	0.94	0.23	NS
PA (% DM)	0.34^{Aa}	0.09^{Bbc}	0.01^{c}	0.02^{bc}	0.19^{ab}	0.03	*
BA (% DM)	ND	ND	ND	ND	ND	-	-
LA/AA	2.63^{A}	2.86^{A}	2.43	2.03	3.36	0.08	NS

LAB, lactic acid bacteria; IVDMD, *in vitro* dry matter digestibility; SEM, pooled standard error of means; p, probability; TDN, total digestible nutrients; DM, dry matter; OM, organic matter; NCF, non-fiber carbohydrates; TC, total carbohydrates; ME, metabolisable energy; FP, fleig point; WCS, water soluble carbohydrate; LA, lactic acid; AA, acetic acid; PA, propionic acid; BA, butyric acid; ND, not detected.

[1] NF, non-frozen; F, frozen; FM, frozen+% 5 molasses; FLAB, frozen+LAB; FLEN, frozen+LAB+enzyme.

* $p<0.05$; ** $p<0.01$; NS, not significant.

[a,b,c] Values with different superscript in a line differ significantly between treatment groups.

[A, B] Values with different superscript in a line differ significantly between freezing treatment.

104.18 (Table 2). The quality of sunflower silages (104.18 and 121.80) was assessed as "good quality" (Akyildiz, 1986). The WSC content of group NF (2.65%) was higher than group F (2.43%). Freezing and subsequent thawing result in severe water loss and seepage of water-soluble substances. Particularly, loss of carbohydrates after thawing causes decrease in TDN of the feeds (Martino and Zaritzky, 1988). The propionic acid (PA) concentration of group NF (0.337%) was higher than that of group F (0.09%). The propionic acid content is an important parameter for silage quality. Propionic acid may improve the aerobic stability of corn silages (Britt and Tuber, 1975).

Addition of LAB caused an increase in CF ratio in supplemented groups compared to NF and FM groups ($p<0.05$). The NDF and ADF ratios were higher in FLAB and FLEN groups than in NF and FM groups. Hemicellulose and ADL ratios of treatment groups were not significantly different ($p>0.05$). Martino and Zaritzky (1988) noted that freezing and thawing may alter cellulose content of feeds. According to Tan et al. (2015), the NDF and ADF values of sunflower silage may vary between 55.85%-57.62% and 37.81%-43.09%, respectively. In this study, these values varied between 35.70%-42.14% and 30.81%-36.41%, respectively. Fennoma and Powrie (1963) reported that freezing may change the cellulose content of foods. Koc et al. (2009) indicated that LAB inoculation decreased cellulose content of sunflower silages compared to the control group. Some studies showed that LAB+E mixture inoculation reduced the cell wall contents of silages (Nadeau et al., 2000). Ozduven et al. (2009) reported that LAB+E mixture addition decreased the NDF ratio of silage compared to control and LAB inoculant groups. However, the *in vitro* dry and organic matter digestibility of the silages was not affected by the treatments. The freezing process caused a decrease in IVDMD value of the treatment groups compared to NF group. Within the freezing groups, only in the FM group, IVDMD value was similar to the NF group and this group's IVDMD value was higher than those of the FLAB and FLEN groups ($p<0.01$). There is not enough experimental data about frozen silage material and the effect of LAB inoculation in the literature. Demirel et al. (2006) reported that OM digestibility of LAB+E silages could be increased by decreasing NDF in silage materials. In this experiment, lower NDF and ADF content of groups (NF, FM groups) correlated with an increase in *in vitro* DM digestibility of silage. There is a positive correlation between *in vitro* dry matter digestibility and low cell wall components of forages. It is expected that inoculation of cell wall degrading enzymes in silage may improve concentration of WSC available to LAB, and after appropriate fermentation, increase the digestibility of OM and fiber (Xing et al., 2009). On the other hand, some other studies showed that LAB and LAB+enzyme inoculants provided an improvement in digestibility or degradability of silage DM (Kamarloiy and Teimouri Yansari, 2008).

The highest TDN (80.93) and NFC (39.28) contents were determined in the FM group. Silage OM significantly differed among groups ($p<0.05$) and the highest OM was found in the FLAB group (89.95). The OM of feed is related to crude ash content. In this experiment, in the FLAB group, DM was high and CA was low, therefore OM content was higher than in other treatment groups. The FP was the highest in the NF group (121.80) and the lowest in the FLAB group (99.28) ($p<0.001$). The FP of silages were calculated based on the pH and dry matter content of silages. This value coincides with the "excellent" quality according to the Fleig scoring system.

WSC content of the FM group was significantly higher than those of the other groups ($p<0.01$). PA concentration of the NF group was higher than those of the other groups ($p<0.05$). There were no significant differences among the groups in terms of LA, AA, and LA/AA. The BA was not detected in silage samples. It is generally reported that microbial inoculation of silage had a positive effect on silage fermentation. When forages are inoculated with LAB and LAB+E before ensiling, resulting silage usually has a lower pH and a higher concentration of LA, but lower concentrations of acetic acid than control silage (Kung et al., 1987). Ozduven et al. (2009) reported that sunflower silage treated with LAB and LAB+E mixture inoculants had lower pH and acetic acid contents than that of control silage and a higher LA concentration and LA/acetic acid ratio than that of control silages.

CONCLUSION

The freezing treatments caused significant changes in nutrient composition of sunflower silage. The freezing caused an increase in dry matter content of silage due to loss of water after thawing the silage material and decreased the WSC contents of samples. In this case, this resulted in reduction of TDN of the feed (OM, TDN, NFC, and IVDMD are decreased) and subsequently resulted in energy and carbohydrates losses (ME is reduced). In conclusion, while molasses supplementation improved some quality traits of frozen silage, LAB and LAB+enzyme inoculations were not found to be effective to compensate for the negative impacts of freezing on sunflower silage.

REFERENCES

Akyildiz, A. R. 1986. Feed Science and Technology. Agricultural Faculty Publications (Ankara University, Publication no:895), Ankara, Turkey. 974 p.

AOAC. 1996. Official Methods of Analysis of AOAC International. 16th edition. Gaithersburg, MD, USA.

AOAC. 2000. Official Methods of Analysis of AOAC International. 17th edition Gaithersburg, MD, USA.

Barker, S. B. and W. H. Summerson. 1941. The colorimetric determination of lactic acid in biological material. J. Biol. Chem. 138:535-554.

Britt, D. G. and J. T. Huber. 1975. Fungal growth during fermentation and refermentation of nonprotein nitrogen treated corn silage. J. Dairy. Sci. 58:1666-1671.

Chandler, P. 1990. Energy prediction of feeds by forage testing explorer. Feedstuffs 62:12.

Deinum, B. and A. Maassen. 1994. Effects of drying temperature on chemical composition and *in vitro* digestibility of forages. Anim. Feed Sci. Technol. 46:75-86.

Demirel, M., D. Bolat, S. Celik, Y. Bakici, and A. Tekeli. 2006. Evaluation of fermentation qualities and digestibility of silages made from sorghum and sunflower alone and the mixtures of sorghum-sunflower. J. Biol. Sci. 6:926-930.

Dubois, M., K. A. Gilles, J. K. Hamilton, P. A. Rebers, and F. Smith. 1956. Calorimetric method for determination of sugars and related substances. Anal. Chem. 28:350-356.

Fennoma, O. and W. D. Powrie. 1963. Fundamentals of low-temperature food preservation. Adv. Food Res. 13:219-347.

Fussell, R. J. and D. V. McCalley. 1987. Determination of volatile fatty acid (C_2–C_5) and lactic acid in silage by gas chromatography. Analyst 122:1213-1216.

George, R. M. 1993. Freezing proceseses used in the food industry. Trends Food Sci. Technol. 4:134-138.

Gleadow, R. M., M. E. Møldrup, N. H. O'Donnell, and P. N. Stuart. 2012. Drying and processing protocols affect the quantification of cyanogenic glucosides in forage sorghum. J. Sci. Food Agric. 92:2234-2238.

Kamarloiy, M. and A. Teimouri Yansari. 2008. Effect of microbial inoculants on the nutritive value of corn silage for beef cattle. Pakistan J. Biol. Sci. 11:1137-1141.

Koc, F., M. L. Ozduven, L. Coskuntuna, and C. Polant. 2009. The effects of inoculant lactic acid bacteria on the fermentation and aerobic stability of sunflower silage. Poljoprivreda 15:47-52.

Kohn, R. A. and M. S. Allen. 1992. Storage of fresh and ensiled forages by freezing affects fibre and crude protein fractions. J. Sci. Food Agric. 58:215-220.

Kung, L., L. D. Satter, B. A. Jones, K. W. Genin, A. L. Sudoma, G. L. Enders, and H. S. Kim. 1987. Microbial inoculation of low moisture alfalfa silage. J. Dairy Sci. 70:2069-2077.

Martino, M. N. and N. E. Zaritzky. 1988. Ice crystal size modifications during frozen beef storage. J. Food Sci. 53:1631-1637.

Meeske, R., G. Ashbell, Z. G. Weinberg, and T. Kipnis. 1993. Ensiling forage sorghum at two stages of maturity with the addition of lactic acid bacterial inoculants. Anim. Feed Sci. Technol. 43:165-175.

Mehmet, A. B. 2006. Effects of hybrid type, stage of maturity, and fermentation length on whole plant com silage quality. Turk. J. Vet. Anim. Sci. 30: 331-336.

Nadeau, E. M. G., D. R. Buxton, J. R. Russell, M. J. Allison, and J. W. Young. 2000. Enzyme, bacterial inoculant, and formic acid effects on silage composition of orchardgrass and alfalfa. J. Dairy Sci. 83:1487-1502.

Ozduven, M. L., F. Koc, C. Polat, and L. Coskuntuna. 2009. The effects of lactic acid bacteria and enzyme mixture inoculants on fermentation and nutrient digestibility of sunflower silage. Kafkas Univ. J. Fac. Vet. Med. Vet. Fak. Derg. 15:195-199.

Pahlow, G., R. E. Muck, F. Driehuis, S. J. O. Elferink, and S. F. Spoelstra. 2003. 2 Microbiology of Ensiling. Silage Sci. Tech. 42:31.

Park, R. S., R. E. Agnew, and D. J. Kilpatrick. 2002. The effect of freezing and thawing on grass silage quality predictions based on near infrared reflectance spectroscopy. Anim. Feed Sci. Technol. 102:151-167.

Robinson, P. H., D. I. Givens, and G. Getachew. 2004. Evaluation of NRC, UC Davis and ADAS approaches to estimate the metabolizable energy values of feeds at maintenance energy intake from equations utilizing chemical assays and *in vitro* determinations. Anim. Feed Sci. Technol. 114:75-90.

Sniffen, C. J., J. D. O'connor, P. J. Van Soest, D. G. Fox, and J. B. Russell. 1992. A net carbohydrate and protein system for evaluating cattle diets: II. Carbohydrate and protein availability. J. Anim. Sci. 70:3562-3577.

Speck, M. L. and B. Ray. 1977. Effects of freezing and storage on microorganisms in frozen foods: A review. J. Food Protec. 40: 333-336.

SPSS. 1998. Version 17.00 for Windows. SPSS Inc., Chicago, IL, USA.

Tan, M., H. Yolcu, and Z. D. Gül. 2015. Nutritive value of sunflower silages ensiled with corn or alfalfa at different rate. J. Agric. Sci. 21:184-191.

Tilley, J. M. A. and R. A. Terry. 1963. A two-stage technique for the *in vitro* digestion of forage crops. J. Br. Grassl. Soc. 18: 104-111.

Van Soest, P. H., J. B. Robertson, and B. A. Lewis. 1991. Methods for dietary fiber, neutral detergent fiber, and nonstarch polysaccharides in relation to animal nutrition. J. Dairy Sci. 74: 3583-3597.

Weiss, W. P., H. R. Conrad, and N. R. Pierre. 1992. A theoretically-based model for predicting total digestible nutrient values of forages and concentrates. Anim. Feed Sci. Technol. 39:95-110.

Xing, L., J. Chen, and L. J. Han. 2009. The effect of an inoculant and enzymes on fermentation and nutritive value of sorghum straw silages. Bioresour. Technol. 100:488-491.

13

Performance Responses, Nutrient Digestibility, Blood Characteristics, and Measures of Gastrointestinal Health in Weanling Pigs Fed Protease Enzyme

Glenmer B. Tactacan[1], Seung-Yeol Cho[2], Jin H. Cho[3], and In H. Kim*

Department of Animal Resources and Science, Dankook University, Cheonan 330-714, Korea

ABSTRACT: Although exogenous protease enzymes have been used in poultry diets quite extensively, this has not been the case for pig diets. In general, due to their better gut fermentative capacity and longer transit time, pigs have greater capacity to digest dietary proteins than poultry. However, in early-weaned piglets, the stress brought about by weaning adversely affects the digestion of dietary proteins. Therefore, a study was conducted to determine the effects of a commercial protease enzyme in weanling pigs. Indices of growth, nutrient digestibility, blood profiles, fecal microflora, fecal gas emission and fecal scores were measured during the study. A total of 50 weanling pigs (6.42±0.12 kg) at 28 d of age were randomly assigned to receive 1 of 2 dietary treatments: i) control diet (corn-soy based) with no supplemental protease (CON), and ii) control diet+200 g/ton protease (PROT) for 42 d. A completely randomized design consisting of 2 treatments, 5 replicates, and 5 pigs in each replicate was used. Growth performance in terms of body weight (27.04±0.38 kg vs 25.75±0.39 kg; $p<0.05$) and average daily gain (491±7.40 g vs 460±7.46 g; $p<0.05$) in PROT fed pigs were increased significantly, but gain per feed (0.700±0.01 vs 0.678±0.01; $p>0.05$) was similar between treatments at d 42. Relative to CON pigs, PROT fed pigs had increased ($p<0.05$) apparent total tract digestibility (84.66%±0.65% vs 81.21%±1.13% dry matter and 84.02%±0.52% vs 80.47%±1.22% nitrogen) and decreased ($p<0.05$) NH_3 emission (2.0±0.16 ppm vs 1.2±0.12 ppm) in the feces at d 42. Except for a decreased ($p<0.05$) in blood creatinine level, no differences were observed in red blood cell, white blood cell, lymphocyte, urea nitrogen, and IgG concentrations between treatments. Fecal score and fecal microflora (*Lactobacillus* and *E. coli*) were also similar between CON and PROT groups. Overall, the supplementation of protease enzyme in weanling pigs resulted in improved growth rate and nutrient digestibility. Exogenous protease enzyme reduced fecal NH_3 emission, thus, potentially serving as a tool in lowering noxious gas contribution of livestock production in the environment. (**Key Words:** Enzyme, Growth Performance, Nutrient Digestibility, Protease, Weanling Pig)

INTRODUCTION

Weaning is a critical period in the pig rearing process. At this time, the piglets are exposed to different types of stressors (environmental, nutritional, psychological, and social), causing physiological changes in the structure and function of the piglet's digestive tract. Pluske et al. (1997) and Boudry et al. (2004) reported the shortening of the villi and the elongation of the crypt depth immediately after weaning. On the other hand, Hedemann and Jensen (2004) reported a significant reduction in the activity of pancreatic trypsin and chymotrypsin in newly weaned pigs. As a result, the digestive and absorptive capacity of the piglet is impaired, leading to poor performance such as slow growth rate, poor feed conversion, and high incidence of diarrhea.

The diets for weanling pig usually contain high levels of protein to promote early growth and muscle deposition. Therefore, coupled with piglet's poor capacity to digest and absorb dietary nutrients like proteins, high levels of undigested protein may accumulate and promote microbial fermentation and proliferation of pathogenic bacteria in the gastrointestinal tract. Microbial fermentation of undigested protein produces toxic substances such as ammonia and amines (Htoo et al., 2007), both of which can reduce growth

* Corresponding Author: In H. Kim.
E-mail: inhokim@dankook.ac.kr

[1] JEFO Innovation and Development, JEFO Nutrition Inc., Saint-Hyacinthe, Quebec, J2S 7B6, Canada.

[2] Eugenebio, Suwon 443-380, Korea.

[3] Department of Animal Science, Chungbuk National University, Cheongju 361-763, Korea.

(Gaskins, 2000) and cause diarrhea in piglets (Porter and Kentworthy, 1969, Dong et al., 1996). In addition, increased ammonia excretion in piggery farms poses a major environmental problem.

Recently, interests on the use of exogenous protease enzyme to improve protein utilization in livestock animals led to the conduct of studies related to this field (Yu et al., 2006; Wang et al., 2011; Guggenbuhl et al., 2012; Adebiyi and Olukosi, 2015; Opoku et al., 2015). However, to date, most of the available information on the application of protease enzyme has been generated from poultry studies with limited studies being done in pigs. Given the need for a nutritional tool to support piglets in processing dietary proteins at the time of weaning, the current study was conducted to determine the effects of exogenous protease enzyme on growth performance, nutrient digestibility, blood characteristics, and measures of gastrointestinal health in weanling pigs.

MATERIALS AND METHODS

The experimental protocol describing the management and care of animals were reviewed and approved by the Animal Care and Use Committee of Dankook University. The protease enzyme was provided by JEFO Nutrition Inc. of Saint-Hyacinthe, QC, Canada. The enzyme is an alkaline serine endopeptidase produced from bacterial fermentation product with an optimal pH of 8.5.

Experimental design, animals, and housing

A total of 50 weanling pigs ([Yorkshire×Landrace]× Duroc) with an average body weight (BW) of 6.42±0.12 kg (28 d of age) were used in a 42-d experiment. Pigs were randomly allotted to 2 experimental diets according to their initial BW. There were 5 replicate pens per treatment with 5 pigs per pen. The dietary treatment groups were i) control (CON), basal diet; and ii) basal diet+0.02% exogenous protease enzyme (PROT). All nutrients in diets were formulated to meet or exceed the recommendation of NRC (2012) for weanling pigs and fed in a crumble form (Table 1). All the pigs were housed in an environmentally controlled room with a slatted plastic floor. Each pen was equipped with a 1-sided, stainless steel self-feeder and a nipple drinker to allow the pig *ad libitum* access to feed and water throughout the experimental period.

Table 1. Composition of experimental diets, as-fed basis

Items	D 1 to 7	D 8 to 21	D 22 to 42
Ingredients (%)			
Extruded corn	29.18	44.49	61.97
Dehulled soybean meal (48% CP)	6.94	16.20	25.30
Fermented soybean meal (45% CP)	10.00	5.00	2.50
Fish meal (66% CP)	5.00	3.50	-
Soy oil	3.65	2.55	1.05
Lactose	15.30	8.30	-
Whey	15.00	10.00	5.00
Monocalcium phosphate	1.45	-	-
Dicalcium phosphate	-	1.50	-
Sugar	5.00	3.00	-
Plasma powder	6.00	3.00	-
L-lysine-HCl, 78%	0.29	0.39	0.46
DL-methionine, 50%	0.32	0.30	0.24
L-threonine, 89%	0.13	0.19	0.20
Choline chloride, 25%	0.20	0.10	0.10
Vitamin premix[1]	0.10	0.10	0.10
Trace mineral premix[2]	0.20	0.20	0.20
Limestone	1.24	0.98	1.13
Salt	-	0.20	0.25
Total	100.00	100.00	100.00
Calculated composition			
ME (kcal/kg)	3,640	3,540	3,410
Analyzed composition (%)			
Crude protein	21.12	19.87	19.07
Lysine	1.65	1.53	1.32
Methionine	0.61	0.60	0.55
Calcium	0.97	0.91	0.87
Phosphorus	0.82	0.73	0.67
Crude fiber	1.39	1.79	2.49

[1] Provided per kg of complete diet: 11,025 IU of vitamin A, 1,103 IU of vitamin D_3, 44 IU of vitamin E, 4.4 mg of vitamin K, 8.3 mg of riboflavin, 50 mg of niacin, 4 mg of thiamine, 29 mg of d-pantothenic, 60 mg of biotin, 166 mg of choline, and 33 μg of vitamin B_{12}.

[2] Provided per kg of complete diet: 80 mg of Fe (as $FeSO_4 \cdot 7H_2O$), 12 mg of Cu (as $CuSO_4 \cdot 5H_2O$), 85 mg of Zn (as $ZnSO_4$), 8 mg of Mn (as MnO_2), 0.28 mg of I (as KI), and 0.15 mg of Se (as $Na_2SeO_3 \cdot 5H_2O$).

Sampling and measurements

Individual pig BW and feed consumption on a per pen basis were recorded on d 1, 7, 21, and 42 to calculate average daily gain (ADG), average daily feed intake (ADFI), and gain per feed (G/F). Apparent total tract digestibility (ATTD) of dry matter (DM), gross energy (GE), and nitrogen (N) was determined using chromic oxide (0.2%) as an inert indicator (Kong and Adeola, 2014). Pigs were fed diets mixed with chromic oxide from d 14 to 21 and from d 35 to 42. Fresh fecal grab samples collected from 2 pigs per pen (d 21 and 42) were mixed and pooled, and a representative sample was stored in a freezer at –20°C until analyzed. Before chemical analysis, the fecal samples were thawed and dried at 60°C for 72 h, after which they were finely ground to pass through a 1-mm screen. All feed and fecal samples were analyzed for DM, GE, and N following the procedures outlined by the AOAC (2000). Chromium was analyzed via UV absorption spectrophotometry (Shimadzu UV-1201, Shimadzu, Kyoto, Japan) following the method described by Williams et al.

(1962).

The digestibility was calculated using the following formula:

$$\text{Digestibility (\%)} = \{1 - [(N_f \times C_d)/(N_d \times C_f)]\} \times 100$$

where N_f = nutrient concentration in feces (% DM), N_d = nutrient concentration in diet (% DM), C_d = chromium concentration in diet (% DM), and C_f = chromium concentration in feces (% DM).

For the blood profile analysis, 5 pigs from each treatment were randomly selected and 5-mL blood samples were collected via anterior vena cava puncture on d 42. At the time of collection, blood samples were collected into non-heparinized tubes and vacuum tubes containing K_3EDTA (Becton, Dickinson and Co., Franklin Lakes, NJ, USA) to obtain serum and whole blood, respectively. After collection, serum samples were centrifuged (3,000×g) for 15 min at 4°C. The white blood cells (WBC), red blood cells (RBC), lymphocyte concentration in the blood were determined using an automatic blood analyzer (ADVIA 120, Bayer, Tarrytown, NY, USA). The blood urea nitrogen (BUN) and creatinine concentrations in the serum were determined using an auto-chemical analyzer (HITACHI 747, Hitachi, Tokyo, Japan). Serum IgG was analyzed using nephelometry (Dade Behring, Marburg, Germany).

Fecal scores were determined at 08:00 and 20:00 h using the following fecal scoring system: 1 hard, dry pellet; 2 firm, formed stool; 3 soft, moist stool that retains shape; 4 soft, unformed stool that assumes shape of container; 5 watery liquid that can be poured. Feces and urine were collected on d 14, and 42. A total of 300 g fresh feces samples were stored in 2.6 L plastic boxes for replicates. The samples were fermented for 7 d at room temperature (28°C). After the fermentation period, a Gastec (model GV-100) gas-sampling pump was utilized for gas detection (Gastec Corp., Gastec detector tube No. 3M and 3La for NH_3; No. 70 and 70L for R.SH (total mercaptan), Gastec Corp, detector tube, Japan). The adhesive plasters were punctured, and 100 mL of headspace air was sampled approximately 2.0 cm above the feces surface.

Procedures of microbial shedding

Fecal samples were collected directly via massaging the rectum of 5 pigs in each treatment on d 42. Samples were pooled and immediately placed on ice for transportation to the lab. One gram of the composite fecal sample from each pen was diluted with 9 mL of 1% peptone broth (Becton, Dickinson and Co., Franklin Lakes, NJ, USA) and was subsequently homogenized. The bacterial count in fecal samples was conducted by plating a serial 10-fold dilution (in 1% peptone solution) into MacConkey agar plates (Difco Laboratories, Detroit, MI, USA) and *Lactobacilli* medium III agar plates (Medium 638, DSMZ, Braunschweig, Germany) for isolation of *E. coli* and *Lactobacillus*, respectively. The *Lactobacilli* medium III agar plates were then incubated for 48 h at 39°C under anaerobic conditions. The MacConkey agar plates were incubated for 24 h at 37°C. The *E. coli* and *Lactobacillus* colonies were counted immediately after removal from the incubator.

Statistical analysis

All experimental data were analyzed using the GLM Procedure as a completely randomized design (SAS Inst. Inc., Cary, NC, USA). The pen was used as the experimental unit. Differences among treatment means were determined using t-test. A probability level of $p \leq 0.05$ was used as the criterion for statistical significance.

RESULTS

Piglet performance

Body weight, ADG, ADFI, and G/F are shown in Table 2. The initial BW was similar between treatments at 6.42±0.12 kg (mean±standard error). At d 7 and 21, no differences in BW were observed. However, final BW (d 42) was increased ($p<0.05$) in piglets fed the PROT diet more than those fed the CON diet. The ADG from d 1 to 7 and d 1

Table 2. Performance of weanling pigs fed diets supplemented with and without protease enzyme

Item	CON[1]	PROT[1]	SEM[2]	p-value[3]
BW (kg)				
d 1	6.42	6.42	0.12	0.9889
d 7	8.35	8.61	0.16	0.2510
d 21	14.52	15.17	0.25	0.0758
d 42	25.75	27.04	0.38	0.0213
ADG (g/d)				
d 1 to 7	277	313	8.7	0.0068
d 8 to 21	441	469	12	0.1023
d 22 to 42	535	565	11.3	0.0898
d 1 to 42	460	491	7.4	0.0052
ADFI (g/d)				
d 1 to 7	326	358	5.1	0.0002
d 8 to 21	596	626	5.3	0.0004
d 22 to 42	855	866	6.6	0.0865
d 1 to 42	678	701	9.2	0.0907
G/F				
d 1 to 7	0.850	0.874	0.020	0.3554
d 8 to 21	0.740	0.749	0.019	0.7380
d 22 to 42	0.626	0.652	0.011	0.1509
d 1 to 42	0.678	0.700	0.009	0.1140

SEM, standard error of the mean; BW, body weight; ADG, average daily gain; ADFI, average daily feed intake; G/F, gain per feed.

[1] CON, control diet; PROT, protease diet (0.02% of the diet).

[2] n = 25/treatment.

[3] p-value for the effect of diet.

to 42, and ADFI from d 1 to 7 and d 8 to 21 were increased significantly in piglets fed the PROT diet compared to those fed the CON diet. However, the G/F in the CON and PROT group was similar throughout the study.

Apparent total tract digestibility

The ATTD of DM, N, and GE at d 21 and 42 are shown in Table 3. The ATTD of DM and N were increased ($p<0.05$) in pigs fed the PROT diet at d 21 and 42 compared with pigs fed the CON diet. However, no effect was observed on the ATTD of GE at d 21 and d 42.

Blood profile measurements

There were no differences in concentrations of RBC, WBC, lymphocyte, BUN, and IgG between treatment groups (Table 4). However, the creatinine concentration in blood was decreased ($p<0.05$) in piglets belonging to the PROT group compared to the CON group.

Fecal score, fecal gas emission, and fecal microflora

There was no difference in fecal score between treatments (Table 5). The NH_3 emission from the feces was reduced ($p<0.05$) in piglets fed the PROT diet relative to those fed the CON diet. However, R.SH emission (total mercaptan) was the same in both groups. Fecal *Lactobacillus* and *E. coli* populations were not affected by protease enzyme supplementation.

DISCUSSION

The supplementation of diets with exogenous protease enzyme to improve protein utilization has generated interest within the modern pig industry. However, in relation to their application in poultry nutrition, the use of protease in swine diets has not been explored as extensively. This is because overall, pigs have greater capacity to digest dietary proteins than chickens due to their better gut fermentative capacity and longer digesta transit time. However, in early-weaned piglets, the stress of weaning adversely affects the digestion of feed nutrients including proteins and amino acids (Pluske et al., 1997; Hedemann and Jensen, 2004). Therefore, it is worth investigating to determine the effects of exogenous protease supplementation in weanling pigs.

In the current study, PROT diet increased the final BW and the overall ADG of newly weaned piglets but did not affect the overall G/F. Previous study conducted by Wang et al. (2011) reported an increase in ADG when nursery pigs were fed diet supplemented with a protease enzyme (keratinase). Similarly, the growth performance of weanling pigs was improved when protease enzyme was added in a piglet diet based on low digestible protein sources (Zuo et al., 2015). Improvements in the utilization of dietary proteins and amino acids were anticipated with protease supplementation and may have accounted for the observed improvements in growth performance. Indeed, DM and N digestibility were increased significantly in weanling pigs fed with PROT diets. The ATTD of DM and N components of the PROT diets were increased consistently relative to the control when measured at d 21 and d 42. Guggenbuhl et al. (2012) and Zuo et al. (2015) also reported an increased in crude protein digestibility in a corn-soy based diet

Table 3. Nutrient digestibility in weanling pigs fed diets supplemented with and without protease enzyme

Item	CON[1]	PROT[1]	SEM[2]	p-value[3]
3 week (%)				
Dry matter	82.65	86.03	0.86	0.0266
Nitrogen	81.34	85.30	0.90	0.0155
Energy	83.04	83.70	0.74	0.5951
6 week (%)				
Dry matter	81.21	84.66	0.89	0.0245
Nitrogen	80.47	84.02	0.87	0.0229
Energy	82.81	83.04	0.74	0.8462

SEM, standard error of the mean.
[1] CON, control diet; PROT, protease diet (0.02% of the diet).
[2] n = 6/treatment.
[3] p-value for the effect of diet.

Table 4. Blood profiles of weanling pigs fed diets supplemented with and without protease enzyme

Item	CON[1]	PROT[1]	SEM[2]	p-value[3]
RBC ($10^6/\mu L$)	5.68	6.44	0.73	0.5861
WBC ($10^3/\mu L$)	17.4	15.87	2.08	0.6274
Lymphocyte (%)	6.95	10.33	3.18	0.5149
Creatinine (mg/dL)	1.14	0.81	0.04	0.0013
BUN (mg/dL)	7.83	7.88	0.79	0.9658
IgG (mg/dL)	475.8	490.3	27.5	0.7402

SEM, standard error of the mean; RBC, red blood cells; WBC, white blood cells; BUN, blood urea nitrogen; IgG, immunoglobulin G.
[1] CON, control diet; PROT, protease diet (0.02% of the diet).
[2] n = 4/treatment.
[3] p-value for the effect of diet.

Table 5. Fecal microflora, fecal score, and fecal gas emission in weanling pigs fed diets supplemented with and without protease enzyme

Item	CON[1]	PROT[1]	SEM[2]	p-value[3]
R.SH (ppm)	2.2	1.8	0.2	0.1097
NH_3 (ppm)	2.0	1.2	0.1	0.0039
Lactobacillus (log_{10}cfu/g)	7.38	7.54	0.08	0.2218
E. coli (log_{10}cfu/g)	6.22	6.14	0.04	0.1852
Fecal score[4]	3.62	3.60	0.01	0.8976

SEM, standard error of the mean.
[1] CON, control diet; PROT, protease diet (0.02% of the diet).
[2] n = 5/treatment.
[3] p-value for the effect of diet.
[4] Fecal score: 1 = hard, dry pellet, 2 = firm, formed stool, 3 = soft, moist stool that retains shape, 4 = soft, unformed stool that assumes shape of container, 5 = watery liquid that can be poured.

supplemented with a protease when fed to weanling pigs. Huo et al. (1993) explained that protease enzyme may neutralize anti-nutritive factors such as protease inhibitors and thus could account for its positive effect on protein digestibility. An *in vitro* study utilizing gel electrophoresis demonstrated the degradation of purified lectin, Kunitz and Bowman Birk inhibitors after being incubated in a solution with a protease enzyme (Nielsen et al., 2013). In terms of energy digestibility, PROT diet did not impact the ATTD of GE in the current study. This is in contrast with the findings of McAllister (1993) who reported an increase in the digestion of cornstarch following the application of a serine protease in a rumen *in vitro* model. In this study, the improvement in GE digestibility was attributed to the disruption of the protein matrix surrounding the starch granules. In parallel, energy digestibility values of protease-supplemented feed ingredients selected to represent both vegetable (soybean meal, rapeseed meal, and peas) and animal proteins (meat and bone meal) were increased significantly in adult cockerels when compared to the un-supplemented group (Gauthier, unpublished data).

In the current study, blood creatinine concentration was decreased significantly in nursery pigs fed PROT diet but RBC, WBC, lymphocyte, IgG, and BUN levels were similar to the CON group. The degradation of creatine and creatine phosphate in muscle is the main source of creatinine fluxed in the blood. Creatinine levels in the blood can be elevated during periods of diarrhea as a result of the increase mobilization of protein reservoirs from muscle to compensate for the decreased nutrient intake and/or absorption. However, because the piglets from the current study did not suffer from diarrhea (as indicated by fecal score), the lower creatinine levels in the PROT supplemented diet may indicate an overall better amino acid status relative to the CON pigs. Brosnan et al. (2011) reported that the replacement of degraded creatine in the muscle imposes an appreciable burden on the metabolism of methionine and arginine, and thus, negatively impacting the total body amino acid balance.

The high protein levels in weaning diets coupled with the impaired ability of the weanling pigs to digest dietary proteins as a result of weaning may result in the proliferation of pathogenic bacteria in the gut and subsequently increased predisposition to diarrhea (Porter and Kentworthy, 1969; Dong et al., 1996). Therefore, it was hypothesized that feeding weanling pigs with protease enzyme would reduce enteric pathogens and lower the incidence of diarrhea. However, in the current study, the PROT diet did not affect the fecal microflora (*Lactobacillus* and *E. coli*) and fecal score of weanling pigs. The observed fecal score was also not indicative of any incidence of diarrhea for both groups. In contrast, Zhang et al. (2014) found that dietary supplementation of an exogenous multi-enzyme consisting of xylanase, amylase, and protease, increased the counts of *Lactobacilli* spp. and reduced the populations of *Escherichia coli* spp. in the feces, resulting in reduced diarrhea index in weanling piglets. The reason for the differences in between experiments could have been due to the differences in types and levels of enzymes used, the nature of the diets (Mirzaie et al., 2012; Wen et al., 2012) and the sanitary condition of the facility where the study was conducted. In the current study, the piglets were housed in a relatively clean research facility. It is well recognized that the level of bacterial population in the gastrointestinal tract can be dictated by sanitary conditions in the environment where pigs are raised.

The emission of odorous gas such as NH_3 from pig production facilities contributes to pollution in the environment and may affect the performance of pigs. In the current study, the fecal NH_3 concentration was significantly decreased by PROT diet. On the contrary, McAlpine et al. (2012) reported that finisher pigs offered protease-supplemented diets alone had significantly higher NH_3 emissions compared to basal fed pigs. However, NH_3 emissions were reduced when protease was combined with xylanase. Although the large differences in age and diet might have accounted for the observed differences in results across experiments, further research is necessary to elucidate the important factors involved. Microbial fermentation of undigested proteins and amino acids in the hindgut produces NH_3 and contributes to the NH_3 output in the manure (Gaskins, 2000). Therefore, in the current study, the lower fecal NH_3 concentration in the PROT pigs may indicate better digestion of dietary proteins and amino acids. This is through limiting the availability of non-digested proteins, which then serve as substrate for NH_3 production in the large intestine.

Overall, it can be concluded that protease supplementation in weanling pigs can effectively improved growth performance and nutrient digestibility, particularly of dietary proteins. Furthermore, protease supplementation can significantly reduce fecal NH_3 emission, thus, potentially serving as a tool in lowering noxious gas contribution of livestock production in the environment.

REFERENCES

Adebiyi, A. O. and O. A. Olukosi. 2015. Metabolizable energy content of wheat distillers' dried grains with solubles supplemented with or without a mixture of carbohydrases and protease for broilers and turkeys. Poult. Sci. 94:1270-1276.

AOAC. 2000. Official Methods of Analysis. 17th ed. Assoc. Off. Anal. Chem. Washington, DC, USA.

Boudry, G., V. Péron, I. ILe Huërou-Luron, J. P. Lallès, and B. Sève. 2004. Weaning induces both transient and long-lasting modifications of absorptive, secretory, and barrier properties of piglet intestine. J. Nutr. 134:2256-2262.

Brosnan, J. T., R. P. da Silva, and M. E. Brosnan. 2011. The metabolic burden of creatine synthesis. A Review. Amino Acids 40:1325-1331.

Dong, G., A. Zhou, F. Yang, K. Chen, K. Wang, and D. Dao. 1996. Effect of dietary protein levels on the bacterial breakdown of protein in the large intestine, and diarrhoea in early-weaned piglets. Acta Vet. Zootec. Sin. 27:293-302.

Gaskins, H. R. 2000. Intestinal bacteria and their influence on swine growth. In Swine Nutrition (Eds. A. J. Lewis and L. L. Southern). 2nd ed. CRC Press, New York, NY, USA. pp. 585-608.

Guggenbuhl, P., Y. Waché, and J. W. Wilson. 2012. Effects of dietary supplementation with a protease on the apparent ileal digestibility of the weaned piglet. J. Anim. Sci. 90:152-154.

Hedemann, M. S. and B. B. Jensen. 2004. Variations in enzyme activity in stomach and pancreatic tissue and digesta in piglets around weaning. Arch. Anim. Nutr. 58:47-59.

Htoo, J. K., B. A. Araiza, W. C. Sauer, M. Rademacher, Y. Zhang, M. Cervantes, and R. T. Zijlstra. 2007. Effect of dietary protein content on ileal amino acid digestibility, growth performance, and formation of microbial metabolites in ileal and cecal digesta of early-weaned pigs. J. Anim. Sci. 85:3303-3312.

Huo, G. C., V. R. Fowler, and M. Bedford. 1993. The use of enzymes to denature antinutritive factors in soybean. In: Recent Advances of Research in Antinutritional Factors in Legume Seeds (Eds. A. F. B. van der Poel, J. Huisman, and H. S. Saini). Proceedings of the Second International Workshop on 'Antinutritional factors (ANFs) in Legume seeds'. Wageningen Pers, Wageningen, The Netherlands. pp. 517-521.

Kong, C. and O. Adeola. 2014. Evaluation of amino acid and energy utilization in feedstuff for swine and poultry diets. Asian Australas. J. Anim. Sci. 27:917-925.

McAllister, T. A., R. C. Phillippe, L. M. Rode, and K. J. Cheng. 1993. Effect of the protein matrix on the digestion of cereal grains by ruminal microorganisms. J. Anim. Sci. 71:205-212.

McAlpine, P. O., C. J. O'Shea, P. F. Varley, P. Solan, T. Curran, and J. V. O'Doherty. 2012. The effect of protease and nonstarch polysaccharide enzymes on manure odor and ammonia emissions from finisher pigs. J. Anim. Sci. 90:369-371.

Mirzaie, S., M. Zaghari, S. Aminzadeh, M. Shivazad, and G. G. Mateos. 2012. Effects of wheat inclusion and xylanase supplementation of the diet on productive performance, nutrient retention, and endogenous intestinal enzyme activity of laying hens. Poult. Sci. 91:413-425.

Nielsen, P., K. Pontoppidan, M. U. Faruk, J. Broz, and I. Knap. 2013. *In vitro* degradation of soybean anti-nutritional factors by a mono component protease. Int. Poult. Sci. Forum. M80. (Abstr.).

NRC. 2012. Nutrient Requirements of Swine. 11th Rev. Ed. National Academies Press, Washington, DC, USA.

Opoku, E. Y., H. L. Classen, and T. A. Scott. 2015. Evaluation of inclusion level of wheat distillers dried grains with solubles with and without protease or β-mannanase on performance and water intake of turkey hens. Poult. Sci. 94:1600-1610.

Pluske, J. R., D. J. Hampson, and I. H. Williams. 1997. Factors influencing the structure and function of the small intestine in the weaned pig: A Review. Livest. Prod. Sci. 51:215-236.

Porter, P. and R. Kentworthy. 1969. A study of intestinal and urinary amines in pigs in relation to weaning. Res. Vet. Sci. 10:440-447.

Wang, D., X. S. Piao, Z. K. Zeng, T. Lu, Q. Zhang, P. F. Li, L. F. Xue, and S. W. Kim. 2011. Effects of keratinase on performance, nutrient utilization, intestinal morphology, intestinal ecology and inflammatory response of weaned piglets fed diets with different levels of crude protein. Asian Australas. J. Anim. Sci. 24:1718-1728.

Wen, C., L. C. Wang, Y. M. Zhou, Z. Y. Jiang, and T. Wang. 2012. Effect of enzyme preparation on egg production, nutrient retention, digestive enzyme activities, and pancreatic enzyme messenger RNA expression of late-phase laying hens. Anim. Feed Sci. Technol. 172:180-186.

Williams, C. H., D. J. David, and O. Iismaa. 1962. The determination of chromic oxide in faeces samples by atomic absorption spectrophotometry. J. Agric. Sci. 59:381-385.

Yu, B., S. T. Wu, C. C. Liu, R. Gauthier, and P. W. S. Chiou. 2006. Effects of enzyme inclusion in a maize-soybean diet on broiler performance. Anim. Feed Sci. Technol. 134:283-294.

Zhang, G. G., Z. B. Yang, Y. Wang, W. R. Yang, and H. J. Zhou. 2014. Effects of dietary supplementation of multi- enzyme on growth performance, nutrient digestibility, small intestinal digestive enzyme activities, and large intestinal selected microbiota in weanling pigs. J. Anim. Sci. 92:2063-2069.

Zuo, J., B. Ling, L. Long, T. Li, L. Lahaye, C. Yang, and D. Feng. 2015. Effect of dietary supplementation with protease on growth performance, nutrient digestibility, intestinal morphology, digestive enzymes and gene expression of weaned piglets. Anim. Nutr. 1:276-282.

14

Effects of Condensed Tannins in Mao (*Antidesma thwaitesianum* Muell. Arg.) Seed Meal on Rumen Fermentation Characteristics and Nitrogen Utilization in Goats

P. Gunun*, M. Wanapat[1], N. Gunun[2], A. Cherdthong[1], S. Sirilaophaisan, and W. Kaewwongsa[2]

Department of Animal Science, Faculty of Natural Resources, Rajamangala University of Technology-Isan, Sakon Nakhon Campus, Phangkhon, Sakon Nakhon 47160, Thailand

ABSTRACT: Mao seed is a by-product of the wine and juice industry, which could be used in animal nutrition. The current study was designed to determine the effect of supplementation of mao (*Antidesma thwaitesianum* Muell. Arg.) seed meal (MOSM) containing condensed tannins (CT) on rumen fermentation, nitrogen (N) utilization and microbial protein synthesis in goats. Four crossbred (Thai Native×Anglo Nubian) goats with initial body weight (BW) 20±2 kg were randomly assigned to a 4×4 Latin square design. The four dietary treatments were MOSM supplementation at 0%, 0.8%, 1.6%, and 2.4% of total dry matter (DM) intake, respectively. During the experimental periods, all goats were fed a diet containing roughage to concentrate ratio of 60:40 at 3.0% BW/d and pangola grass hay was used as a roughage source. Results showed that supplementation with MOSM did not affect feed intake, nutrient intakes and apparent nutrient digestibility ($p>0.05$). In addition, ruminal pH and ammonia nitrogen (NH_3-N) were not influenced by MOSM supplementation, whilst blood urea nitrogen was decreased quadraticly ($p<0.05$) in goats supplemented with MOSM at 2.4% of total DM intake. Propionate was increased linearly with MOSM supplementation, whereas acetate and butyrate were remained the same. Moreover, estimated ruminal methane (CH_4) was decreased linearly ($p<0.05$) when goats were fed with MOSM at 1.6% and 2.4% of total DM intake. Numbers of bacteria and protozoa were similar among treatments ($p>0.05$). There were linear decreases in urinary N ($p<0.01$) and total N excretion ($p<0.01$) by MOSM supplementation. Furthermore, N retention was increased linearly ($p<0.05$) when goats were fed with MOSM supplementation at 1.6% and 2.4% of total DM intake. Microbial protein synthesis were not significantly different among treatments ($p>0.05$). From the current study, it can be concluded that supplementation of MOSM at 1.6% to 2.4% of total DM intake can be used to modify ruminal fermentation, especially propionate and N utilization in goats, without affecting the nutrient digestibility, microbial populations and microbial protein synthesis. (**Key Words:** Mao Seed Meal, Goats, Rumen Fermentation, Nitrogen Utilization, Microbial Protein Synthesis)

INTRODUCTION

Many feed additives have been developed to improve the efficiency of nutrient use by decreasing methane (CH_4) production and reducing nitrogen (N) excretion, among which ionophore antibiotics have been successful. However, ionophore antibiotics have been banned in Europe since 2006, and many countries outside the European Union are also considering a ban (Jayanegara et al., 2009). For this reason, attention has recently shifted to natural antimicrobials as a safe means of modifying ruminal fermentation. Currently, the use of plant secondary compounds, including tannins and saponins, that are potent modifiers of ruminal fermentation and intraruminal recycling of microbial protein aims at improving the efficiency of dietary N utilization and mitigating N losses in ruminants (Hristov and Jouany, 2005).

* Corresponding Author: P. Gunun.
E-mail: pongsatorng@hotmail.com

[1] Tropical Feed Resources Research and Development Center (TROFREC), Department of Animal Science, Faculty of Agriculture, Khon Kaen University, Khon Kaen 40002, Thailand.

[2] Program in Animal Production Technology, Faculty of Technology, Udon Thani Rajabhat University, Udon Thani 41000, Thailand.

Mao or Mamao *(Antidesma thwaitesianum* Muell. Arg.) is favored by consumers in the local market in Thailand because of its good color and taste. Mao grows very well over a variety of soil types and is naturalised in Africa, Australia, islands in the Pacific Ocean and tropical Asia. Currently, Mao juice and mao wine industry produce very large amounts of seed and skin pulp, which is considered as an environmental problem. Mao seed contains a large amount of plant secondary compounds, especially condensed tannins (CT) (Gunun et al., 2014). Our previous study (Gunun et al., 2014) illustrated that supplementation with mao seed has the potential to manipulate rumen fermentation, by decreasing protozoa. However, mao seed meal (MOSM) have not been investigated for use as feed additives in *in vivo.* Therefore, the objective of this study was to investigate the effect of MOSM supplementation on rumen fermentation, microbial populations, N utilization and microbial protein synthesis in goats.

MATERIALS AND METHODS

Animals, treatments and experimental design

Four crossbred (Thai Native×Anglo Nubian) goats with initial body weight (BW) 20±2 kg were randomly assigned to a 4×4 Latin square design. The dietary treatments were as follows: Supplementation with MOSM at 0%, 0.8%, 1.6%, and 2.4% of total dry matter (DM) intake. Fresh mao seed was provided from Department of Food Science and Technology, Faculty of Natural Resources, Rajamangala University of Technology-Isan, Sakon Nakhon Campus, Thailand; and sundried for 2 to 3 days, then ground to pass a 1 mm sieve. All animals were fed a diet containing roughage to concentrate ratio (R:C) of 60:40 at 3.0% BW/d and pangola grass (*Digitaria eriantha* Steud., synonym *D. decumbens*) hay was used as a roughage source. The diet of R:C was offered to the animals twice per day in the morning (07:00) and afternoon (16:00). Goats were housed individually in ventilated pens with wooden slotted flooring in an open goat barn raised above the ground. Clean fresh water and feed blocks were available at all times. The experiment was conducted over four periods, each lasting for 21 days: The first 14 days were used for feed intake measurements and the remaining 7 days for total urine and fecal collection. Chemical compositions of concentrate and pangola grass hay are presented in Table 1.

Data collection and sampling procedures

Feeds offered and refusals samples were collected during the last 7 days of each period in the morning and afternoon feedings. Fecal samples were collected from the total collection of each individual goat on each treatment during the last 7 days of each period in the morning and afternoon feeding. Feed, refusals and fecal samples were

Table 1. Ingredients and chemical composition of concentrate, pangola grass hay and mao seed meal (MOSM) used in the experiment

Item	Concentrate	Pangola grass hay	MOSM
Ingredient (% dry matter)			
Cassava chip	54.1	-	-
Rice bran	15.7	-	-
Palm kernel meal	12.0	-	-
Soybean meal	11.7	-	-
Urea	2.2	-	-
Molasses	2.5	-	-
Mineral and vitamin mixture[1]	1.0	-	-
Salt	0.5	-	-
Sulfur	0.3	-	-
Chemical composition			
Dry matter (%)	92.2	87.5	95.7
	---------- % of dry matter --------		
Organic matter	90.0	91.7	95.2
Crude protein	16.1	7.9	10.7
Neutral detergent fiber	28.1	67.6	65.0
Acid detergent fiber	16.8	35.9	47.8
Ash	10.0	8.3	4.8
Condensed tannins	-	-	9.6
Saponins	-	-	9.2

MOSM, mao seed meal.

[1] Minerals and vitamins (each kg contains): Vitamin A, 10,000,000 IU; Vitamin E, 70,000 IU; Vitamin D, 1,600,000 IU; Fe, 50 g; Zn, 40 g; Mn, 40 g; Co, 0.1 g; Cu, 10 g; Se, 0.1 g; I, 0.5 g.

dried at 60°C, ground (1 mm screen using a Cyclotech Mill, Tecator, Hoganas, Sweden) and analysed using the standard methods of AOAC (1995) for DM, crude protein (CP), and ash, while neutral detergent fiber (NDF) and acid detergent fiber (ADF) were analysed according to Van Soest et al. (1991). Content of CT in MOSM was analysed using the modified vanillin-HCl method based on Burns (1971). Crude saponin concentrations were measured using methanol extraction following the method of Kwon et al. (2003) and modified by Poungchompu et al. (2009).

Urine samples were analyzed for total N according to AOAC (1995), and allantoin in urine was determined by high-performance liquid chromatography (HPLC) (Instruments by controller water model 600E, Milford, MA, USA; water model 484 UV detector; column novapak C18; column size 3.9 mm×300 mm; mobile phase 10 mM H_2PO_4 [pH 2.5]) as described by Chen et al. (1993). The amount of microbial purines absorbed (X, mmol/d) presumably proportional to the purine derivatives (PD) excreted (Y, mmol/d), was estimated based with the following equation described by Chen et al. (1990) for sheep as: $Y = 0.84X+(0.150BW^{0.75}e^{-0.25X})$. The supply of microbial N (MN) was estimated by urinary excretion of PD according to Chen and Gomes (1995): MN (g/d) =

$70X/(0.116 \times 0.83 \times 1,000) = 0.727X$; where X is PD absorption in mmol/d, digestibility of microbial purine is 0.83, the N content of purines is 70 mg N/mmol, and the ratio of purine-N:total N in mixed rumen microbes is 11.6:100. The efficiency of microbial N synthesis (EMNS) to denote the microbial N supplied to the animal per unit of digestible organic matter apparently fermented in the rumen (DOMR) was calculated using the following formula: EMNS = MN (g/d)/DOMR (assuming that rumen digestion was 65% organic matter of digestion in total tract, DOMR = DOMI×0.65; DOMI = digestible organic matter intake).

At the end of each period, rumen fluid was collected at 0, 3, and 6 h post-feeding. Rumen fluid was taken from the rumen by a stomach tube connected to a vacuum pump. Ruminal pH was determined using a portable pH meter (HANNA Instruments HI 8424 microcomputer, Singapore). Rumen fluid samples were then filtered through 4 layers of cheesecloth and divided into two portions. The first portion was used for ammonia nitrogen (NH_3-N) analysis, comprised 5 mL of 1 M H_2SO_4 and 50 mL of rumen fluid. It was centrifuged at 16,000 g for 15 min and the supernatant stored at –20°C. Ruminal NH_3-N concentrations were analysed using a Kjeltech Auto 1030 Analyzer, Tecator, Hoganiis, Sweden (AOAC, 1995) and volatile fatty acids (VFA) analysis was performed using HPLC (Instruments by controller water model 600E; water model 484 UV detector; column novapak C18; column size 3.9 mm×300 mm; mobile phase 10 mM H_2PO_4 [pH 2.5]) (Samuel et al., 1997). CH_4 production was calculated by standard equations according to CH_4 = 0.45 (acetate)–0.275 (propionate)+0.40 (butyrate) (Moss et al., 2000). A second portion was fixed with 10% formalin solution in sterilized 0.9% saline solution. The total direct counts of bacteria and protozoa were made by the methods of Galyean (1989) based on the use of a haemocytometer (Boeco, Hamburg, Germany).

A blood sample (about 10 mL) was collected from the jugular vein at the same time as rumen fluid sampling into tubes containing 12 mg of ethylene diaminetetraacetic acid, and plasma was separated by centrifugation at 500 g for 10 min at 4°C and stored at –20°C until analysis of blood urea nitrogen (BUN) according to Crocker (1967).

Statistical analysis

All data were subjected to analysis of variance according to a 4×4 Latin square design using the general linear models procedures (SAS, 1996). Data were analyzed using the model: $Yijk = \mu + Di + Pj + gk + eijk$; where $Yijk$ = the dependent variable, μ = overall mean, Di = fixed effect of diet, Pj = fixed effect of period, gk = random effect of goat, and $eijk$ = residual error. Orthogonal polynomial contrasts (linear, quadratic and cubic) were used to estimate the effect of MOSM supplementation. Significant effects were identified at $p<0.05$.

RESULTS

Chemical composition of diet

Table 1 shows feed ingredients and chemical compositions of concentrate, pangola grass hay and MOSM. The concentrate diet was formulated by using available local feed resources and contained CP and NDF at 16.1% and 28.1%, respectively. Moreover, pangola grass hay contained CP and NDF at 7.9% and 67.6%, respectively. MOSM contained CP, NDF, and CT at 10.7%, 65.0%, and 9.6%, respectively.

Feed intake and nutrient digestibility

Results of feed intakes and apparent nutrient digestibilities as influenced by MOSM supplementation are presented in Table 2. The results show that feed intake, nutrient intakes and apparent nutrient digestibility were not affected by MOSM supplementation ($p>0.05$).

Characteristics of rumen fermentation and blood metabolites

The result of ruminal pH, NH_3-N, and BUN affected by MOSM supplementation are presented in Table 3. Supplementation of MOSM did not affect ruminal pH and NH_3-N concentrations ($p>0.05$), while BUN concentrations were decreased quadraticly ($p<0.05$) when goats were fed MOSM with 2.4% of total DM intake. The effect of MOSM supplementation on ruminal VFA concentrations is presented in Table 4. The concentrations of total VFA, acetate and butyrate were similar among treatments ($p>0.05$), while propionate was increased linearly by MOSM supplementation ($p<0.05$). Furthermore, the calculated production of ruminal CH_4 was decreased linearly ($p<0.05$) when goats were fed MOSM at 1.6% and 2.4% of total DM intake.

Rumen microorganism populations

Table 5 presents information on the rumen microorganism population as influenced by MOSM supplementation. The population of bacteria, holotrich protozoa, entodiniomorph protozoa and total protozoa counts were unaltered by dietary treatments ($p>0.05$)

Nitrogen utilization and microbial protein synthesis

Table 6 presents the results of N utilization and microbial CP synthesis of goats as influenced by MOSM supplementation. Total N intake was not different among treatments ($p>0.05$) and ranged from 10.5 to 11.1 g/d. Fecal N excretion, and N absorption were not affected by MOSM supplementation ($p>0.05$). The amount of N excreted in the

Table 2. Effect of mao seed meal supplementation on feed intake and nutrients digestibility in goats

Item	MOSM (% of total DM intake)				SEM	Contrast[1]		
	0	0.8	1.6	2.4		L	Q	C
DM intake								
Pangola grass hay								
g/d	322.9	345.4	322.8	346.6	12.38	ns	ns	ns
g/kg $BW^{0.75}$	31.8	31.1	31.7	33.5	0.96	ns	ns	ns
Concentrate								
g/d	250.8	231.9	252.5	257.7	14.84	ns	ns	ns
g/kg $BW^{0.75}$	24.4	22.1	24.3	24.5	0.71	ns	ns	ns
MOSM								
g/d	0.00	4.73	8.93	14.3	0.49	**	ns	ns
g/kg $BW^{0.75}$	0.00	0.45	0.88	1.38	0.04	**	ns	ns
Total intake								
g/d	573.7	582.0	584.2	618.6	24.85	ns	ns	ns
g/kg $BW^{0.75}$	56.2	55.7	56.8	59.3	1.15	ns	ns	ns
Nutrients intake (g/d)								
Organic matter	521.8	530.0	531.8	563.4	22.55	ns	ns	ns
Crude protein	65.9	65.1	67.1	70.3	3.14	ns	ns	ns
Neutral detergent fiber	282.8	301.7	294.9	315.6	11.66	ns	ns	ns
Acid detergent fiber	158.1	165.2	162.5	174.2	6.47	ns	ns	ns
Apparent digestibility (%)								
Dry matter	71.2	70.8	70.2	70.8	0.47	ns	ns	ns
Organic matter	71.9	72.2	72.0	72.3	0.41	ns	ns	ns
Crude protein	80.3	78.3	80.9	78.2	0.74	ns	ns	ns
Neutral detergent fiber	65.4	64.2	63.9	62.1	0.59	ns	ns	ns
Acid detergent fiber	54.0	53.4	53.1	53.5	0.90	ns	ns	ns

MOSM, mao seed meal, DM, dry matter; SEM, standard error of the mean; BW, body weight.
[1] L, linear; Q, quadratic; C, cubic.
** $p<0.01$, ns, Non-significant ($p>0.05$).

urine and total N excretion were decreased linearly ($p<0.01$) with MOSM supplementation at 1.6% and 2.4% of total DM intake. Moreover, N retention and N retention to N intake ratios were increased linearly ($p<0.05$) when goats were fed with MOSM at 1.6% and 2.4% of total DM intake. PD excretion, microbial N supply, microbial CP synthesis

Table 3. Effect of mao seed meal supplementation on rumen pH, NH_3-N and BUN in goats

Item	MOSM (% of total DM intake)				SEM	Contrast[1]		
	0	0.8	1.6	2.4		L	Q	C
pH								
0 h-post feeding	6.8	6.8	6.9	6.8	0.05	ns	ns	ns
3	6.5	6.5	6.5	6.5	0.03	ns	ns	ns
6	6.3	6.6	6.7	6.8	0.05	*	ns	ns
Mean	6.5	6.6	6.7	6.7	0.08	ns	ns	ns
NH_3-N (mg/dL)								
0 h-post feeding	16.9	19.7	17.6	18.4	0.91	ns	ns	ns
3	23.6	24.9	20.6	20.1	0.92	ns	ns	ns
6	18.4	17.9	17.9	19.8	1.17	ns	ns	ns
Mean	19.6	20.8	18.7	19.4	0.57	ns	ns	ns
BUN (mg/dL)								
0 h-post feeding	13.3	14.2	14.0	13.5	0.34	ns	ns	ns
3	18.0	18.3	19.5	15.5	0.47	ns	*	ns
6	14.3	16.4	16.8	12.8	0.42	ns	**	ns
Mean	15.2	16.2	16.8	13.9	0.35	ns	*	ns

BUN, blood urea nitrogen; MOSM, mao seed meal; DM, dry matter; SEM, standard error of the mean.
[1] L, linear; Q, quadratic; C, cubic.
* $p<0.05$, ** $p<0.01$; ns, non-significant ($p>0.05$).

Table 4. Effect of mao seed meal supplementation on VFA concentration and CH_4 production in goats

Item	MOSM (% of total DM intake)				SEM	Contrast[1]		
	0	0.8	1.6	2.4		L	Q	C
Total VFA (mmol/d)								
0 h-post feeding	107.1	96.1	102.1	95.2	1.36	*	ns	*
3	118.8	115.5	110.7	106.8	2.37	ns	ns	ns
6	98.0	96.3	99.9	100.4	1.97	ns	ns	ns
Mean	108.0	102.6	104.2	100.8	1.67	ns	ns	ns
VFA (% TVFA)								
Acetate (C2)								
0 h-post feeding	66.4	66.4	65.8	63.7	1.24	ns	ns	ns
3	65.2	63.9	62.4	63.8	1.27	ns	ns	ns
6	67.1	64.7	63.2	66.2	1.54	ns	ns	ns
Mean	66.2	65.0	63.8	64.6	0.71	ns	ns	ns
Propionate (C3)								
0 h-post feeding	22.5	23.4	22.5	25.9	0.63	*	ns	ns
3	24.7	25.3	27.8	27.1	0.93	ns	ns	ns
6	23.1	24.0	25.4	26.6	0.98	ns	ns	ns
Mean	23.4	24.2	25.2	26.5	0.64	*	ns	ns
Butyrate (C4)								
0 h-post feeding	11.1	10.2	11.7	10.4	0.69	ns	ns	ns
3	10.1	10.8	9.8	9.1	0.58	ns	ns	ns
6	9.8	11.3	11.4	7.2	0.54	ns	*	ns
Mean	10.3	10.8	11.0	8.9	0.46	ns	ns	ns
C2:C3								
0 h-post feeding	3.0	2.8	2.9	2.5	0.11	ns	ns	ns
3	2.6	2.5	2.2	2.4	0.10	ns	ns	ns
6	2.9	2.7	2.5	2.5	0.12	ns	ns	ns
Mean	2.8	2.7	2.5	2.4	0.14	ns	ns	ns
CH_4 estimation[2] (mol/100 mol)								
0 h-post feeding	28.1	27.5	28.1	25.7	0.71	ns	ns	ns
3	26.6	26.1	24.4	24.9	0.58	ns	ns	ns
6	27.8	27.0	26.0	25.4	0.68	ns	ns	ns
Mean	27.5	26.9	26.2	25.3	0.40	*	ns	ns

VFA, volatile fatty acids; MOSM, mao seed meal; DM, dry matter; SEM, standard error of the mean.
[1] L, linear; Q, quadratic; C, cubic. [2] CH_4 gas production was calculated.
* $p<0.05$, ns, non-significant ($p>0.05$).

and EMNS were not significantly affected by dietary treatments ($p>0.05$).

DISCUSSION

Supplementation of tannin containing plants to ruminant diets usually reduces feed intake because of reduced palatability, decreased rate of digestion and development of conditioned aversion (Mueller-Harvey, 2006). Beauchemin et al. (2008) reported that feed intake was reduced by high doses (>50 g/kg DM) of CT uptake. The negative effect of CT on feed intake was caused by astringency of CT and short-term post-ingestive malaise (Landau et al., 2000). However, in the present study, dietary tannins sources had no effect on total DM intake and nutrient intakes when used at a suitable level (<50 g/kg DM) as a supplement.

Significant differences in ruminal pH between control and treatments occurred at 6 h-post feeding. The ruminal pH in control was significantly lower than those of treatment group. However, mean ruminal pH was not altered by dietary treatments and was in optimal range for the ecology and fermentation by microbes as reported by Gunun et al. (2013). Rumen NH_3-N concentrations varied in response to feeding, especially at the peaks occurring at 3 h post-feeding. Concentration of NH_3-N in the rumen fluid is the net result of NH_3-N production from the feed, fermentation of protein, absorption through the rumen wall and passage out of the rumen and utilization by microbes. Rumen NH_3-N concentration averaged 19.6 mg/dL for dietary treatments, a value that is close the optimal concentration of 15 to 30 mg/dL of NH_3-N as an optimal level for microbial growth in tropical conditions (Wanapat and Pimpa, 1999). When a CT containing plant is masticated, insoluble CT-protein complexes are formed;

Table 5. Effect of mao seed meal supplementation on ruminal microbes in goats

Item	MOSM (% of total DM intake)				SEM	Contrast[1]		
	0	0.8	1.6	2.4		L	Q	C
Direct count (cell/mL)								
Bacteria ($\times 10^9$)								
0 h-post feeding	2.3	2.5	2.5	2.2	0.79	ns	ns	ns
3	2.4	2.3	2.1	2.3	0.69	ns	ns	ns
6	2.5	2.4	2.3	2.4	0.92	ns	ns	ns
Mean	2.5	2.4	2.3	2.3	0.85	ns	ns	ns
Protozoa ($\times 10^5$)								
Horotrich								
0 h-post feeding	1.6	1.7	1.8	1.2	0.11	ns	ns	ns
3	3.2	2.2	2.8	1.7	0.35	ns	ns	ns
6	1.4	2.5	2.2	1.9	0.16	ns	ns	ns
Mean	2.0	2.1	2.2	1.6	0.10	ns	ns	ns
Entodiniomorph								
0 h-post feeding	0.4	0.6	0.6	0.9	0.11	ns	ns	ns
3	0.5	0.4	1.1	0.8	0.34	ns	ns	ns
6	1.0	0.6	0.6	0.8	0.16	ns	ns	ns
Mean	0.6	0.5	0.8	0.8	0.10	ns	ns	ns
Total protozoa								
0 h-post feeding	2.0	2.3	2.4	2.1	0.26	ns	ns	ns
3	3.7	2.6	2.9	2.5	0.59	ns	ns	ns
6	2.4	3.1	2.8	2.7	0.20	ns	ns	ns
Mean	2.7	2.7	3.0	2.4	0.26	ns	ns	ns

MOSM, mao seed meal; DM, dry matter; SEM, standard error of the mean; ns, non-significant (p>0.05).
[1] L, linear; Q, quadratic; C, cubic.

these are stable over the pH range 3.5 to 7.0 but dissociate in the abomasum and duodenum, hence decreasing ruminal CP degradation and NH_3-N concentrations (Makkar, 2003). However, reactivity between CT and proteins depends

Table 6. Effect of mao seed meal supplementation on N utilization and microbial protein synthesis in goats

Item	MOSM (% of total DM intake)				SEM	Contrast[1]		
	0	0.8	1.6	2.4		L	Q	C
N utilization								
N intake (g/d)								
Pangola grass hay	4.0	4.4	4.3	4.4	1.15	ns	ns	ns
Concentrate	6.5	6.5	6.5	6.5	0.33	ns	ns	ns
MOSM	0.00	0.08	0.15	0.23	0.009	**	ns	ns
Total	10.5	11.0	11.0	11.1	0.45	ns	ns	ns
N excretion (g/d)								
Fecal	2.1	2.4	2.0	2.4	0.11	ns	ns	ns
Urine	3.2	3.2	1.0	1.0	0.19	**	ns	ns
Total	5.3	5.6	3.0	3.4	0.17	**	ns	ns
N balance (g/d)								
Absorption	8.5	8.6	9.0	8.7	0.37	ns	ns	ns
Retention	5.2	5.4	8.0	7.7	0.47	*	ns	ns
% of N retention to N intake	49.5	46.5	71.9	69.9	2.29	**	ns	*
PD excretion (mmol/d)	15.4	13.4	12.8	16.7	1.91	ns	ns	ns
MN (g N/d)	11.2	9.7	9.3	12.2	1.39	ns	ns	ns
MCP (g CP/d)	70.0	60.8	58.3	75.9	8.69	ns	ns	ns
EMNS (g N/kg OMDR)	41.6	33.5	44.4	52.4	4.98	ns	ns	ns

MOSM, mao seed meal; DM, dry matter; SEM, standard error of the mean; PD, purine derivatives; MN, microbial nitrogen; MCP, microbial crude protein; EMNS, efficiency of microbial N synthesis, g of N/kg of OMDR; OMDR, organic matter digested in the rumen.
[1] L, linear; Q, quadratic; C, cubic.
* p<0.05, ** p<0.01; ns, non-significant (p>0.05).

partly on the molecular weight, type of tertiary structure, and amino acid content of proteins. In the present study, ruminal NH_3-N concentrations and microbial CP synthesis were not significantly affected by dietary treatments. A possible explanation for this is that CT in MOSM may not affect the protein degradability. This is in agreement with Animut et al. (2008) who reported that rumen NH_3-N concentrations were not affected by varying source of CT in goats. Moreover, BUN is an indicator of protein degradation in the rumen (Pathak et al., 2013). In the current study, BUN was decreased when goats were fed with MOSM at 2.4% of total DM intake. Higher CT in MOSM may have further reduced availability of degradable protein in the rumen, resulting in a further decrease in BUN concentrations. The results from the present study were similar to those of Dey et al. (2008), who reported that BUN concentrations was decreased by supplementation of CT from *Ficus infectoria* leaves at 1.5% and 2.0% in lambs.

Propionate formation can be considered as a competitive pathway for CH_4 production (Moss et al., 2000). The increase of propionate in the supplemented groups led to the reduction of calculated CH_4 production in this study. The CT suppressed CH_4 production by shifting hydrogen (H_2) from the CH_4 pathway to produce propionate (Wanapat et al., 2014). Simailarly, Animut et al. (2008) reported that propionate was increased, while CH_4 was decreased by supplementation with CT containing forage, Kobe lespedeza, in goats. Our previous study (Gunun et al., unpublished) showed that supplementation of mao pomace containing CT could reduce CH_4 production up to 40% in *in vitro*. The present study showed the reduction of CH_4 by MOSM without affecting total DM intake. However, total VFA, acetate and butyrate were similar among treatments. The results from the present study were similar to the report of Wanapat et al. (2014) that VFA production and C2:C3 were not affected by mangosteen peel containing CT, possibly due to lack of effect of supplemented CT on bacteria or to adaptation of rumen microorganisms to CT (Patra and Saxena, 2011).

The total direct count is a traditional method and may not reflect accurately bacteria in the rumen, especially bacteria attached to plant particles. In future studies, quantitative polymerase chain reaction (qPCR) method will need to be used to more accurately measure the number of total bacteria. Effects of CT on rumen protozoa are variable and mostly depend on the type of CT, their origin and supplementation levels (Patra and Saxena, 2011). In the present study, total bacteria, total protozoa, horotrich and entodiniomorph counts were not affected by MOSM supplemention. Similarly, Animut et al. (2008) established that CT in Kobe lespedeza was not responsible for the anti-protozoal activity in goat. In contrast, Gunun et al. (2014) reported that protozoal population was dramatically decreased with MOSM supplementation in *in vitro*.

In the present study, supplementation with MOSM did not affect digestibility of CP. These results contrast with those of Soltan et al. (2013) who reported a decrease in apparent N digestibility due to higher fecal N excretion in the presence of CT. These authors suggested that higher faecal N excretion in the presence of active tannins was matched by lower urinary N excretion due to the lower degradability of protein in the rumen and subsequent lower NH_3-N absorption. This shift in N excretion routes was not relevant in our experiment since no significant differences were observed in fecal N excretion. However, urinary N excretion was lower, which resulted in higher N retention when goats were fed with MOSM at 1.6% and 2.4% of total DM intake. The reduction in urinary N in our study was 68.8% by MOSM compared to control. A possible explanation for this is that CT in MOSM was increased rate of recycling N into the rumen, presumable as saliva urea. Similar to the present study, increased N retention in goat given plants-containing CT due to lowered nitrogen excretion through urine has been reported earlier by several workers (Pathak et al., 2013).

Urinary PD excretion is often used to estimate ruminal microbial protein synthesis in ruminants. The principle is that duodenal purine bases, as a microbial marker, are efficiently absorbed at the small intestine and the majority of their metabolites excreted via kidney with a specific urinary recovery in each animal species. Allantoin was the major PD, accounting for 78% to 85% of total PD (Hernandez et al., 2014). In the present study, PD excretion was not affected by MOSM supplementation. Microbial CP synthesis in the rumen provides the majority of protein supplied to small intestine of ruminants, accounting for 50% to 80% of total absorbable protein. Supplementation with MOSM did not affect microbial CP synthesis. Moreover, total protozoa counts were not affected by MOSM supplementation compared with the control. Thus it seems unlikely that the protozoa substantively contributed to the estimated increase in duodenal flow of microbial N as estimated from the urinary excretion of PD. The results from the present study were similar to those of Dentinho et al. (2014), who reported that microbial N supply and protozoa counts were unaffected by *Salix babylonica* and *Leucaena leucocephala* leaf extracts in lambs. The microbial CP flows from the rumen as calculated from PD excretion using the equation of Chen and Gomes (1995) were 70.0, 60.8, 58.3, and 75.9 g CP/d when goats were fed with MOSM at 0%, 0.8%, 1.6%, and 2.4% of total DM intake, respectively.

CONCLUSIONS

Supplementation of MOSM at 1.6% to 2.4% of total

DM intake resulted in increased VFA production especially propionate, N balance through reducing N losses in urine and reduced CH_4 production, while it did not adversely affect feed intake, digestibility, microbial populations and microbial protein synthesis in goats.

ACKNOWLEDGMENTS

The authors would like to express their most sincere gratitude and appreciation to the Program in Animal Production Technology, Faculty of Technology, Udon Thani Rajabhat University for the use of the research facilities. The cooperation of J. Jantapong, S. Sangkaew, and P. Thattamlay whom participated in this study and the Department of Animal Science, Faculty of Natural Resources, Rajamangala University of Technology-Isan, Sakon Nakhon Campus was highly appreciated.

REFERENCES

Animut, G., R. Puchala, A. L. Goetsch, A. K. Patra, T. Sahlu, V. H. Varel, and J. Wells. 2008. Methane emission by goats consuming different sources of condensed tannins. Anim. Feed Sci. Technol. 144:228-241.

AOAC. 1995. Official Method of Analysis, 16th ed. Animal Feeds: Association of Official Analytical Chemists, Arlington, VA, USA.

Beauchemin, K. A., M. Kreuzer, F. O'Mara, and T. A. McAllister. 2008. Nutritional management for enteric methane abatement: a review. Aust. J. Exp. Agric. 48:21-27.

Burns, R. E. 1971. Method for estimation of tannin in the grain sorghum. Agron. J. 63:511-512.

Chen, X. B. and M. J. Gomes. 1995. Estimation of microbial protein supply to sheep and cattle based on urinary excretion of purine derivative-an overview of the technique details. Occasional Publication 1992. International Feed Resources Unit, Rowett Research Institute, Aberdeen, UK.

Chen, X. B., F. D. DeB. Hovell, E. R. Ørskov, and D. S. Brown. 1990. Excretion of purine derivatives by ruminants: effect of exogenous nucleic acid supply on purine derivative excretion by sheep. Br. J. Nutr. 63:131-142.

Chen, X. B., D. J. Kyle, and E. R. Ørskov. 1993. Measurement of allantoin in urine and plasma by high-performance liquid chromatography with pre-column derivatization. J. Chromatogr. B Biomed. Sci. Appl. 617:241-247.

Crocker, C. L. 1967. Rapid determination of urea nitrogen in serum or plasma without deproteinization. Am. J. Med. Technol. 33:361-365.

Dentinho, M. T. P., A. T. Belo, and R. J. B. Bessa. 2014. Digestion, ruminal fermentation and microbial nitrogen supply in sheep fed soybean meal treated with *Cistus ladanifer* L. tannins. Small Rumin. Res. 119:57-64.

Dey, A., N. Dutta, K. Sharma, and A. K. Pattanaik. 2008. Effect of dietary inclusion of *Ficus infectoria* leaves as a protectant of proteins on the performance of lambs. Small Rumin. Res. 75:105-114.

Galyean, M. 1989. Laboratory Procedure in Animal Nutrition Research. Department of Animal and Food Sciences, Texas Tech University, Lubbock TX, USA.

Gunun, P., M. Wanapat, and N. Anantasook. 2013. Rumen fermentation and performance of lactating dairy cows affected by physical forms and urea treatment of rice straw. Asian Australas. J. Anim. Sci. 26:1295-1303.

Gunun, P., N. Anantasook, M. Wanapat, S. Sirilaophaisan, A. Cherdthong, C. Wachirapakorn, and C. Yuangklang. 2014. Effect of Mao (*Antidesma thwaitesianum* Mull. Arg.) seed supplementation on *in vitro* rumen protozoal population and digestibility using a gas production technique. Khon Kaen Agr. J. 42(Suppl. 4):47-53.

Hernandez, P., A. Z. M. Salem, S. Lopez, X. Z. Sun, R. Rojo, L. M. Camacho, M. M. Y. Elghandour, and M. Gonzalez-Ronquillo. 2014. Influence of *Salix babylonica* and *Leucaena Leucocephala* leaf extracts on ruminal fermentation characteristics, urinary purine derivative excretion and microbial protein synthesis. Livest. Sci. 163:80-84.

Hristov, A. N. and J. -P. Jouany. 2005. Factors affecting the efficiency of nitrogen utilization in the rumen. In: Nitrogen and Phosphorus Nutrition of Cattle: Reducing the Environmental Impact of Cattle Operations (Eds. E. Pfeffer and A. N. Hristov) CAB International, Wallingford, UK. pp. 117-166.

Jayanegara, A., N. Togtokhbayar, H. P. S. Makkar, and K. Becker. 2009. Tannins determined by various methods as predictors of methane production reduction potential of plants by an *in vitro* rumen fermentation system. Anim. Feed Sci. Technol. 150:230-237.

Kwon, J. H., J. M. R. Belanger, J. R. J. Pare, and V. A. Yaylayan. 2003. Application of the microwave-assisted process (MAPTM) to the fast excretion of ginseng saponins. Food Res. Int. 36:491-498.

Landau, S., N. Silanikove, Z. Nitsan, D. Barkai, H. Baram, F. D. Provenza, and A. Perevolotsky. 2000. Short-term changes in eating patterns explain the effects of condensed tannins on feed intake in heifers. Appl. Anim. Behav. Sci. 69:199-213.

Makkar, H. P. S. 2003. Effects and fate of tannins in ruminant animals, adaptation to tannins, and strategies to overcome detrimental effects of feeding tannin rich feeds. Small Rumin. Res. 49:241-256.

Moss, A. R., J. P. Jouany, and J. Newbold. 2000. Methane production by ruminants: its contribution to global warming. Anim. Res. 49:231-253.

Mueller-Harvey, I. 2006. Unravelling the conundrum of tannins in animal nutrition and health. J. Sci. Food Agric. 86:2010-2037.

Pathak, A. K., N. Dutta, P. S. Banerjee, A. K. Pattanaik, and K. Sharma. 2013. Influence of dietary supplementation of condensed tannins through leaf meal mixture on intake, nutrient utilization and performance of *Haemonchus contortus* infected sheep. Asian Australas. J. Anim. Sci. 26:1446-1458.

Patra, A. K. and J. Saxena. 2011. Exploitation of dietary tannins to improve rumen metabolism and ruminant nutrition. J. Sci. Food Agric. 91:24-37.

Poungchompu, O., M. Wanapat, C. Wachirapakorn, S. Wanapat, and A. Cherdthong. 2009. Manipulation of ruminal fermentation and methane production by dietary saponins and tannins from mangosteen peel and soapberry fruit. Arch. Anim. Nutr. 63:389-400.

Samuel, M., S. Sagathewan, J. Thomus, and G. Mathen. 1997. An HPLC method for estimation of volatile fatty acids of rumen fluid. Indian J. Anim. Sci. 67:805-807.

SAS. 1996. User's Guide: Statistic, Version 5. edn. SAS. Institute, Cary, NC, USA.

Soltan, Y. A., A. S. Morsy, S. M. Sallam, R. C. Lucas, H. Louvandini, M. Kreuzer, and A. L. Abdalla. 2013. Contribution of condensed tannins and mimosine to the methane mitigation caused by feeding *Leucaena leucocephala*. Arch. Anim. Nutr. 67:169-184.

Van Soest, P. J., J. B. Robertson, and B. A. Lewis. 1991. Method for dietary fiber, neutral detergent fiber and non-starch polysaccharide in relation to animal nutrition. J. Dairy Sci. 74:3583-3597.

Wanapat, M. and O. Pimpa. 1999. Effect of ruminal NH_3-N levels on ruminal fermentation, purine derivatives, digestibility and rice straw intake in swamp buffaloes. Asian Australas. J. Anim. Sci. 12:904-907.

Wanapat, M., V. Chanthakhoun, K. Phesatcha, and S. Kang. 2014. Influence of mangosteen peel powder as a source of plant secondary compounds on rumen microorganisms, volatile fatty acids, methane and microbial protein synthesis. Livest. Sci. 162:126-133.

15

Improvement of Nutritive Value and *In vitro* Ruminal Fermentation of *Leucaena* Silage by Molasses and Urea Supplementation

K. Phesatcha and M. Wanapat*

Tropical Feed Resources Research and Development Center (TROFREC), Department of Animal Science, Faculty of Agriculture, Khon Kaen University, Khon Kaen 40002, Thailand

ABSTRACT: *Leucaena* silage was supplemented with different levels of molasses and urea to study its nutritive value and *in vitro* rumen fermentation efficiency. The ensiling study was randomly assigned according to a 3×3 factorial arrangement in which the first factor was molasses (M) supplement at 0%, 1%, and 2% of crop dry matter (DM) and the second was urea (U) supplement as 0%, 0.5%, and 1% of the crop DM, respectively. After 28 days of ensiling, the silage samples were collected and analyzed for chemical composition. All the nine *Leucaena* silages were kept for study of rumen fermentation efficiency using *in vitro* gas production techniques. The present result shows that supplementation of U or M did not affect DM, organic matter, neutral detergent fiber, and acid detergent fiber content in the silage. However, increasing level of U supplementation increased crude protein content while M level did not show any effect. Moreover, the combination of U and M supplement decreased the content of mimosine concentration especially with M2U1 (molasses 2% and urea 1%) silage. The result of the *in vitro* study shows that gas production kinetics, cumulation gas at 96 h and *in vitro* true digestibility increased with the increasing level of U and M supplementation especially in the combination treatments. Supplementation of M and U resulted in increasing propionic acid and total volatile fatty acid whereas, acetic acid, butyric acid concentrations and methane production were not changed. In addition, increasing U level supplementation increased NH_3-N concentration. Result from real-time polymerase chain reaction revealed a significant effect on total bacteria, whereas *F. succinogenes* and *R. flavefaciens* population while *R. albus* was not affected by the M and U supplementation. Based on this study, it could be concluded that M and urea U supplementation could improve the nutritive value of *Leucaena* silage and enhance *in vitro* rumen fermentation efficiency. This study also suggested that the combination use of M and U supplementation level was at 2% and 1%, respectively. (**Key Words:** *Leucaena*, Silage, Rumen Fermentation, *In vitro* Gas Production, Urea, Molasses)

INTRODUCTION

Tropical legume forages are a rich source of crude protein (CP) and minerals for animal nutrition, in addition to their contribution to a sustainable agro ecosystem (Bansi et al., 2014). The identification of alternative sources of dietary protein for ruminants is driven by the desire to reduce feeding cost and to ensure profitability and sustainability of livestock production systems. Ensiling is a method for preserving moist crops which ensures animal feed availability throughout the year. Silage processing is based on lactic acid fermentation under anaerobiosis and preserves the nutritive and sanitary qualities of the crops (Cazzato et al., 2011). Legume silages are better accepted by the animal than grass silages, with a tendency to higher animal performance. *Leucaena* (*Leucaena leucocephala*) is high in palatability, digestibility and digestible protein and is often recommended as ruminant feed (Barros Rodriguez et al., 2013). Ensiling may be an appropriate method for preservation and toxic reduction because *Leucaena* is harvested during the rainy season when drying is rather difficult. Sunagawa et al. (1989) reported that around 90% of mimosine is destroyed after 14 to 21 days of ensiling. Silage is widely used in farms and has a substantial role in animal production systems. High silage quality is a key factor in minimizing the cost of production and sustaining animal health. Increasing use of silage has resulted in

* Corresponding Author: M. Wanapat.
E-mail: metha@kku.ac.th

continuing efforts to minimize the quality losses. The main aim of ensiling is to preserve fodder under anaerobic condition, where anaerobic microbes build up organic acids, mainly lactic acid, by using fermentable carbohydrates, and aerobic stability describes the length of time that silage remains stable. As a result, the pH decreases, and the forage is preserved. A good additive increases the nutrient recovery, decreases heating of the silage and fungi development during the storage or feed out period and results in increased gas production and fermentation (Salem et al., 2013). Ensiling with additional carbon and nitrogen sources could improve the quality of silage. Therefore, the aim of this study was to investigate the effect of molasses and urea supplementation on *Leucaena* silage quality and *in vitro* gas production and ruminal fermentation profiles.

MATERIALS AND METHODS

Dietary substrate, animals, experimental design and treatments

Leucaena was harvested and immediately chopped in 2 to 3 cm lengths and ensiled to the respective supplementation treatments according to a 3×3 factorial arrangement in a completely randomized design (CRD). Factor A was molasses (M) supplementation at 0%, 1%, and 2%, and factor B was urea (U) supplementation at 0%, 0.5%, and 1.0% of the *Leucaena* dry matter (DM). A mixture of M and U was dissolved in water, sprayed onto the *Leucaena* which was then packed into plastic bags. The silage bags were kept in room temperature (about 25°C to 30°C). All treatments were done in triplicates at 1 kg each. After 28 days of ensiling, 200 g of *Leucaena* silage were sampled for analysis of DM, organic matter (OM) and CP (AOAC, 1990), and acid detergent fiber and neutral detergent fiber (NDF) (Van Soest et al., 1991). In addition, mimosine was analysed by the modified methods of Dalzell et al. (2012). Feed ingredients and chemical compositions of concentrate, rice straw and *Leucaena* silage are shown in Table 1 and 2.

Silage samples were prepared and weighed (total substrate mixture 200 mg of DM) into 60 mL glass bottles for various times of incubation to study the rumen fermentation using *in vitro* gas techniques. All treatments were assigned according to a 3×3 factorial arrangement in a CRD with 3 bottles per treatment including triplicates of blank (medium only) in an incubation for 3 runs.

Rumen inoculums

Strict anaerobic techniques were used in all steps during the rumen fluid transferring and incubation periods. Rumen fluid samples were removed from swamp buffaloes (1 liter per animal) before morning feeding (7:00 h) under vacuum via the rumen fistula into a 2 liter plastic flask and transferred into 2 pre-warmed thermos flasks (1 liter) (Menke et al., 1979; Makkar et al., 1995). The fluid was then transported to the laboratory.

Table 1. Feed ingredients and chemical composition of concentrate, rice straw, and *Leucaena* leaf

	Concentrate	Rice straw	Fresh *Leucaena*
Ingredients (g/kg DM)			
Cassava chip	750		
Rice bran	60		
Palm kernel meal	50		
Coconut meal	80		
Urea	15		
Molasses	15		
Tallow	10		
Salt	10		
Sulfur	10		
Mineral premix	10		
Chemical composition			
Dry matter (g/kg DM)	924	911	324
Organic matter	916	895	936
Crude protein	121	23	21
Neutral detergent fiber	206	760	316
Acid detergent fiber	134	594	173
Condensed tannins	-	-	36
Mimosine	-	-	84

Medium solution preparation

In the present study, the medium was prepared for determination of gas production and fermented material during various incubation times. Therefore, the medium preparation was as described by Makkar et al. (1995). The mixture was kept stirring under CO_2 at 39°C using a magnetic stirrer fitted with a hot plate. A portion (30 mL) of the rumen-fluid medium was transferred into each bottle and incubated in a water bath at 39°C.

Substrate incubation

The method used for *in vitro* fermentation was based on the technique described by Menke et al. (1979). The sets of sample incubations for the determination of fermentation end-products and gas production were prepared each time. The bottles with the mixture of substrate treatments were pre-warmed in a water bath at 39°C for 1 h before filling with 30 mL of rumen inoculums mixture. The bottles were then sealed with rubber stoppers and aluminum caps and incubated in a water bath set at 39°C.

Sample collection and analysis

Gas production kinetics: During the incubation, the gas production was recorded at 1, 2, 4, 6, 8, 12, 24, 36, 48, 72, and 96 h. Cumulative gas production data was fitted to the

Table 2. Chemical composition of *Leucaena* silage for all treatments

Treatment[1]	DM	Ash	OM	CP	NDF	ADF	Mimosine
	g/kg DM						
T1 (Control)	332	76	924	215	387	224	27.1
T2 (M1%)	324	83	917	214	371	221	26.5
T3 (M2%)	322	93	907	218	372	210	25.6
T4 (U0.5%)	335	78	922	223	374	220	26.7
T5 (U1.0%)	337	79	921	235	376	218	27.0
T6 (M1.0% U0.5%)	321	84	916	222	374	217	26.6
T7 (M1.0% U1.0%)	328	94	906	237	367	212	16.4
T8 (M2.0% U0.5%)	319	96	904	232	368	209	15.5
T9 (M2.0% U1.0%)	325	97	903	248	363	207	15.3
SEM	0.87	5.53	1.02	3.14	2.98	1.05	8.72
Contrast							
Con vs Supp	ns	*	ns	*	ns	ns	*
Con vs M	ns	ns	ns	ns	ns	ns	ns
Con vs U	ns	ns	ns	0.06	ns	ns	ns
Con vs MU	ns	ns	ns	*	ns	ns	*
M1.0% vs M2.0%	ns	ns	ns	ns	ns	ns	ns
U0.5% vs U1.0%	ns	ns	ns	*	ns	ns	ns

DM, dry matter; OM, organic matter; CP, crude protein; NDF, neutral detergent fiber; ADF, acid detergent fiber; SEM, standard error of the mean; ns, non significant.

[1] M1% and M2% were molasses supplementation at 1% and 2% of *Leucaena* DM, respectively and U0.5% and U1.0% were urea supplementation at 0.5% and 1.0% of *Leucaena* DM, respectively.

model of Ørskov and McDonald (1979) as follows:

$$y = a+b(1-e^{(-ct)})$$

Where a = the gas production from the immediately soluble fraction, b = the gas production from the insoluble fraction, c = the gas production rate constant for the insoluble fraction (b), t = incubation time, (a+b) = the potential extent of gas production. y = gas produced at time "t".

Determination of fermentation parameters: The rumen inoculum mixtures were sampled at 0, 4, 8, and 12 h of fermenting post inoculation. Ruminal inoculum fluids were collected at 0, 4, 8, and 12 h post inoculation. Rumen fluid samples were then filtered through four layers of cheesecloth. Samples were divided into 2 portions. The first portion, around twenty milliliters of rumen inoculum, was put into plastic bottles for ammonia nitrogen (NH_3-N) and volatile fatty acid (VFA) analysis. The sample was centrifuged at 16,000×g for 15 min, and the supernatant was stored at –20°C before NH_3-N analysis by using the micro Kjeldahl methods (AOAC, 1990) and VFA analysis by high performance liquid chromatography (HPLC; Instruments by controller water model 1525, Waters Corporation, Milford MA, USA) water model 2707 auto sampler; water model 2489 UV detector and BREEZE software; column novapak C18; column size 3.9 mm×300 mm; mobile phase 10 mM H_2PO_4 [pH 2.5]) (Samuel et al., 1997). The second portion was fixed with 10% formalin solution in sterilized 0.9% saline solution. The total direct count was made by the methods of Galyen (1989) based on the use of a haemocytometer (Boeco, Hamburg, Germany). The last portion was stored at –20°C for DNA extraction (Yu and Morrison, 2004).

Extraction of genomic DNA and real-time polymerase chain reaction: Community DNA was extracted from 0.5 g of rumen content (fluid and digesta) by the RBB+C method (Yu and Morrison, 2004). In brief, the RBB+C method employs two rounds of bead beating in the presence of NaCl and sodium dodecyl sulphate, followed by sequential ammonium acetate and isopropanal precipitations. The precipitated nucleic acids were then treated with RNase A and proteinase K, and the DNA was purified using columns from QIAGEN DNA Mini Kit (QIAGEN, Valencia, CA, USA), according to manufacturer's recommendations. The targeted bacteria were total bacteria, the three predominant cellulolytic bacteria (*F. succinogenes*, *R. flavefaciens*, and *R. albus*) and protozoa. Primers for *F. succinogenes*, Fs219f (5′-GGTATGGGATGAGCTTGC-3′) and Fs654r (5′-GCCTGCCCCTGAACTATC-3′) were selected to allow amplification of all 10 *F. succinogenes* strains deposited in Gene Bank. For *R. albus* primers, Ra1281f (5′-CCCTAAAAGCAGTCTTAGTTCG-3′) and Ra1439r (5′-CCTCCTTGCGGTTAGAA CA-3′) and for *R. flavefaciens* primers, Rf154f (5′-TCTGGAAACGGATGGTA-3′) and Rf425r (5′-CCTTTAAGACAGGAGTTTACAA-3′) were also selected to allow species-specific amplification of all seven *R. flavefaciens* strains deposited in Gene Bank.

Primers for total bacteria and protozoa were F (5'-GC-clamp-CCTACGGGAGGCAGCAG3'), R (5'GWATTAC CGCGGCKGCTG3') and F (5'-GCTTTCGWTGGTA GTGTTT-3'), R (5'-ACTTGCCCTCYAATCGTWCT-3'). These primers were chosen from previously published sequences that demonstrated species specific amplification (Koike and Kobayashi, 2001). The conditions of the real-time polymerase chain reaction (PCR) for *F. succinogenes* were as follows: 30 s at 94°C for denaturing, 30 s at 60°C for annealing and 30 s at 72°C for extension (48 cycles), except for 9 min denaturation in the first cycle and 10 min extension in the last cycle. Amplification of 16s rDNA for *R. albus* and *R. flavefaciens* was carried out similarly except an annealing temperature of 55°C.

To establish a quantitative assay, amplified target 16s rDNA of each species by using specific primers and PCR conditions as described previously, the purified DNA were quantified by spectrophotometry with multiple dilutions. The target DNA was quantified by using serial 10-fold dilutions from 10^1 to 10^8 DNA copies of the previously quantified DNA standards. Real-time PCR amplification and detection were performed in a Chromo 4TM system (Bio-Rad, Hercules, CA, USA). In brief, Biostools QuantiMix Easy SYG Kit was used for PCR amplification and samples were assayed in duplicate in a 20 μL reaction mixture contained 4 to 6 mM $MgCl_2$, 10 μL of Mastermix (including; Taq DNA polymerase, reaction buffer, dNTP mixture, $MgCl_2$ and SybrGreen), 2 μL of DNA template and 0.8 μL of each primer (10 μM/μL).

Digestibility: At 12 and 24 h post inoculation, the *in vitro* true digestibility (IVTD) of a set of samples was determined. In brief, the content of the bottle was transferred quantitatively to a spout-less beaker by repeated washing with 100 mL neutral detergent solution. The content was refluxed for 1 h and filtered through pre weighed Gooch crucibles. The DM of the residue was weighed and IVTD of feed was calculated based on the following equation:

IVTD = {(DM of feed taken for incubation – NDF residue) ×100}/DM of feed taken for incubation

Statistical analysis

Data used for the statistical analyses consisted of 3 levels of molasses supplementation, 3 levels of urea supplementation, 3 replications, and runs making a total of 27 observations. All obtained data were subjected to the general linear models procedures of the Statistical Analysis System Institute (SAS, 1998) according to a 3×3 factorial arrangement in CRD. The statistical model including molasses level, urea level and interaction effects were: $Y_{ij} = \mu + A_i + B_j + AB_{ij} + \varepsilon_{ij}$; where Y_{ijk} is an observation, μ is the overall mean, A is molasses level effect (i = 1, 2, 3), B is urea level effect (j = 1, 2, 3), AB is interaction effect of molasses level and urea level, and ε_{ij} the residual effect. Multiple comparisons among treatment means were performed by Duncan's New Multiple Range Test (DMRT) and orthogonal contrast (Steel and Torrie, 1980).

RESULTS AND DISCUSSION

Feed ingredients and chemical compositions

The feed ingredients and chemical composition of concentrate, rice straw and fresh *Leucaena* leaf are shown in Table 1. Rice straw was used as roughage source. However, the CP of rice straw was low (23 g/kg DM) and high in NDF (760 g/kg DM). *Leucaena* silage (Table 2) contained CP 215 to 248 g/kg of DM and mimosine 15.3 to 27.1 g/kg of DM. Increasing U supplementation level increased the CP content of the silage and this was similar to the finding of Wanapat et al. (2013) who investigated whole crop rice silage. Energy is usually the limiting factor for growth of anaerobic microbes and provision of U and M might have increased the microbial mass that lead to increased CP (Staples et al., 1981). The provision of carbon skeleton and energy for microbial growth might have synchronized with ammonia released from urea hydrolysis, consequently increasing the CP content of forages ensiled (Salem et al., 2013). Furthermore, fermentation decreased 85% to 90% of mimosine. The result was in agreement with that reported by Sunagawa et al. (1989) who found mimosine reduction over 90% in *Leucaena silage* either with or without additives. The reduction of mimosine by ensiling being higher than by sun drying (14.5% to 51.1% of the original samples) was reported by Wee and Wang (1987). These results indicated that *Leucaena* silage is an interesting alternative for feed preservation.

Gas production kinetics and *in vitro* digestibility

Cumulative gas production for each of the substrate treatments presented as gas production and values for kinetics of gas production models for substrates studied are given in Table 3. The values for the estimated parameters obtained from the kinetics of gas production models for supplements studied revealed that the intercept value (a) for the different treatments representing gas production from soluble fractions and gas production rate constants for the insoluble fraction (c) ranged from –4.72 to –1.82 and 0.04 to 0.12, respectively. Gas production from the insoluble fraction (b), potential extent of gas production (a+b) and cumulative gas production at 96 h were significantly different with U supplementation ($p<0.01$). The treatments with U supplementation showed the higher gas production. Effect of M and U supplementation on digestibility from *in vitro* incubation are shown in Table 3. There was no interaction effect between M and U on DM digestibility at

Table 3. Effect of *Leucaena* silage on gas production kinetics and degradability from *in vitro* incubation with swamp buffalo rumen fluid

Treatment[1]	Gas kinetics[2]				Gas[3]	*IVDMD*	
	a	b	c	a+b		12 h	24 h
T1 (Control)	−2.57	26.7	0.12	24.1	23.6	23.7	41.2
T2 (M1%)	−1.82	38.5	0.04	36.7	36.0	47.4	45.2
T3 (M2%)	−1.83	34.0	0.04	32.2	31.6	48.8	54.9
T4 (U0.5%)	−3.12	35.5	0.04	32.4	31.9	60.5	54.9
T5 (U1.0%)	−2.25	37.6	0.04	35.4	34.6	59.2	58.3
T6 (M1.0% U0.5%)	−2.59	38.6	0.04	36.0	35.8	63.0	58.4
T7 (M1.0% U1.0%)	−2.08	36.7	0.04	34.6	33.9	57.9	62.8
T8 (M2.0% U0.5%)	−2.66	36.9	0.04	34.2	33.7	61.6	64.8
T9 (M2.0% U1.0%)	−4.72	49.0	0.04	44.3	42.9	64.2	66.1
SEM	0.63	4.52	0.03	4.14	4.02	5.87	10.4
Contrast							
Con vs Supp	ns	*	*	*	*	*	*
Con vs M	ns	ns	*	0.05	0.05	**	ns
Con vs U	ns	ns	*	0.07	0.06	ns	ns
Con vs MU	ns	*	ns	*	0.09	*	*
M1.0% vs M2.0%	ns	ns	ns	ns	ns	*	ns
U0.5% vs U1.0%	ns	ns	ns	ns	ns	ns	ns

DM, dry matter; IVDMD, *in vitro* dry matter digestibility; SEM, standard error of the mean; ns, non significant.

[1] M1% and M2% were molasses supplementation at 1% and 2% of *Leucaena* DM, respectively and U0.5% and U1.0% were urea supplementation at 0.5% and 1.0% of *Leucaena* DM, respectively.

[2] a, the gas production from the immediately soluble fraction; b, the gas production from the insoluble fraction; c, the gas production rate constant for the insoluble fraction (b); a+b, the gas potential extent of gas production.

[3] Cumulative gas production at 96 h (mL/0.2 g DM substrate).

* $p<0.05$; ** $p<0.01$; *** $p<0.001$.

hours 12 and 24 of incubation. According to Cone and van Gelder (1999), comparison of gas production data of samples differing widely in CP content can lead to misinterpretations. Generally, low gas production would indicate low degradability, but feedstuffs high in CP normally produce less gas during fermentation, even if their extent of degradation is high, because protein fermentation produces ammonia, which influences the carbonate buffer equilibrium by neutralizing H^+ ions from VFA without release of carbon dioxide. In the present study *Leucaena* silage contained high CP but produced more gas. The addition of M and U to *Leucaena* silage increased digestibility after 12 and 24 h of incubation, suggesting that during the ensiling process molasses might have removed some chemical linkages of hemicelluloses and thus enhanced their solubility in detergent solutions and also possibly due to the ability of rumen microorganism to degrade the plant secondary metabolites like alkaloids and saponins (Hart et al., 2008) and utilize them as an energy source.

Rumen fermentation

In the *in vitro* gas production technique, NH_3-N concentration was highest in T9 (M2.0% U1.0%). While NH_3-N was found to be the lowest in control. The concentration of NH_3-N was increased dramatically based on the time incubation. This result could be due to the effects of tannins contained in *Leucaena* silage which protect CP from degradation by the formation of tannin-protein complexes in the rumen, thereby increasing metabolizable protein supply to the duodenum (Waghorn, 2008). NH_3-N concentration was found higher in the treatments with high level of U supplementation. There was no effect of M supplementation on NH_3-N. Availability of NH_3 is an important determinant of microbial protein production as the majority of rumen bacteria use NH_3 as a nitrogen source. It is essential to know what concentration of NH_3 will support maximal microbial growth in order to make judgments regarding utilization of non-protein N. The NH_3–N concentration of all treatments ranged from 16.8 to 22.8 mg/dL (Table 4). However, Satter and Slyter (1974) suggested NH_3 concentrations from 3 to 5 mg/dL as optimal to produce ruminal microorganism growth, which was relatively less than those observed in this work. It appears that, once NH_3 starts to accumulate, the growth of bacteria utilizing NH_3 is not enhanced by increasing NH_3 concentration (Satter and Slyter, 1974).

Interaction between M and U supplementation affected the proportion of propionic acid (Table 4). Total VFA concentrations in M2.0% U1.0% was higher than other treatments ($p<0.05$). In addition, supplementation of M2.0% U0.5% and M2.0% U1.0% resulted in a higher

Table 4. Effect of*Leucaena* silage on ammonia nitrogen, volatile fatty acid and methane production from *in vitro* incubation with swamp buffalo rumen fluid

Treatment[1]	NH_3-N (mg/dL)	TVFA (mM/L)	C_2	C_3 (mol/100mol)	C_4	C_2:C_3	CH_4[2]
T1 (Control)	16.8	86.4	68.7	22.6	8.7	1.7	20.2
T2 (M1%)	17.1	93.0	61.3	30.1	8.6	1.8	20.8
T3 (M2%)	17.6	92.9	62.2	30.4	7.4	1.8	20.6
T4 (U0.5%)	18.7	87.7	60.3	32.3	7.4	1.6	18.6
T5 (U1.0%)	20.7	92.7	58.8	32.6	8.6	1.7	19.6
T6 (M1.0% U0.5%)	18.6	94.0	59.6	32.2	8.2	1.8	21.3
T7 (M1.0% U1.0%)	20.9	99.0	57.8	32.7	9.5	1.8	20.8
T8 (M2.0% U0.5%)	19.4	95.3	57.4	34.1	8.5	1.8	21.2
T9 (M2.0% U1.0%)	22.8	100.4	56.6	34.5	8.9	1.8	21.1
SEM	0.30	4.01	0.84	1.07	0.50	0.09	0.77
Contrast							
Con vs Supp	*	*	ns	*	ns	ns	ns
Con vs M	ns	*	ns	ns	ns	ns	ns
Con vs U	*	*	ns	ns	ns	ns	ns
Con vs MU	ns	*	ns	*	ns	ns	ns
M1.0% vs M2.0%	0.08	ns	*	ns	ns	ns	ns
U0.5% vs U1.0%	ns	ns	ns	ns	ns	ns	ns

NH_3-N, ammonia nitrogen; TVFA, total volatile fatty acid; C_2, acetic acid; C_3, propionic acid; C_4, butyric acid; C_2:C_3, acetic acid:propionic acid ratio; SEM, standard error of the mean; ns, non significant; DM, dry matter.

[1] M1% and M2% were molasses supplementation at 1% and 2% of *Leucaena* DM, respectively and U0.5% and U1.0% were urea supplementation at 0.5% and 1.0% of *Leucaena* DM, respectively.

[2] Methane production (mM/L) calculated by Moss et al. (2000) = 0.45 (C_2)–0.275 (C_3)+0.4 (C_4).

* p<0.05; ** p<0.01; *** p<0.001.

(p<0.05) proportion of propionic acid (34.4 and 34.1 mole/100 mole, respectively) and a lower (p<0.01) proportion of acetic acid which was highest in the control. While the proportion of butyric acid was not affected by M and U supplementation. Calculation of ruminal methane (CH_4) production using VFA proportions according to Moss et al. (2000) showed that methane production was not influenced by interaction of U and M supplementation. In contrast, Anantasook and Wanapat (2012) reported that a high proportion of propionic acid was caused by a decreased methane production due to tannins contained in rain tree pot meal. Effects of tannins on increased propionic acid and reduced acetic to propionic ratio have been found to vary with diets and applications.

Rumen microbes

In the present study, effects of M and U supplementation on microbes from *in vitro* incubation are shown in Table 5 and 6. The results revealed a significant effect on bacterial populations by M and U supplementation, while protozoa and fungi zoospores were not effect by supplementation. As compared with the control group, e supplementation resulted in a larger bacteria population (p<0.05). This effect may be due to the cause of *Leucaena* silage supplemented with U an M which contains high level of nitrogen and carbon source. The additional protein provided by the *Leucaena* would have increased availability of ammonia for rumen micro flora, stimulating microbial growth and increasing rate of breakdown of the forage (Barros-Rodriguez et al., 2013).

The effect of M and U supplementation on microbes from *in vitro* incubation with swamp buffalo rumen fluid is shown in Table 6. The real-time PCR for quantification of ruminal microbes with specific targets (total bacterial, *R. albus*, *F. succinogenes*, *R. flavefaciens*, and protozoa) are reported in Table 6. The total bacteria and three dominant cellulolytic bacteria were found affected by M and U supplementation. Predominant cellulolytic bacteria in *in vitro* incubation were affected by the M and U supplementation, except for *R. albus*. Supplementation of M and U decreased the population of protozoa (p<0.05). Koike and Kobayashi (2001) reported that *F. succinogenes* was the most dominant bacteria among the three species of cellulolytic bacteria. Changes of the population size or the proportion of cellulolytic bacterial numbers in the rumen may be due to some effect of tannins in *Leucaena*. Goel et al. (2008) reported that the *F. succinogenes* population was increased when supplementation with *S. sesban* leaves and *Fenugreek* seeds, while the *R. flavefaciens* population increased with *Carduus* leaves and *fenugreek* supplementation. Moreover, McSweeney et al. (2001) reported that the protein-tannin complexes reduce the

Table 5. Effect of *Leucaena* silage on microorganisms from *in vitro* incubation with swamp buffalo rumen fluid

Treatment[1]	Protozoa ($\times 10^5$cell/mL)			Fungi ($\times 10^6$ cell/mL)			Bacteria ($\times 10^8$ cell/mL)		
	4 h	12 h	Mean	4 h	12 h	Mean	4 h	12 h	Mean
T1 (Control)	1.4	1.8	1.6	2.3	2.8	2.6	15.6	14.2	14.9
T2 (M1%)	1.6	1.4	1.5	3.4	3.2	3.3	14.2	18.8	16.5
T3 (M2%)	2.3	2	2.2	5.1	4.7	4.9	17.5	20.6	19.1
T4 (U0.5%)	1.7	2.2	2.0	5.3	5.6	5.5	20.4	22.4	21.4
T5 (U1.0%)	3.2	1.8	2.5	6.2	4.3	5.3	26.8	21.5	24.2
T6 (M1.0% U0.5%)	2.5	1.5	2.0	6	5.1	5.6	24.6	26.3	25.5
T7 (M1.0% U1.0%)	2.2	1.9	2.1	5.4	6.2	5.8	30.2	28.7	29.5
T8 (M2.0% U0.5%)	3.4	1.7	2.6	5.2	6.1	5.7	33.5	30.2	31.9
T9 (M2.0% U1.0%)	3.2	2.1	2.7	5.6	6.3	6.0	35.2	40.3	37.8
SEM	0.43	0.22	0.51	0.34	0.60	1.12	0.31	1.08	2.01
Contrast									
Con vs Supp	ns	ns	ns	ns	ns	ns	*	*	*
Con vs M	ns	ns	ns	ns	ns	ns	ns	ns	ns
Con vs U	ns	ns	ns	ns	ns	ns	ns	ns	ns
Con vs MU	ns	ns	ns	ns	ns	ns	*	*	*
M1.0% vs M2.0%	ns	ns	ns	ns	ns	ns	ns	ns	ns
U0.5% vs U1.0%	ns	ns	ns	ns	ns	ns	ns	ns	ns

SEM, standard error of the mean; ns, non significant; DM, dry matter.

[1] M1% and M2% were molasses supplementation at 1% and 2% of *Leucaena* DM, respectively and U0.5% and U1.0% were urea supplementation at 0.5% and 1.0% of *Leucaena* DM, respectively.

* $p<0.05$; ** $p<0.01$; *** $p<0.001$.

availability of fermentable N for microbial activity in the rumen. Kumar and Singh (1984) reported that tannins in tree leaves inhibited proteolysis of casein and subsequent ammonia production *in vitro*. Therefore, U addition would provide fermentable N for stimulating microbial fermentation in the rumen. Wanapat and Cherdthong (2009)

Table 6. Effect of *Leucaena* silage on cellulolytic bacteria and protozoa populations in *in vitro* incubation in swamp buffalo fluid as determined by real-time PCR

Treatment[1]	Real-time PCR technique, copies/mL of incubation				
	Total bacteria ($\times 10^9$ cell/mL)	*F.succinogenes* ($\times 10^7$ cell/mL)	*F. flavefaciens* ($\times 10^6$ cell/mL)	*R. albus* ($\times 10^6$ cell/mL)	Protozoa ($\times 10^4$ cell/mL)
T1 (Control)	2.44	3.42	1.21	2.14	3.32
T2 (M1%)	3.15	3.17	1.32	2.21	2.16
T3 (M2%)	4.24	3.56	1.47	1.97	2.31
T4 (U0.5%)	5.87	3.61	1.84	2.35	2.55
T5 (U1.0%)	6.32	3.80	1.68	2.66	2.49
T6 (M1.0% U0.5%)	6.01	3.72	2.03	3.34	2.03
T7 (M1.0% U1.0%)	6.55	3.88	2.15	3.41	2.18
T8 (M2.0% U0.5%)	6.83	4.01	2.65	3.32	1.97
T9 (M2.0% U1.0%)	7.35	4.22	2.43	3.17	1.89
SEM	0.07	0.42	0.06	0.14	0.12
Contrast					
Con vs Supp	*	*	*	ns	*
Con vs M	*	ns	ns	ns	ns
Con vs U	*	ns	ns	ns	ns
Con vs MU	**	*	*	ns	*
M1.0% vs M2.0%	*	ns	ns	ns	ns
U0.5% vs U1.0%	*	ns	ns	ns	ns

PCR, polymerase chain reaction; SEM, standard error of the mean; ns, non significant; DM, dry matter.

[1] M1% and M2% were molasses supplementation at 1% and 2% of *Leucaena* DM, respectively and U0.5% and U1.0% were urea supplementation at 0.5% and 1.0% of *Leucaena* DM, respectively.

* $p<0.05$; ** $p<0.01$; *** $p<0.001$.

reported that increasing protein supplementation for ruminants lead to a higher population of ruminal microbes. The reason could be due to a nitrogen source which could support microbial production in the rumen. The population of *F. succinogenes* was higher than those of *R. albus*. The finding of this study was similar to that of Wanapat and Cherdthong (2009), who studied rumen cellulolytic bacteria population using real-time PCR. They found that the population of *F. Succinogenes* was more abundant than *R. albus* (3.0×10^9 vs 2.93×10^6 copies/mL of rumen fluid).

CONCLUSION

Based on this study, it could be concluded that supplementation of molasses and urea could efficiently improve *Leucaena* silage quality in terms of the chemical composition being high in protein and low in NDF contents. The present results suggest that supplementation of urea and molasses to *Leucaena* silage enhanced *in vitro* rumen fermentation efficiency, especially by the addition of urea at 1% and molasses at 2% of crop DM. However, further study using *Leucaena* silage supplemented with urea and molasses in feeding trials emphasizing lactating dairy cows and fattening beef cattle should be investigated.

ACKNOWLEDGMENTS

The authors would like to express their most sincere gratitude and appreciation to the Thailand Research Fund (TRF) via "The Royal Golden Jubilee Ph.D. Program", Tropical Feed Resources Research and Development Center (TROFREC), and the Department of Animal Science, Faculty of Agriculture, Khon Kaen University, Thailand, for their kind financial support and provision of research facilities.

REFERENCES

Anantasook, N. and M. Wanapat. 2012. Influence of rain tree pod meal supplementation on rice straw based diets using *in vitro* gas fermentation technique. Asian Australas. J. Anim. Sci. 25:325-334.

AOAC. 1990. Official Methods of Analyses, 15th edn. Association of Official Analytical Chemists, Arlington, VA, USA.

Bansi, H., E. Wina, P. R. Matitaputy, and V. Tufarelli. 2014. Evaluation of Zapoteca tetragona forage as alternative protein source in ruminants' feeding. Ital. J. Anim. Sci. 13:147-150.

Barros-Rodríguez, M., J. Solorio-Sánchez, C. Sandoval-Castro, A. V. Klieve, E. B. Briceno-Poot,L.Ramirez-Aviles, and R. Rojas-Herrera. 2013. Effect of two intake levels of *Leucaena leucocephala* on rumen function sheep. Trop. Grasslands-Forrajes Tropicales 1:55-57.

Cazzato, E., V. Laudadio, A. Corleto, and V. Tufarelli. 2011. Effects of harvest date, wilting and inoculation on yield and forage quality of ensiling safflower (*Carthamus tinctoriusL.*) biomass. J. Sci. Food Agric. 91:2298-2302.

Cone, J. W. and A. H. Van Gelder. 1999. Influence of protein fermentation on gas production profiles. Anim. Feed Sci. Technol. 76:251-264.

Cudjoe, N. and V. Mlambo. 2014. Buffer nitrogen solubility, *in vitro* ruminal partitioning of nitrogen and *in vitro* ruminal biological activity of tannins in leaves of four tree species. J. Anim. Physiol. Anim. Nutr. 98:722-730.

Dalzell, S. A., D. J. Burnett, J. E. Dowsett, V. E. Forbes, and H. M. Shelton. 2012. Prevalence of mimosine and DHP toxicity in cattle grazing *Leucaena leucocephala* pastures in Queensland, Australia. Anim. Prod. Sci. 52:365-372.

Galyen, M. 1989. Laboratory procedures in animal nutrition research. New Mexico State University, Las Cruces, NM, USA.

Goel, G., H. P. S. Makkar, and K. Becker. 2008. Changes in microbial community structure, methanogenesis and rumen fermentation in response to saponin-rich fractions from different plant materials. J. Appl. Microbiol.105:770-777.

Hart, K. J., D. R. Yáñez-Ruiz, S. M. Duval, N. R. McEwan, and C. J. Newbold. 2008. Plant extracts to manipulate rumen fermentation. Anim. Feed Sci. Technol. 147:8-35.

Koike, S. and Y. Kobayashi. 2001. Develop and use of competitive PCR assays for the rumen cellulolytic bacteria: *Fibrobactor succinogenes*, *Ruminococcus albus*, and *Ruminococcus flavefaciens*. FEMS Microbiol. Lett. 204:361-366.

Kumar, R., M. Singh. 1984. Tannins: their adverse role in ruminant nutrition. J. Agr. Food Chem. 32:447-453.

Makkar, H. P. S., M. Blummel, and K. Becker. 1995. Formation of complexes between polyvinyl pyrrolidones or polyethylene glycols and tannins, and their implication in gas production and true digestibility in *in vitro* techniques. Br. J. Nutr. 73:897-913.

McSweeney, C. S., B. Palmer, D. M. McNeil, and D. O. Krause. 2001. Microbial interactions with tannins: Nutritional consequences for ruminants. Anim. Feed Sci. Technol. 91:83-93.

Menke, K. H., L. Raab, A. Salewski, H. Steingass, D. Fritz, and W. Schneider. 1979. The estimation of the digestibility and metabolizable energy content of ruminant feedstuffs from the gas production when they are incubated with rumen liquor *in vitro*. J. Agric. Sci. 93:217-222.

Moss, A. R., J. P. Jouany, and J. Newbold. 2000. Methane production by ruminants: its contribution to global warming. Anim. Res. 49:231-253.

Mathew, S., S. Sagathevan, J. Thomas, and G. Mathen. 1997. An HPLC method for estimation of volatile fatty acids of rumen fluid. Indian J. Anim. Sci. 67:805-807.

Orskov, E. R. and I. McDonal. 1979. The estimation of protein degradability in the rumen from incubation measurements weighted according to rate of passage. J. Agric. Sci. 92:499-503.

Salem, A. Z. M., C. S. Zhou, Z. L. Tan, M. Mellado, M. C. Salazar, M. M. M. Y. Elghandopur, and N. E. Odongo. 2013. *In vitro* ruminal gas production kinetics of four fodder trees ensiled with or without molasses and urea. J. Integr. Agric. 12:1234-1242.

SAS. 1998. User's Guide: Statistic, Version 6, 12th edn. SAS Inst. Inc., Cary, NC, USA.

Satter, L. D. and L. L. Slyter. 1974. Effect of ammonia concentration on ruminal microbial protein production *in vitro*. Br. J. Nutr. 32:199-208.

Staples, C. R., G. C. Fahey Jr, L. L. Berger, and R. B. Rindsig. 1981. Evaluation of dairy waste fiber as a roughage source for ruminants. J. Dairy Sci. 64:662-671.

Steel, R. G. D. and J. H. Torrie. 1980. Principles and Procedures of Statistics. McGraw Hill Book Co., New York, NY, USA.

Sunagawa, K., F. Hongo, Y. Kawashima, and S. Tawata. 1989. The effect of mimosine reduced *Leucaena* feed on sheep. JPN. J. Zootech. Sci. 60:133-140.

Van Soest, P. J., J. B. Robertson, and B. A. Lewis. 1991. Methods for dietary fiber neutral detergent fiber, and nonstarch polysaccharides in relation to animal nutrition. J. Dairy Sci. 74:3583-3597.

Waghorn, G. 2008. Beneficial and detrimental effects of dietary condensed tannins for sustainable sheep and goat production-progress and challenges. Anim. Feed Sci. Technol. 147:116-139.

Wanapat, M. and A. Cherdthong. 2009. Use of real-time PCR technique in studying rumen cellulolytic bacteria population as affected by level of roughage in Swamp buffalo. Curr.Microbiol. 58:294-299.

Wanapat, M., S. Kang, P. Khejornsart, and R. Pilajun. 2013. Improvement of whole crop rice silage nutritive value and rumen degradability by molasses and urea supplementation. Trop. Anim. Health Prod. 45:1777-1781.

Wee, K. L. and S. S. Wang. 1987. Effect of postharvest treatment on the degradation of mimosine in *Leucaena leucocephala* leaves. J. Sci. Food. Agric. 39:195-201.

Yu, Z. and M. Morrison. 2004. Improved extraction of PCR-quality community DNA from digesta and fecal samples. BioTechniques 36:808-812.

16

Evaluation of Relative Bioavailability of 25-Hydroxycholecalciferol to Cholecalciferol for Broiler Chickens

J. C. Han[1,]*, G. H. Chen[1,2], J. G. Wang[1,2], J. L. Zhang[1,2], H. X. Qu[1], C. M. Zhang[1], Y. F. Yan[1,]*, and Y. H. Cheng[3]

[1] Department of Animal Science, College of Life Science, Shangqiu Normal University, Shangqiu, Henan 476000, China

ABSTRACT: This study was conducted to evaluate the relative bioavailability (RBV) of 25-hydroxycholecalciferol (25-OH-D_3) to cholecalciferol (vitamin D_3) in 1- to 21-d-old broiler chickens fed with calcium (Ca)- and phosphorus (P)-deficient diets. On the day of hatch, 450 female Ross 308 broiler chickens were assigned to nine treatments, with five replicates of ten birds each. The basal diet contained 0.50% Ca and 0.25% non-phytate phosphorus (NPP) and was not supplemented with vitamin D. Vitamin D_3 was fed at 0, 2.5, 5.0, 10.0, and 20.0 μg/kg, and 25-OH-D_3 was fed at 1.25, 2.5, 5.0, and 10.0 μg/kg. The RBV of 25-OH-D_3 was determined using vitamin D_3 as the standard source by the slope ratio method. Vitamin D_3 and 25-OH-D_3 intake was used as the independent variable for regression analysis. The linear relationships between the level of vitamin D_3 or 25-OH-D_3 and body weight gain (BWG) and the weight, length, ash weight, and the percentage of ash, Ca, and P in femur, tibia, and metatarsus of broiler chickens were observed. Using BWG as the criterion, the RBV value of 25-OH-D_3 to vitamin D_3 was 1.85. Using the mineralization of the femur, tibia, and metatarsus as criteria, the RBV of 25-OH-D_3 to vitamin D_3 ranged from 1.82 to 2.45, 1.86 to 2.52, and 1.65 to 2.05, respectively. These data indicate that 25-OH-D_3 is approximately 2.03 times as active as vitamin D_3 in promoting growth performance and bone mineralization in broiler chicken diets. (**Key Words:** 25-Hydroxycholecalciferol, Cholecalciferol, Relative Bioavailability, Growth, Bone, Broiler Chicken)

INTRODUCTION

Cholecalciferol (vitamin D_3) has been used as a feed additive to regulate calcium (Ca) and phosphorus (P) metabolism and bone development in animals for many years. Vitamin D_3 undergoes 25-hydroxylation in animal livers to transform 25-hydroxycholecalciferol (25-OH-D_3). The commercial 25-OH-D_3 has been produced and approved for use in poultry and swine feed in China in 2014.

Previous research has shown that body weight and feed efficiency of broiler chickens fed with 25-OH-D_3 were greater than those of the birds fed with vitamin D_3 (Yarger et al., 1995; Fritts and Waldroup, 2003). Replacing vitamin D_3 by 25-OH-D_3 at 50% has a beneficial effect on the growth performance of broilers (Koreleski and Swiatkiewicz, 2005). These data indicate that the relative bioavailability (RBV) of 25-OH-D_3 is higher than that of vitamin D_3.

However, no consistent results have been obtained in the RBV of 25-OH-D_3 to vitamin D_3. Soares et al. (1995) reviewed that the RBV of 25-OH-D_3 to vitamin D_3 ranged from 1.0 to 4.0 when Ca absorption, plasma Ca, bone ash, bone strength, and tibial dyschondroplasia were used as the criteria. Atencio et al. (2005) found that the RBV values of 25-OH-D_3 to vitamin D_3 were 1.38, 1.33, 1.28, and 1.11 for egg production, hatchability, late embryo mortality, and body ash of the progeny in broiler breeder hen diets, respectively. These data reveal the differences in the RBV of 25-OH-D_3 to vitamin D_3 among the studies. The RBV of 25-OH-D_3 to vitamin D_3 in poultry diets should be further clarified.

Researchers usually use the tibia to evaluate the RBV of

* Corresponding Authors: J. C. Han.
E-mail: j.c.han@hotmail.com / Y. F. Yan.
E-mail: yanyf01@sina.com
[2] College of Animal Husbandry and Veterinary Science, Henan Agricultural University, Zhengzhou, Henan 450002, China.
[3] Department of Biotechnology and Animal Science, National Ilan University, I-Lan 26047, Taiwan.

vitamin D derivatives. In fact, the differences in growth and mineralization among the femur, tibia, and metatarsus in poultry have been observed (Applegate and Lilburn, 2002; Goetting-Fuchs et al., 2012; Han et al., 2015). The femur and metatarsus should also be used as criteria to evaluate the RBV of 25-OH-D_3 to vitamin D_3.

Therefore, the present study was conducted to investigate the effects of 25-OH-D_3 and vitamin D_3 on the growth performance and development of the femur, tibia, and metatarsus and to re-evaluate the RBV of 25-OH-D_3 to vitamin D_3 in broilers fed with Ca- and P-deficient diets.

MATERIALS AND METHODS

Birds, diets, and management

All of the procedures adopted in this study were approved by the Animal Care Committee of Shangqiu Normal University.

On the day of hatch, 450 female Ross 308 broiler chickens were assigned to nine treatments, with five replicates of ten birds each. The initial body weight of broiler chickens was 46.5±1.9 g. Birds from 1 to 13 d of age were reared in stainless steel starter cages (70 cm×70 cm×30 cm). At 14 d, the broilers were transferred to stainless steel grower cages (190 cm×50 cm×35 cm). The basal diet contained 0.50% Ca and 0.25% non-phytate phosphorus (NPP) and was not supplemented with vitamin D. Vitamin D_3 was fed at 0, 2.5, 5.0, 10.0, and 20.0 μg/kg and 25-OH-D_3 was fed at 1.25, 2.5, 5.0, and 10.0 μg/kg. The birds were provided mash diet (Table 1) and water *ad libitum*. The lighting system consisted of 23 h of light from 1 d to 3 d and 20 h of light from 4 d to 21 d. Room temperature was controlled at 33°C from 0 d to 3 d, 30°C from 4 d to 7 d, 27°C from 8 d to 14 d, and 24°C from 15 d to 21 d.

Table 1. Ingredients and nutrient composition of the basal diet

Items	Basal diet
Ingredient (%)	
Corn	60.73
Soya bean meal (43% CP)	32.00
Soya bean oil	1.60
Soya bean protein isolate (65% CP)	3.47
Limestone	0.67
Dicalcium phosphate	0.71
L-lysine·HCl (98%)	0.14
DL-methionine (98%)	0.14
Trace mineral premix[1]	0.01
Vitamin premix[2]	0.03
Choline chloride (50%)	0.20
Sodium chloride	0.30
Nutrient composition	
Metabolizable energy (kcal/kg)	2,975.20
Crude protein (CP, %)	21.24
Analyzed calcium (Ca, %)	0.52
Analyzed total phosphorus (TP, %)	0.49
Non-phytate phosphorus (NPP, %)	0.25

[1] The trace mineral premix provided the following (per kg of diet): 80 mg iron; 40 mg zinc; 8 mg copper; 60 mg manganese; 0.35 mg iodine; and 0.15 mg selenium.

[2] The vitamin premix provided the following (per kg of diet): 8,000 IU vitamin A; 20 IU vitamin E; 0.5 mg menadione; 2.0 mg thiamine; 8.0 mg riboflavin; 35 mg niacin; 3.5 mg pyridoxine; 0.01 mg vitamin B_{12}; 10.0 mg pantothenic acid; 0.55 mg folic acid; and 0.18 mg biotin.

25-OH-D_3 and vitamin D_3

Crystalline 25-OH-D_3 (98%) and vitamin D_3 (99%) were supplied by Changzhou Book Chemical Co., Ltd. (Changzhou, China) and Jiaxing Tianhecheng Biological Technology Co., Ltd. (Jiaxing, China), respectively. The 25-OH-D_3 and vitamin D_3 solutions were prepared by the method of Biehl and Baker (1997). Crystalline 25-OH-D_3 and vitamin D_3 were weighed and dissolved in ethanol. Then, they were diluted to a final concentration of 10 mg/L of 25-OH-D_3 or vitamin D_3 in a solution of 5% ethanol and 95% propylene glycol. After the preparation, the solution of 25-OH-D_3 or vitamin D_3 was added to the diets.

Sample collection

The birds were weighed on d 21 after 12 h of fasting. Ten chicks per treatment were randomly selected for the collection of blood, femur, tibia, and metatarsus. Plasma samples (5 mL) were collected through cardiac puncture and centrifuged for 10 min at 3,000×g at 20°C. The birds were killed after collecting the blood samples. The femur, tibia, and metatarsus of the individual birds were excised and frozen at –20°C for analysis.

Sample analysis

Plasma Ca and inorganic phosphorus (Pi) were determined using a Shimadzu CL-8000 analyzer (Shimadzu Corp., Kyoto, Japan) following the instructions of the manufacturer.

The left femur, tibia, and metatarsus were boiled for 5 min to loosen the muscle tissues using the method of Hall et al. (2003). The meat, connective tissue, and fibula bone were completely removed using scissors and forceps. The bones were placed in a container of ethanol for 24 h (removing water and polar lipids) after cleaning. Afterward, the bones were further extracted in anhydrous ether for 24 h (removing non-polar lipids). The bones were dried at 105°C for 24 h before weighing. The bone ash content was determined by ashing the bone in a muffle furnace for 48 h at 600°C.

The right tibia was utilized to analyze the breaking-strength, which was determined using an all-digital electronic universal testing machine (Shenzhen Hengen Instrument Co. Ltd., Shenzhen, China). The tibias were cradled on two support points measuring 4 cm apart. Force was applied to the midpoint of the same face of each tibia using a 50 kg load cell with a crosshead speed of 10 mm/min (Jendral et al., 2008).

Dietary and bone Ca were determined by the ethylene diamine tetraacetic acid titration method, and P was determined by photometric methods after reaction with ammonium molybdate and ammonium metavanadate (Han et al., 2013).

Statistical analysis

Replicate means are the experimental units in the statistical analysis. Data were analyzed with a general linear model of the SAS software (SAS Institute, 2002). The RBV of 25-OH-D_3 was determined using vitamin D_3 as the standard source by the slope ratio method (Littell et al., 1997). Feed intake differed for the dietary treatments so that vitamin D intake rather than vitamin D content was used as the independent variable for regression analysis. The model is as follows: $y = a+b_1x_1+b_2x_2$, where y is the response, x_1 is vitamin D_3 intake, x_2 is 25-OH-D_3 intake, a is the intercept, and b_1 and b_2 are the slope of vitamin D_3 and 25-OH-D_3, respectively. Orthogonal comparisons were performed to determine the linear and quadratic effects of the 25-OH-D_3 or vitamin D_3 levels on growth performance and bone mineralization. The basal diet treatment was included for both vitamin D_3 and 25-OH-D_3 when conducting orthogonal polynomial contrast test. Means were compared by conducting Tukey test when probability values were significant ($p<0.05$).

RESULTS

Growth performance

Dietary 25-OH-D_3 linearly affected body weight gain (BWG), feed intake (FI), feed efficiency, and mortality in 1- to 21-d-old broiler chickens ($p<0.05$, Table 2). Vitamin D_3 levels also influenced the above parameters ($p<0.05$). 25-OH-D_3 or vitamin D_3 did not affect the plasma Ca or Pi concentration ($p>0.05$).

Bone mineralization

The femur, tibia, and metatarsus are three leg bones in poultry. They reflect the body bone quality of birds. The linear relationships between the level of 25-OH-D_3 or vitamin D_3 and the weight, length, ash weight, and the percentage of ash, Ca, and P of the femur in broiler chickens were observed ($p<0.05$, Table 3).

Dietary 25-OH-D_3 or vitamin D_3 linearly improved the tibia breaking-strength ($p<0.05$, Table 4). Similar results were observed in the relationship between the level of 25-OH-D_3 or vitamin D_3 and the weight, length, ash weight, and the percentage of ash, Ca, and P of the tibia.

Increasing the 25-OH-D_3 or vitamin D_3 level linearly increased the weight, length, ash weight, and the percentage of ash, Ca, and P of the metatarsus ($p<0.05$, Table 5).

Relative bioavailability of 25-OH-D_3 to vitamin D_3

The slope ratio method was used to evaluate the RBV of 25-OH-D_3 to vitamin D_3 in broiler chickens (Table 6). Vitamin D_3 and 25-OH-D_3 intake was used as the independent variable for regression analysis. The slopes of

Table 2. Effects of vitamin D_3 and 25-OH-D_3 on growth performance and plasma mineral concentration in 1- to 21-d-old broiler chickens

Vitamin D_3 (μg/kg)	25-OH-D_3 (μg/kg)	Growth				Plasma	
		BWG (g/bird)	FI (g/bird)	FE (BWG/FI)	Mortality (%)	Ca (mg/dL)	Pi (mg/dL)
0	0	233^{d}	431^{d}	0.542^{c}	26^{a}	5.45	3.07
2.5		338^{c}	621^{bc}	0.548^{bc}	12^{b}	6.69	3.27
5.0		433^{b}	709^{b}	0.612^{a}	0^{b}	6.90	3.83
10.0		584^{a}	932^{a}	0.627^{a}	0^{b}	6.93	4.82
20.0		602^{a}	955^{a}	0.631^{a}	0^{b}	6.98	4.64
	1.25	321^{c}	585^{c}	0.549^{bc}	10^{b}	5.72	3.24
	2.5	426^{b}	702^{b}	0.606^{ab}	0^{b}	5.70	3.60
	5.0	556^{a}	866^{a}	0.642^{a}	0^{b}	6.70	3.66
	10.0	574^{a}	927^{a}	0.619^{a}	0^{b}	6.73	4.17
SEM		19	27	0.007	2	0.16	0.18
p value							
Vitamin D_3	Linear	<0.001	<0.001	<0.001	<0.001	0.043	0.022
	Quadratic	0.002	0.006	0.363	0.010	0.171	0.890
25-OH-D_3	Linear	<0.001	<0.001	<0.001	<0.001	0.013	0.038
	Quadratic	0.036	0.089	0.067	<0.001	0.732	0.797

25-OH-D_3, 25-hydroxycholecalciferol; BWG, body weight gain; FI, feed intake; FE, feed efficiency; Ca, calcium; Pi, inorganic phosphorus; SEM, standard error of the mean.

[a–d] Means in the same column without a common superscript differ significantly ($p<0.05$).

Table 3. Effects of vitamin D_3 and 25-OH-D_3 on femur mineralization in 1- to 21-d-old broiler chickens

Vitamin D_3 (μg/kg)	25-OH-D_3 (μg/kg)	Weight (g)	Length (cm)	Ash (g)	Ash (%)	Ca (%)	P (%)
0	0	0.56^{d}	3.43^{f}	0.17^{d}	26.64^{e}	10.30^{e}	4.92^{f}
2.5		0.63^{cd}	3.75^{ef}	0.20^{d}	33.89^{cd}	12.70^{cd}	6.04^{de}
5.0		0.78^{b}	4.34^{bcd}	0.29^{c}	37.55^{bc}	13.01^{cd}	6.78^{cde}
10.0		1.01^{a}	4.69^{ab}	0.40^{ab}	39.55^{b}	14.02^{abc}	7.13^{bc}
20.0		1.07^{a}	4.87^{a}	0.45^{a}	43.71^{a}	15.20^{ab}	7.88^{ab}
	1.25	0.69^{bc}	3.85^{def}	0.23^{cd}	33.07^{d}	11.31^{de}	5.91^{e}
	2.5	0.80^{b}	4.10^{cde}	0.29^{c}	36.37^{bcd}	13.48^{bc}	6.68^{cde}
	5.0	0.98^{a}	4.58^{abc}	0.38^{b}	38.64^{b}	14.52^{abc}	6.95^{cd}
	10.0	1.02^{a}	4.77^{ab}	0.45^{a}	44.14^{a}	15.86^{a}	8.22^{a}
SEM		0.03	0.08	0.02	0.81	0.28	0.16
p value							
Vitamin D_3	Linear	<0.001	<0.001	<0.001	<0.001	<0.001	<0.001
	Quadratic	0.352	0.268	0.288	0.033	0.318	0.207
25-OH-D_3	Linear	<0.001	<0.001	<0.001	<0.001	<0.001	<0.001
	Quadratic	0.424	0.590	0.457	0.368	0.761	0.936

25-OH-D_3, 25-hydroxycholecalciferol; SEM, standard error of the mean.
[a–f] Means in the same column without a common superscript differ significantly ($p<0.05$).

vitamin D_3 and 25-OH-D_3 were 16.706 and 30.862, respectively, when BWG was used as the criterion. Thus, the RBV value of 25-OH-D_3 to vitamin D_3 was 1.85 (namely 185%, 30.862 divided by 16.706).

Using BWG as the criterion, the RBV value of 25-OH-D_3 to vitamin D_3 was 1.85. When the weight, length, ash weight, and the percentage of ash, Ca, and P of the femur were used as the criteria, the RBV of 25-OH-D_3 to vitamin D_3 were 1.88, 1.82, 2.00, 2.03, 2.45, and 2.22, respectively. Using the same parameters of the tibia as the criteria, the RBV of 25-OH-D_3 to vitamin D_3 were 2.12, 1.86, 2.17, 2.13, 2.52, and 2.52, respectively. Metatarsus mineralization was also used as a criterion. The above RBV values were 2.05, 1.89, 2.00, 1.76, 1.73, and 1.65, respectively.

Generally, the bioavailability of 25-OH-D_3 is higher than that of vitamin D_3 in broilers. The average RBV of 25-OH-D_3 to vitamin D_3 is approximately 2.03 (namely 203%) in promoting growth performance and bone mineralization in 1- to 21-d-old broiler chicken diets.

DISCUSSION

Previous research has shown that vitamin D_3 levels linearly improve growth and bone quality when broilers are fed with Ca- and NPP-deficient diets; by contrast, growth is quadratically or not significantly affected by vitamin D_3 levels when the Ca and NPP contents are sufficient (Aburto et al., 1998; Baker et al., 1998; Rao et al., 2009). Thus, the

Table 4. Effects of vitamin D_3 and 25-OH-D_3 on tibia mineralization in 1- to 21-d-old broiler chickens

Vitamin D_3 (μg/kg)	25-OH-D_3 (μg/kg)	BS (N)	Weight (g)	Length (cm)	Ash (g)	Ash (%)	Ca (%)	P (%)
0	0	20.28^{d}	0.66^{d}	4.71^{d}	0.18^{e}	26.82^{f}	9.76^{e}	5.08^{d}
2.5		24.94^{d}	0.80^{d}	5.26^{c}	0.26^{d}	32.70^{de}	11.51^{d}	6.00^{cd}
5.0		39.87^{c}	1.01^{bc}	5.71^{bc}	0.37^{c}	36.70^{cd}	13.18^{bc}	6.83^{bc}
10.0		54.10^{b}	1.32^{a}	6.23^{a}	0.51^{b}	39.61^{abc}	13.81^{bc}	7.14^{ab}
20.0		68.91^{a}	1.38^{a}	6.38^{a}	0.57^{ab}	41.65^{ab}	14.13^{ab}	7.15^{ab}
	1.25	21.51^{d}	0.83^{cd}	5.38^{c}	0.26^{d}	31.81^{e}	11.23^{de}	5.53^{d}
	2.5	39.83^{c}	1.06^{b}	5.52^{c}	0.39^{c}	36.88^{cd}	12.44^{cd}	6.77^{bc}
	5.0	49.32^{bc}	1.32^{a}	6.12^{ab}	0.50^{b}	37.76^{bc}	13.14^{bc}	6.91^{abc}
	10.0	71.18^{a}	1.39^{a}	6.30^{a}	0.59^{a}	42.93^{a}	15.59^{a}	7.77^{a}
SEM		2.88	0.04	0.09	0.02	0.78	0.27	0.14
p value								
Vitamin D_3	Linear	<0.001	<0.001	<0.001	<0.001	<0.001	<0.001	<0.001
	Quadratic	0.088	0.633	0.125	0.765	0.059	0.003	0.015
25-OH-D_3	Linear	<0.001	<0.001	<0.001	<0.001	<0.001	<0.001	<0.001
	Quadratic	0.009	0.282	0.195	0.581	0.269	0.288	0.715

25-OH-D_3, 25-hydroxycholecalciferol; BS, breaking-strength; SEM, standard error of the mean.
[a–f] Means in the same column without a common superscript differ significantly ($p<0.05$).

Table 5. Effects of vitamin D_3 and 25-OH-D_3 on metatarsus mineralization in 1- to 21-d-old broiler chickens

Vitamin D_3 (μg/kg)	25-OH-D_3 (μg/kg)	Weight (g)	Length (cm)	Ash (g)	Ash (%)	Ca (%)	P (%)
0	0	0.50^{c}	3.56^{f}	0.12^{c}	22.20^{e}	7.68^{e}	3.67^{f}
2.5		0.58^{c}	3.84^{ef}	0.17^{c}	29.02^{cd}	10.23^{d}	4.94^{de}
5.0		0.80^{b}	4.29^{bc}	0.25^{b}	31.69^{bcd}	11.77^{bc}	5.81^{abcd}
10.0		0.91^{ab}	4.61^{ab}	0.30^{ab}	33.06^{abc}	12.01^{abc}	5.93^{abc}
20.0		0.97^{a}	4.65^{a}	0.35^{a}	36.71^{a}	13.20^{a}	6.68^{a}
	1.25	0.59^{c}	3.92^{de}	0.16^{c}	27.41^{d}	10.39^{d}	4.92^{e}
	2.5	0.80^{b}	4.23^{cd}	0.25^{b}	29.50^{cd}	10.84^{cd}	5.40^{cde}
	5.0	0.94^{ab}	4.55^{abc}	0.30^{ab}	31.65^{bcd}	11.32^{cd}	5.67^{bcde}
	10.0	0.94^{ab}	4.58^{abc}	0.34^{a}	35.86^{ab}	12.88^{ab}	6.35^{ab}
SEM		0.03	0.06	0.01	0.69	0.25	0.14
p value							
Vitamin D_3	Linear	<0.001	<0.001	<0.001	<0.001	<0.001	<0.001
	Quadratic	0.133	0.052	0.582	0.029	<0.001	0.007
25-OH-D_3	Linear	<0.001	<0.001	<0.001	<0.001	<0.001	<0.001
	Quadratic	0.071	0.024	0.519	0.593	0.076	0.104

25-OH-D_3, 25-hydroxycholecalciferol; SEM, standard error of the mean.
[a–f] Means in the same column without a common superscript differ significantly ($p<0.05$).

Ca- and NPP-deficient diet was designed in the present study. The optimal dietary Ca to NPP ratio is 2.0 to promote growth performance and bone mineralization in broiler chickens (Bar et al., 2003; Rao et al., 2007). Therefore, the Ca and NPP levels were 0.50% and 0.25%, respectively.

In the comparison of 25-OH-D_3 to vitamin D_3 at the 3.125 μg/kg level, the RBV of 25-OH-D_3 to vitamin D_3 ranged from 1.08 to 4.00 in broiler breeder hens (Atencio et al., 2005). However, no statistical differences in the performance between 25-OH-D_3 and vitamin D_3 were observed when their level reached 12.5 μg/kg in hens (Atencio et al., 2005) or 50 μg/kg in broiler chickens (Fritts

Table 6. Relative bioavailability (RBV) of 25-OH-D_3 to vitamin D_3 based on vitamin D intake (μg/bird) in 1- to 21-d-old broiler chickens with the slope ratio method

Criteria	Intercept	Slope±SE		p value	R^2	RBV±SE
		Vitamin D_3	25-OH-D_3			
Growth performance						
Weight gain	334.545	16.706±1.799	30.862±3.726	<0.001	0.73	1.85±0.23
Femur mineralization						
Weight	0.666	0.024±0.003	0.045±0.005	<0.001	0.73	1.88±0.24
Length	3.813	0.065±0.008	0.118±0.017	<0.001	0.66	1.82±0.29
Ash weight	0.216	0.014±0.001	0.028±0.003	<0.001	0.81	2.00±0.20
Ash percentage	32.147	0.669±0.075	1.356±0.156	<0.001	0.73	2.03±0.21
Ca percentage	11.776	0.198±0.030	0.485±0.061	<0.001	0.66	2.45±0.41
P percentage	5.797	0.121±0.015	0.269±0.031	<0.001	0.71	2.22±0.29
Tibia mineralization						
Weight	0.838	0.033±0.004	0.070±0.008	<0.001	0.70	2.12±0.31
Length	5.242	0.070±0.009	0.130±0.019	<0.001	0.64	1.86±0.29
Ash weight	0.267	0.018±0.002	0.039±0.004	<0.001	0.79	2.17±0.23
Ash percentage	31.847	0.594±0.083	1.263±0.171	<0.001	0.65	2.13±0.32
Ca percentage	11.216	0.187±0.027	0.472±0.056	<0.001	0.68	2.52±0.40
P percentage	5.876	0.085±0.017	0.214±0.034	<0.001	0.53	2.52±0.59
Metatarsus mineralization						
Weight	0.632	0.020±0.003	0.041±0.006	<0.001	0.59	2.05±0.37
Length	3.919	0.046±0.007	0.087±0.014	<0.001	0.59	1.89±0.34
Ash weight	0.175	0.010±0.001	0.020±0.002	<0.001	0.71	2.00±0.27
Ash percentage	26.868	0.570±0.073	1.004±0.152	<0.001	0.65	1.76±0.26
Ca percentage	9.767	0.202±0.027	0.350±0.057	<0.001	0.62	1.73±0.24
P percentage	4.733	0.113±0.016	0.187±0.033	<0.001	0.60	1.65±0.27

and Waldroup, 2003). These data indicate that the 25-OH-D_3 or vitamin D_3 level should not exceed their requirement when evaluating the RBV of 25-OH-D_3 to vitamin D_3.

The optimal level of 25-OH-D_3 was 10 μg/kg for promoting bone ash in broiler chicken diets (Goodgame et al., 2011). Thus, the level of 25-OH-D_3 ranged from 1.25 to 10 μg/kg in the present study. No significant differences were observed in the tibia ash, breaking-strength, and the contents of Ca and P among the broilers fed with vitamin D_3 ranging from 25 to 1,000 μg/kg (Han et al., 2013). Therefore, vitamin D_3 levels were lower than those of our previous research (Han et al., 2013) and were at 2.5, 5, 10, and 20 μg/kg in the present study. Our results revealed the linear relationship between the 25-OH-D_3 or vitamin D_3 level and the performance or bone mineralization in broiler chickens.

The positive effects of 25-OH-D_3 or vitamin D_3 on growth performance and tibia weight, breaking-strength, and the percentage of ash, Ca, and P in broiler chickens have been observed (Fritts and Waldroup, 2003; Rao et al., 2006; Han et al., 2012). The improvement of vitamin D derivatives on performance and bone of birds was caused by the increase of Ca and P utilization. Research has shown that addition of vitamin D_3 (Qian et al., 1997), 25-OH-D_3 (Ledwaba and Roberson, 2003), 1α-OH-D_3 (Shirley, 2003; Han et al., 2012), or 1,25-$(OH)_2$-D_3 (Edwards, 2002) improves the retention of Ca and P in broiler chicken diets.

Previous research has shown that when bone ash was used as criteria, the RBV of 25-OH-D_3 to vitamin D_3 ranged from 1.0 to 2.5 (Soares et al., 1995). The RBV was from 1.11 to 1.38 in broiler breeder hens (Atencio et al., 2005). The present study showed that the RBV of 25-OH-D_3 to vitamin D_3 is approximately 2.03 for promoting growth performance and bone mineralization in 1- to 21-d-old broiler chickens. The differences in RBV values among the criteria were observed. Tibia gave the highest RBV of 25-OH-D_3 to vitamin D_3 (2.22) and followed by femur (2.07). The BWG and metatarsus criteria yielded the lowest values (1.85).

CONCLUSIONS

The present study indicates that the average RBV of 25-OH-D_3 to vitamin D_3 is approximately 2.03 (namely 203%) for promoting growth performance and bone mineralization in 1- to 21-d-old broiler chickens fed with Ca- and P-deficient diets.

ACKNOWLEDGMENTS

This study was supported by the National Natural Science Foundation of China (31101732), the Innovation Scientists and Technicians Troop Construction Projects of Henan Province, the Foundation of the Education Department of Henan Province (16A230003), and the Shangqiu Normal University Foundation (2013GGJS10).

REFERENCES

Aburto, A., H. M. Edwards Jr., and W. M. Britton. 1998. The influence of vitamin A on the utilization and amelioration of toxicity of cholecalciferol, 25-hydroxycholecalciferol, and 1,25 dihydroxycholecalciferol in young broiler chickens. Poult. Sci. 77:585-593.

Applegate, T. J. and M. S. Lilburn. 2002. Growth of the femur and tibia of a commercial broiler line. Poult. Sci. 81:1289-1294.

Atencio, A., G. M. Pesti, and H. M. Edwards. Jr. 2005. Twenty-five hydroxycholecalciferol as a cholecalciferol substitute in broiler breeder hen diets and its effect on the performance and general health of the progeny. Poult. Sci. 84:1277-1285.

Baker, D. H., R. R. Biehl, and J. L. Emmert. 1998. Vitamin D_3 requirement of young chicks receiving diets varying in calcium and available phosphorus. Br. Poult. Sci. 39:413-417.

Bar, A., D. Shinder, S. Yosefi, E. Vax, and I. Plavnik. 2003. Metabolism and requirements for calcium and phosphorus in the fast-growing chicken as affected by age. Br. J. Nutr. 89:51-60.

Biehl, R. R. and D. H. Baker. 1997. Utilization of phytate and nonphytate phosphorus in chicks as affected by source and amount of vitamin D_3. J. Anim. Sci. 75:2986-2993.

Edwards Jr., H. M. 2002. Studies on the efficacy of cholecalciferol and derivatives for stimulating phytate utilization in broilers. Poult. Sci. 81:1026-1031.

Fritts, C. A. and P. W. Waldroup. 2003. Effect of source and level of vitamin D on live performance and bone development in growing broilers. J. Appl. Poult. Res. 12:45-52.

Goetting-Fuchs, C., R. Günther, V. G. Liesner, B. G. Liesner, M. Beyerbach, and J. Kamphues. 2012. Investigations on skeletal development, bone mineralisation as well as calcium and phosphorus levels in blood of male fattening turkeys. Europ. Poult. Sci. 76:121-130.

Goodgame, S. D., F. J. Mussini, C. Lu, C. D. Bradley, S. E. Watkins, and P. W. Waldroup. 2011. Evaluation of a fermentation source of 25-hydroxycholecalciferol in broiler diets. Int. J. Poult. Sci. 10:295-299.

Hall, L. E., R. B. Shirley, R. I. Bakalli, S. E. Aggrey, G. M. Pesti, and H. M. Edwards. Jr. 2003. Power of two methods for the estimation of bone ash of broilers. Poult. Sci. 82:414-418.

Han, J. C., H. X. Qu, J. G. Wang, G. H. Chen, Y. F. Yan, J. L. Zhang, F. M. Hu, L. Y. You, and Y. H. Cheng. 2015. Comparison of the growth and mineralization of the femur, tibia, and metatarsus of broiler chicks. Braz. J. Poult. Sci. 17:333-339.

Han, J. C., H. X. Qu, J. Q. Wang, J. H. Yao, C. M. Zhang, G. L. Yang, Y. H. Cheng, and X. S. Dong. 2013. The effects of dietary cholecalciferol and 1α-hydroxycholecalciferol levels in a calcium- and phosphorus-deficient diet on growth performance and tibia quality of growing broilers. J. Anim. Feed Sci. 22:158-164.

Han, J. C., Y. Liu, J. H. Yao, J. Q. Wang, H. X. Qu, Y. F. Yan, J. Yue, J. L. Ding, Z. T. Shi, and X. S. Dong. 2012. Dietary calcium levels reduce the efficacy of one alpha-hydroxycholecalciferol in phosphorus-deficient diets of broilers. J. Poult. Sci. 49:34-38.

Jendral, M. J., D. R. Korver, J. S. Church, and J. J. R. Feddes. 2008. Bone mineral density and breaking strength of white leghorns housed in conventional, modified, and commercially available colony battery cages. Poult. Sci. 87:828-837.

Koreleski, J. and S. Swiatkiewicz. 2005. Efficacy of different limestone particle size and 25-hydroxycholecalciferol in broiler diets. J. Anim. Feed Sci. 14:705-714.

Ledwaba, M. F. and K. D. Roberson. 2003. Effectiveness of twenty five hydroxycholecalciferol in the prevention of tibial dyschondroplasia in Ross cockerels depends on dietary calcium level. Poult. Sci. 82:1769-1777.

Littell, R. C., P. R. Henry, A. J. Lewis, and C. B. Ammerman. 1997. Estimation of relative bioavailability of nutrients using SAS procedures. J. Anim. Sci. 75:2672-2683.

Qian, H., E. T. Kornegay, and D. M. Denbow. 1997. Utilization of phytate phosphorus and calcium as influenced by microbial phytase, cholecalciferol, and the calcium: Total phosphorus ratio in broiler diets. Poult. Sci. 76:37-46.

Rao, S. V. R., M. V. L. N. Raju, A. K. Panda, G. S. Sunder, and R. P. Sharma. 2006. Effect of high concentrations of cholecalciferol on growth, bone mineralization, and mineral retention in broiler chicks fed suboptimal concentrations of calcium and nonphytate phosphorus. J. Appl. Poult. Res. 15:493-501.

Rao, S. V. R., M. V. L. N. Raju, A. K. Panda, G. S. Sunder, and R. P. Sharma. 2009. Performance and bone mineralisation in broiler chicks fed on diets with different concentrations of cholecalciferol at a constant ratio of calcium to non-phytate phosphorus. Br. Poult. Sci. 50:528-535.

Rao, S. V. R., M. V. L. N. Raju, and M. R. Reddy. 2007. Performance of broiler chicks fed high levels of cholecalciferol in diets containing sub-optimal levels of calcium and non-phytate phosphorus. Anim. Feed Sci. Tech. 134:77-88.

SAS Institute. 2002. SAS User's Guide. 9th edn. SAS Inst. Inc., Cary, NC, USA.

Shirley, R. B. 2003. Evaluation of Phytase, Vitamin D_3 Derivatives, and Broiler Breed Differences on Nutrient Utilization, Broiler Performance, Leg Disorders, and the Expression of Intestinal Calbindin-28 kd mRNA and Protein. Ph.D. Dissertation. University of Georgia, Athens, GA, USA.

Soares Jr., J. H., J. M. Kerr, and R. W. Gray. 1995. 25-hydroxycholecalciferol in poultry nutrition. Poult. Sci. 74:1919-1934.

Yarger, J. G., C. A. Saunders, J. L. McNaughton, C. L. Quarles, B. W. Hollis, and R. W. Gray. 1995. Comparison of dietary 25-hydroxycholecalciferol and cholecalciferol in broiler chickens. Poult. Sci. 74:1159-1167.

17

Effects of Supplemental Liquid DL-methionine Hydroxy Analog Free Acid in Diet on Growth Performance and Gastrointestinal Functions of Piglets

C. Kaewtapee, N. Krutthai, and C. Bunchasak*

Department of Animal Science, Faculty of Agriculture, Kasetsart University, Bangkok 10900, Thailand

ABSTRACT: This study was conducted to determine the effect of dietary supplementation of liquid DL-methionine hydroxy analog free acid (DL-MHA) on growth performance and gastrointestinal conditions of piglets. One hundred and eighty crossbred barrow piglets (Large White×Landrace, body weight: 12.48±0.33 kg) were divided into three groups with ten replications of six piglets each. Piglets received DL-MHA in diet at a concentration of 0 (control group), 0.15%, or 0.24%. The results indicated that increasing the standardized ileal digestible (SID) of sulfur amino acids (SAA) to lysine (SID SAA:Lys) ratio by supplementation of DL-MHA tended to increase (quadratic; $p<0.10$) weight gain and ADG, and showed slightly greater (linear; $p<0.10$) gain:feed ratio. The pH in the diet and cecum linearly decreased ($p<0.01$), whereas pH in colon had a quadratic response ($p<0.01$) with increasing supplementation of DL-MHA. By greater supplementation of DL-MHA, the population of *Lactobacillus* spp. in rectum was likely to increase (quadratic; $p<0.10$), but *Escherichia coli* population in the diet was reduced (quadratic; $p<0.05$). Acetic acid concentration and total short-chain fatty acids in cecum linearly increased ($p<0.05$), whereas valeric acid in cecum quadratically increased ($p<0.05$) with increasing DL-MHA levels. Moreover, the villous height of the jejunum quadratically increased ($p<0.01$) as the supplementation of DL-MHA was increased. It is concluded that the addition of DL-MHA in diet improved the growth performance and the morphology of gastrointestinal tract of piglets. (**Key Words:** Diet, Gastrointestinal Functions, Growth Performance, Liquid DL-methionine, Piglets, Short-chain Fatty Acids)

INTRODUCTION

Methionine (Met) is considered as the second or third limiting amino acid in diet for modern nursery pigs (Gaines et al., 2005), and it plays several roles such as an initiating amino acid in protein synthesis and as the principal biological methylating agent in the body (Bunchasak, 2014). Commercially, both in powder DL-methionine (DL-Met, 99% powder) and its analogue liquid DL-methionine hydroxy analog free acid (DL-MHA, 88% aqueous solution) are supplemented in diets or drinking water for improving growth responses (Kaewtapee et al., 2010; Krutthai et al., 2015). Although the composition or material used in diets such as practical, semi-purified, or purified diets influence the bioefficacy of DL-MHA compared to DL-Met (Bunchasak, 2014), a smaller effect of DL-MHA and DL-Met can be seen when a practical diet formulated based on corn-soybean or broken rice-soybean is used (Bunchasak, 2014; Krutthai et al., 2015). In addition, in a broken rice-soybean based diet, Kongkaew (2014) concluded that bioefficacy of DL-MHA seemed to be 88% of DL-Met (w/w), and the standardized ileal digestible (SID) of sulfur amino acids (SAA) to Lys (SID SAA:Lys) ratio for maximum growth rate was around 0.60 to 0.63.

Liquid DL-MHA has a low pH (~1.00) due to its monocarboxylic acid with a hydroxyl group on α carbon, and is defined as an organic acid (pKa: 3.86) until it is absorbed and transformed to L-Met in the liver (Dibner and Buttin, 2002). As an organic acid, DL-MHA has broader antimicrobial activities and the minimum inhibitory concentration for *Escherichia coli* (*E. coli*) was 0.24% v/v (Poosuwan et al., 2007). In piglets, *E. coli* is a major problem associated with reducing the growth performance and increasing the volume of slurry, while it is well known that an acidic condition in the intestinal tract can stimulate the growth of lactobacillus bacteria which inhibit *E. coli*

* Corresponding Author: C. Bunchasak.
E-mail: agrchb@ku.ac.th

(Risley et al., 1992), reduce the amount of digestive scouring and produce short chain fatty acids (SCFAs) along the gastrointestinal tract (Sakata et al., 1995). Consequently, DL-MHA supplementation in diet may give benefits as both a Met source and an organic acid for production performance and gut health.

Therefore, the objective of this research was to study the effect of DL-MHA supplementation in diet as a Met source and an organic acid on the growth performance, microorganism in the gastrointestinal tract and diet, SCFAs in the cecum and small intestinal morphology of piglets.

MATERIALS AND METHODS

Experimental animals and management

One hundred eighty 7-wk-old cross bred barrows (Large White×Landrace, 12.48±0.33 kg body weight), were housed in pens (1.00×2.50 m) and the environmental temperature was maintained at an average of 29.18±0.84°C. The duration of the experiment was 6-wk. The water was supplied to each pen by two nipples; feed and water were offered *ad libitum*. The pigs were kept, maintained and treated in accordance with International Guiding Principles for Biomedical Research Involving Animals that are accepted standards for animal welfare.

Dietary treatments

There were three treatments with ten replications of six piglets per pen in a completely randomized design. All nutrients of basal diet were calculated to meet the recommendation of National Research Council (1998), except SAA. Since true ileal digestibility of amino acids in National Research Council (1998) was transformed to SID (Stein et al., 2007), the SID of SAA in basal diet (0.51%) was lower than the recommendation (0.58% for pigs from 10 to 20 kg). The SID of Lys, Thr, and Trp in basal diet was formulated to meet the requirements according to National Research Council (1998). Subsequently, SAA were limiting amino acids in basal diet. The optimum requirement of SID SAA:Lys ratio by supplementation with DL-MHA ranged from 60% to 62%, whereas the 69% of SID SAA:Lys ration was an excessive level, resulting in the negative effect on growth performance (Gaines et al., 2005). Furthermore, the basal diet was supplemented with DL-MHA (based on 88% bioefficacy of DL-Met; w/w; Bunchasak et al., 2014) at a concentration of 0 (basal diet), 0.15% or 0.24%, consequently the SID SAA:Lys ratio was 49%, 61%, and 69%, respectively. Accordingly, the liquid form, DL-MHA was mixed with broken rice as a carrier at ratio of 1:10, and then filtered on a screen (0.5×0.5 mm) to confirm it was well dispersed. Consequently, this product was blended with other feed ingredients. The ingredient composition of the basal diet is shown in Table 1.

Table 1. The ingredient composition of basal diet

Items	
Ingredient (%)	
Broken rice	42.97
Corn	4.99
Raw rice bran	8.13
Soybean meal (48% CP)	3.99
Full fat soybean	27.34
Fish meal (60% CP)	2.00
Soybean oil	1.29
L-Lys-HCl	0.35
L-Thr	0.10
Skim milk (33% CP)	4.99
Monodicalcium phosphate (21% phosphorus)	1.82
Calcium carbonate	1.08
Salt	0.21
Premix[1]	0.50
Corn starch	0.24
Calculated chemical composition	
Metabolizable energy (MJ/kg)[2]	14.00
Crude protein (%)	20.00
Calcium (%)	1.00
Available phosphorus (%)	0.49
Standardized ileal digestible AA (%)	
Lys	1.05
Thr	0.85
Trp	0.25
Met	0.28
Cys	0.23
SAA	0.51
SAA:Lys ratio	49

CP, crude protein; AA, amino acids; SAA, sulfur amino acids.

[1] Premix content; Vitamin A (retinyl acetate) 1,200 mg, vitamin D_3 (cholecalciferal) 16,000 μg, vitamin E (dl-α-tocopheryl acetate) 21,818 mg, K_3 (menadione) 1.4 g, B_1 (thiamin) 0.6 g, B_2 (riboflavin) 0.3 g, B_6 (pyridoxine) 0.75 g, B_{12} (cyanocobalamin) 14 mg, nicotinic acid 20 g, pantothenic acid 10 g, folic acid 0.44 g, D-biotin 0.04 g, choline chloride 60 g, Fe ($FeSO_4 \cdot H_2O$) 45 g, Cu ($CuSO_4 \cdot 5H_2O$) 40 g, Mn (MnO) 15 g, Zn (ZnO) 40 g, Co ($CoCO_3$) 0.2 g, I ($Ca(IO_3)_2$) 0.4 g, Se (Na_2SeO_3) 0.06 g, carrier (grinded corn cob) added to 1 kg.

[2] The energy content was derived from the calculation.

Growth performance records

Initial body weight (d 0) and final body weight (d 41) were measured. Average daily feed intake (ADFI) was recorded daily on a per pen basis. Average daily gain (ADG) and gain:feed ratio (G:F) were calculated on a per pen basis. Due to the sparing effect between Met and Cys and that the bioefficacy of DL-MHA was 88% of DL-Met (Bunchasak, 2014), the supplemental dietary DL-MHA was considered as SAA requirement. Therefore, the SID of SAA was increased initially from 0.51% in basal diet to 0.64% which was calculated by 0.51+(0.15×0.88), and 0.72% which was calculated by 0.51+(0.24×0.88) in diet supplemented with DL-MHA at 0.15% and 0.24%, respectively. Consequently, daily SAA intake was calculated from ADFI.

Diet and gastrointestinal pH

At the end of the trial, ten pigs per treatment were euthanized using CO_2 asphyxiation in an atmosphere of less than 2% oxygen (air displaced by CO_2). The digesta contents from stomach, duodenum, jejunum, ileum, cecum, colon and rectum were immediately collected for the determination of pH. Moreover, diets were sampled for 10 g from each test diet and mixed with 90 mL distilled water (pH 7) for 10 min at room temperature. The pH in diet and digesta samples were directly measured with a pH meter (IQ Scientific Instruments, Inc., Carlsbad, CA, USA).

Bacteria counts in cecum

Five grams of digesta sample was diluted with 45 mL of 1% peptone solution (Oxoid Laboratories, Basingstoke, UK). Ten-fold serial dilution was used to reduce to 1:10 of concentration and 0.1 mL was applied onto duplicate agar plates for each dilution. *Lactobacillus* spp. and *E. coli* grew on De Man, Rogosa and Sharpe agar (Difco, Becton, Dickinson and Company, Sparks, MD, USA) and MacConkey agar (Laboratorios Britania s.a., Buenos Aires, Argentina), respectively. Using the spread plate technique to determine the number of bacteria, all agar plates were incubated for 24 h at 37.0°C.

Analysis of short chain fatty acids

The samples of digesta from the cecum were centrifuged (TOMY model MX-301, TOMY Kogyo Co., Ltd., Tokyo, Japan) in microfuge tubes at 14,000 rpm, 4°C for 10 min, and 1.5 mL of the supernatant was transferred to a microfuge tube. The concentrations of volatile fatty acids were analyzed with gas chromatography (Shimadzu Model GC-2010 High-end, Shimadzu, Kyoto, Japan) and a flame ionization detector (GC-FID, Kyoto, Japan). One μL of the supernatant was injected into a silica capillary column (DB-WAX, 30 m×0.25 mm i.d., film thickness of 0.50 μm, J&W Scientific, Folsom, CA, USA) at a column temperature of 150°C. The carrier gas (He) flow rate was 1.4 mL/min. and the split ratio was 1:20. The temperatures of the injection port and detector were both programmed at 225°C.

Morphology of small intestine

Morphology of the duodenum, jejunum and ileum was evaluated with a light microscope. The tissues were removed and immediately fixed in 10% neutral buffered formalin, and then carefully embedded in paraffin. For each paraffin block, at least 10 sections of 7 μm thickness were prepared, and then stained with hematoxylin-eosin for histological evaluation. In the morphological evaluation of the small intestine, villous height, crypt depth and villous height:crypt depth ratio were measured. The measurement of villous height (from the tip of the villous to the villous-crypt junction) and crypt depth (from the villous-crypt junction to the lower limit of the crypt) were recorded as the mean of 10 fields for each specimen.

Statistical analysis

The data were analyzed as a completely randomized design using the MIXED procedures of SAS (SAS Institute Inc., 2008). The pig or pen was used as the experimental unit. Initial body weight was used as a covariate for analysis of weight gain, final weight and ADG. Orthogonal polynomial contrasts were used to assess for linear and quadratic effects of DL-MHA levels (0, 0.15%, and 0.24%). The contrast coefficients were calculated for unequally spaced treatment levels using IML procedures of SAS (SAS Institute Inc., 2008).

RESULTS AND DISCUSSION

Growth performance

The results of growth performance are presented in Table 2. Increasing the SID SAA:Lys ratio by supplementation of DL-MHA tended to increase (quadratic; $p<0.10$) weight gain and ADG, and showed slightly greater (linear; $p<0.10$) G:F. There was no difference ($p>0.05$) in ADFI, but daily SAA intake was linearly increased ($p<0.01$) as DL-MHA level increased.

The supplementation of DL-MHA significantly improved the growth performance, although feed intake was depressed by about 5% and 6% in 0.15% and 0.24%, respectively. The lower feed intake may explain the decrease in growth promotion of 0.24% group. Although the relative bioefficacy of DL-MHA to DL-Met has been reported to be lower than 88% (Kim et al., 2006; Dilger and Baker, 2008), Bunchasak et al. (2014) and Krutthai et al. (2015) showed that 88% bioefficacy may be an acceptable with regard to SAA requirement. Therefore, based on 88% bioefficacy (w/w) compared to DL-Met, supplementation of DL-MHA at 0.15% and 0.24% of diet result in SID SAA:Lys ratio at 61% and 69%, respectively. This study confirms Gaines et al. (2005) and Peak (2005) who reported that optimal SID SAA:Lys ratio for modern lean-genotype piglets was approximately 60% to 62%, while an excessive Met consumption or excessively high SID SAA:Lys ratio as 0.24% DL-MHA supplementation diet (SID SAA:Lys ratio = 69%) did not give any additional benefit to the growth performance of the piglets.

pH in diet and gastrointestinal tract

The pH in diet and gastrointestinal tract of piglets is presented in Table 3, pH in the diet and cecum linearly

Table 2. Least squares means and SEM of growth performance and methionine intake of piglets receiving different amounts of DL-MHA in diet[1, 2]

Items	Control (Basal diet)	DL-MHA 0.15%	DL-MHA 0.24%	SEM	p-values	
					Linear	Quadratic
Body weight (kg)						
d 0	12.5	12.5	12.5	0.06	NA	NA
d 41	29.0	30.7	30.0	0.41	0.064	0.051
Weight gain (kg)	16.5	18.2	17.5	0.37	0.064	0.051
ADG (g/d)	403	444	428	8.9	0.064	0.051
ADFI (g/d)	1,044	991	981	24.2	0.302	0.795
G:F	0.39	0.45	0.44	0.015	0.094	0.283
Daily SAA intake (g/d)	5.32	6.34	7.06	0.202	<0.001	0.837

SEM, standard error of the mean; DL-MHA, DL-methionine hydroxy analog free acid; NA, not statistical analysis; ADG, average daily gain; ADFI, average daily feed intake; G:F, gain:feed ratio; SAA, sulfur amino acids.

[1] DL-MHA was calculated at 88% bioefficacy of DL-Met.

[2] Values are least squares means of 10 replicates.

decreased ($p<0.01$), whereas pH in colon had a quadratic response ($p<0.01$) with increasing supplementation of DL-MHA. However, inducing this acidity did not impact ($p>0.05$) on pH in the stomach, duodenum, jejunum and rectum, although pH in the ileum tended to linearly decrease ($p<0.10$).

Organic acids probably assist the immature digestive process of young piglets by decreasing the pH of the stomach, and promoting peptic activity and protein digestion (Kirchgessner and Roth, 1982). However, in this study, DL-MHA supplementation in diet did not affect gastric pH, which was similar to Mathew et al. (1991) and Risley et al. (1992) who reported that supplementation of organic acids in diet did not change the pH in the gut. Recently, Krutthai et al. (2015) also reported low potentiality of acidifying by DL-MHA throughout gastrointestinal tract of piglets. Since supplemental DL-MHA induces dietary acidity (Krutthai et al., 2015), the presence of an acidic diet may affect a low pH in stomach, leading to a negative feedback mechanism that inhibits HCl secretion (Yen, 2001). Moreover, an unchanged pH in some part of gastrointestinal tract (stomach, duodenum, jejunum and ileum) may be caused by some factors such as greater feed buffer capacity (Bolduan et al., 1988), digesta transit time (Walsh et al., 2004) and physiological homeostasis in the tract.

Nevertheless, DL-MHA supplementation decreased pH in the cecum and colon where the fermentative segments are located. Generally, the retained feed residues and endogenous materials in the hind gut are fermented by microbes, and the unabsorbed DL-MHA or SAA in the hind gut (cecum and colon) may be utilized by acid bacteria to produce SCFAs or lactic acid (Macfarlane and Gibson, 1995) and result in a decrease of pH. However, the exact mechanism of this observation is unclear.

Microorganisms in gastrointestinal tract

The effect of supplementation of DL-MHA in diet on *Lactobacillus* spp. and *E. coli* in the hind gut of piglets is shown in Table 4. The population of *Lactobacillus* spp. in rectum was likely to increase (quadratic; $p<0.10$), but *E. coli* population in the diet was reduced (quadratic; $p<0.05$) by greater supplementation of DL-MHA.

In practice, feedstuffs and drinking water contaminated

Table 3. Least squares means and SEM for pH in the diet and gastrointestinal tract of piglets receiving different amounts of DL-MHA in diet[1]

Items	Control (Basal diet)	DL-MHA 0.15%	DL-MHA 0.24%	SEM	p-values	
					Linear	Quadratic
Diet	5.81	5.68	5.58	0.029	<0.001	0.743
Stomach	3.33	3.70	3.55	0.312	0.744	0.739
Duodenum	5.32	5.14	5.47	0.183	0.819	0.502
Jejunum	6.14	6.08	6.27	0.086	0.616	0.462
Ileum	6.70	6.57	6.41	0.067	0.087	0.718
Cecum	6.20	6.07	5.86	0.047	0.003	0.318
Colon	6.14	5.82	5.92	0.038	0.002	0.009
Rectum	6.22	6.20	6.17	0.048	0.692	0.917

SEM, standard error of the mean; DL-MHA, DL-methionine hydroxy analog free acid.

[1] Values are least squares means of 10 replicates.

Table 4. Least squares means and SEM for populations of *Lactobacillus* spp. and *E. coli* in the gastrointestinal tract and *E. coli* in diet of piglets receiving different amounts of DL-MHA (Log_{10}cfu/mL)[1]

Items	Control (Basal diet)	DL-MHA 0.15%	DL-MHA 0.24%	SEM	p-values	
					Linear	Quadratic
Lactobacillus spp.						
Cecum	9.47	9.54	9.39	0.086	0.754	0.532
Rectum	9.46	9.62	9.37	0.057	0.700	0.085
E. coli						
Diet	3.97	3.50	3.66	0.090	0.019	0.027
Cecum	10.43	10.62	10.57	0.070	0.363	0.497
Rectum	9.30	9.12	9.27	0.095	0.795	0.434

SEM, standard error of the mean; DL-MHA, DL-methionine hydroxy analog free acid.

[1] Values are least squares means of 10 replicates.

with *E. coli* have been reported to adversely affect animal health (Risley et al., 1992). According to Mathew et al. (1991), organic acid could reduce *E. coli* populations in diet and improved growth performance. Therefore, DL-MHA may be defined as an organic acid that directly inhibits pathogenic bacteria contamination in the diet. However, DL-MHA supplementations have less effect on the populations of this microorganism in the cecum and rectum, although the pH in these segments was decreased by DL-MHA supplementation. Similarly, Risley et al. (1992) failed to depress the population of *E. coli* in the gastrointestinal tract by feeding with fumaric acid. Due to the optimal pH range of *E. coli* is 6.0 to 8.0 (Tan, 2006), it can be postulated that the reduction of pH in diets (from 6.20 to 5.86) by DL-MHA supplementation was not enough to decrease population of *E. coli* in the cecum and rectum.

Short chain fatty acids in cecum

The effects of the DL-MHA supplementation on the concentration of SCFAs in the cecum are presented in Table 5. The addition of DL-MHA in diet had a linear increase ($p<0.05$) in acetic acid and total SCFAs concentration in the cecum, whereas valeric acid in the cecum quadratically increased ($p<0.05$) with increasing DL-MHA levels.

Large intestine is the major site for microbial fermentation resulting in the production of gas, and the highest concentration of SCFAs is found in the cecum (Macfarlane and Gibson, 1995). In this study, an increase in SCFAs production with supplemental dietary DL-MHA may be caused by an extension time of digestion. According to Walsh et al. (2004), a diet supplemented with organic acids had a longer retention time in stomach, thereby enhancing digestibility in piglets. Furthermore, Macfarlane and Macfarlane (1995) showed that a long transit time in the large intestine can have profound effects on bacterial physiology and metabolism, leading to protein breakdown and amino acid fermentation and making an increased contribution to colon SCFA pools.

The varieties of SCFAs in the cecum are in agreement with Wallace (1995) who reported that the concentration of acetate is greater than that of propionate and butyrate. These three SCFAs are produced from the degradation of cysteine (Cys), whereas the main products of Met degradation are propionate and butyrate (Smith and Macfarlane, 1997). Due to the sparing effect between Met and Cys, the supplementation of DL-MHA may lead quantities of residual Cys in the cecum to produce the SCFAs. Surprisingly, the amount of valeric acids, which is produced from branched-chain amino acids (BCAAs; Macfarlane and Macfarlane, 1995), was also significantly increased by DL-MHA supplementation, while the metabolic linkage between SAA and BCAAs is weak.

Small intestinal morphology

The effects of DL-MHA supplementation on morphology of the small intestine are shown in Table 6. The

Table 5. Least squares means and SEM for SCFAs concentrations in the cecum of piglets receiving different amounts of DL-MHA in diet (mmol/L)[1]

Items	Control (Basal diet)	DL-MHA 0.15%	DL-MHA 0.24%	SEM	p-values	
					Linear	Quadratic
Acetic acid	4.83	10.97	10.26	1.125	0.025	0.213
Propionic acid	4.03	6.86	5.92	0.636	0.163	0.223
Butyric acid	2.74	3.70	2.90	0.435	0.785	0.380
Valeric acid	0.04	0.11	0.07	0.011	0.087	0.021
Total SCFAs	11.65	22.52	22.36	2.176	0.022	0.323

SEM, standard error of the mean; SCFAs, short chain fatty acids; DL-MHA, DL-methionine hydroxy analog free acid.

[1] Values are least squares means of 10 replicates.

Table 6. Least squares means and SEM for small intestinal morphology of piglets receiving different amounts of DL-MHA in diet[1]

Items	Control (Basal diet)	DL-MHA 0.15%	DL-MHA 0.24%	SEM	p-values	
					Linear	Quadratic
Villous height (μm)						
Duodenum	438	460	435	7.6	0.995	0.137
Jejunum	376	407	369	5.3	0.893	0.002
Ileum	295	318	316	5.3	0.078	0.399
Crypt depth (μm)						
Duodenum	438	462	444	7.2	0.585	0.180
Jejunum	292	305	293	3.9	0.709	0.128
Ileum	322	332	291	4.9	0.025	0.005
Villous height to crypt depth ratio						
Duodenum	1.06	1.04	1.07	0.023	0.862	0.522
Jejunum	1.35	1.39	1.31	0.024	0.540	0.189
Ileum	0.97	1.01	1.17	0.023	0.001	0.104

SEM, standard error of the mean; DL-MHA, DL-methionine hydroxy analog free acid.
[1] Values are least squares means of 10 replicates.

villous height in segment of the jejunum had a quadratic effect ($p<0.01$), whereas villous height in segment of ileum tended to linearly increase ($p<0.10$) as supplementation of DL-MHA increased. Crypt depths in the duodenum and jejunum were not affected ($p>0.05$) by DL-MHA supplementation, whereas crypt depth in the ileum showed a quadratic effect ($p<0.01$). Subsequently, villous height to crypt depth ratio in the ileum linearly increased ($p<0.01$) with the greater supplementation of DL-MHA.

In general, SCFAs can be absorbed by diffusion in the hindgut wall as the energy substrates for mucosa development and precursors for synthesis of non-essential amino acids, DNA and greater lipids required for intestinal growth (Mroz, 2005). Furthermore, gastrointestinal cell proliferation is also stimulated by SCFAs concentration, particularly butyric acid (Sakata et al., 1995). In current study, however, a significant increase in the amount of butyric acid in the cecum was not observed, while acetic acid and valeric acid contents in the cecum were significantly increased. These findings indicated that at least acetic acid might have closely promoted physiological function of villi, although we could not find any report that indicates the effect of valerate on gastrointestinal morphology of piglets. Moreover, derivatives of DL-MHA such as taurine or glutathione which are antioxidants protect the damage of villi from oxidative stress in the small intestine (Shoveller et al., 2005) may promote the morphology.

Increasing DL-MHA increased (quadratic; $p<0.10$) growth rate, total SCFAs in cecum (linear; $p<0.05$), linearly decreased ($p<0.01$) pH in cecum, and quadratically improved ($p<0.01$) morphology of small intestine of the pigs. Therefore, it can be summarized that three mechanisms are involved in improving intestinal function by DL-MHA supplementation resulting in improvement of growth performance as follows; i) DL-MHA directly stimulates cell elongation as a source of energy or a precursor of protein synthesis, ii) derivatives of DL-MHA such as taurine or glutathione which are antioxidants protect the damage of villi from oxidative stress in the small intestine (Lambert, 2004; Roig-Pérez et al., 2005; Shoveller et al., 2005) and iii) DL-MHA supplementation lowers pH in the cecum and the lower pH condition may increase acid bacteria and produce SCFAs to enhance the physiological status of small intestine villi by the reverse-peristalsis process.

IMPLICATIONS

Liquid DL-MHA can be applied in diet as an acidifier since it inhibited *E. coli* contamination in diet and depressed pH in the large intestine of piglets. Therefore, the dual effect (Met source and organic acid) of 0.15% DL-MHA (61% SID SAA:Lys ratio) could promote growth performance and good conditions of the gastrointestinal tract of piglet.

ACKNOWLEDGMENTS

This research was financially supported by Sumitomo Chemical Co., Ltd. (Tokyo, Japan). The authors would like to thank staff of Department of Animal Science, Kasetsart University Thailand for their skillful technical assistance.

REFERENCES

Bolduan, G., H. Jung, E. Schnabel, and R. Schneider. 1988. Recent advances in the nutrition of weaner pigs. Pig News Info. 9:381-385.

Bunchasak, C. 2014. Comparative bioefficacy of supplemental DL-2-hydroxy-4-[methyl] butanoic acid and DL-methionine on absorption, conversion and production performance. Danex Intercorporation Co., Ltd., Bangkok, Thailand.

Dibner, J. J. and P. Buttin. 2002. Use of organic acids as a model to study the impact of gut microflora on nutrition and metabolism. J. Appl. Poult. Res. 11:453-463.

Dilger, R. N. and D. H. Baker. 2008. Cyst(e)ine imbalance and its effect on methionine precursor utilization in chicks. J. Anim. Sci. 86:1832-1840.

Gaines, A. M., G. F. Yi, B. W. Ratliff, P. Srichana, D. C. Kendall, G. L. Allee, C. D. Knight, and K. R. Perryman. 2005. Estimation of the ideal ratio of true ileal digestible sulfur amino acids:lysine in 8- to 26-kg nursery pigs. J. Anim. Sci. 83:2527-2534.

Kaewtapee, C., N. Krutthai, K. Poosuwan, T. Poeikhampha, S. Koonawootrittriron, and C. Bunchasak. 2010. Effects of adding liquid DL-methionine hydroxy analogue-free acid to drinking water on growth performance and small intestinal morphology of nursery pigs. J. Anim. Physiol. Anim. Nutr. 94:395-404.

Kim, B. G., M. D. Lindemann, M. Rademacher, J. J. Brennan, and G. L. Cromwell. 2006. Efficacy of DL-methionine hydroxy analog free acid and DL-methionine as methionine sources for pigs. J. Anim. Sci. 84:104-111.

Kirchgessner, M. and F. X. Roth. 1982. Fumaric acid as a feed additive in pig nutrition. Pig News Info. 3:259-264.

Kongkaew, P. 2014. Comparative Effect of Methionine Sources and Total Sulfur Amino Acids to Lysine Ratios in Diets on Production Performance and Physiological Status in Intestinal Tract of Nursery Pigs. M.Sc. Thesis, Kasetsart University, Bangkok, Thailand.

Krutthai, N., C. Vajrabukka, K. Markvichitr, A. Choothesa, J. Thiengtham, S. Sawanon, C. Kaewtapee, and C. Bunchasak. 2015. Effect of source of methionine in broken rice-soybean diet on production performance, blood chemistry, and fermentation characteristics in weaned pigs. Czech J. Anim. Sci. 60:123-131.

Lambert, I. H. 2004. Regulation of the cellular content of the organic osmolyte taurine in mammalian cells. Neurochem. Res. 29:27-63.

Macfarlane, G. T. and G. R. Gibson. 1995. Microbiological aspects of the production of short-chain fatty acids in the large bowel. In: Physiological and Clinical Aspects of Short-chain Fatty Acids (Eds. J. H. Cummings, J. L. Rombeau, and T. Sakata). Cambridge University Press, Cambridge, UK. pp. 87-105.

Macfarlane, S. and G. T. Macfarlane. 1995. Proteolysis and amino acid fermentation. In: Human Colonic Bacteria: Role in Nutrition, Physiology and Pathology (Eds. G. R. Gibson and G. T. Macfarlane). CRC Press, Boca Raton, FL, USA. pp. 75-100.

Mathew, A. G., A. L. Sutton, A. B. Scheidt, D. M. Forsyth, J. A. Patterson, and D. T. Kelly. 1991. Effects of a propionic acid containing feed additive on performance and intestinal microbial fermentation of the weanling pig. In: Digestive physiology in Pigs, Proceedings of the fifth international symposium on digestive physiology in the pig. EAAP Publication no. 54, Wageningen, The Netherlands. pp. 464-469.

Mroz, Z. 2005. Organic acids as potential alternatives to antibiotic growth promoters for pigs. Adv. Pork. Prod. 16:169-182.

National Research Council. 1998. Nutrient Requirements of Swine. 10th Ed. National Academy Press, Washington, DC, USA.

Peak, S. 2005. TSAA requirements for nursery and growing pigs. Adv. Pork. Prod. 16:101-107.

Poosuwan, K., C. Bunchasak, K. Prahkarnkaeo, S. Chansawang, and T. Poeikhampha. 2007. Effects of adding methionine hydroxy analog free acid to drinking water on growth performance and gastrointestinal functions of broiler chicks during starter period. In: Proceedings of International Conference on Integration of Science & Technology for Sustainable Development, King Mongkut's Institute of Technology Ladkrabang, Bangkok, Thailand. pp. 90-94.

Risley, C. R., E. T. Kornegay, M. D. Lindemann, C. M. Wood, and W. N. Eigel. 1992. Effect of feeding organic acids on selected intestinal content measurements at varying times postweaning in pigs. J. Anim. Sci. 70:196-206.

Roig-Pérez, S., M. Moretó, and R. Ferrer. 2005. Transepithelial taurine transport in caco-2 cell monolayers. J. Membr. Biol. 204:85-92.

Sakata, T., M. Adachi, M. Hashida, N. Sato, and T. Kojima. 1995. Effect of n-butyric acid on epithelial cell proliferation of pig colonic mucosa in short-term culture. Dtsch. Tierärztl. Wochenschr. 102:163-164.

SAS Institute Inc. 2008. SAS/STAT. User's Guide: Version 9.2. SAS Institute Inc., Cary, NC, USA.

Shoveller, A. K., B. Stoll, R. O. Ball, and D. G. Burrin. 2005. Nutritional and functional importance of intestinal sulfur amino acid metabolism. J. Nutr. 135:1609-1612.

Smith, E. A. and G. T. Macfarlane. 1997. Dissimilatory amino acid metabolism in human colonic bacteria. Anaerobe 3:327-337.

Stein, H. H., B. Séve, M. F. Fuller, P. J. Moughan, and C. F. M. de Lange. 2007. Invited review: Amino acid bioavailability and digestibility in pig feed ingredients: Terminology and application. J. Anim. Sci. 85:172-180.

Tan, H. M. 2006. Acidifiers: Synergy of acids make for better efficacy. Asian. Poul. Mag. 7:30-33.

Wallace, R. J. 1995. Biochemistry and microbiology in the rumen. In: Physiological and clinical aspects of short-chain fatty acids (Eds. J. H. Cummings, J. L. Rombeau, and T. Sakata). Cambridge University Press, Cambridge, UK. pp. 67-71.

Walsh, M. C., L. Peddireddi, and J. S. Radclife. 2004. Acidification of nursery diets and the role of diet buffering capacity. Swine Res. Rep. pp. 25-36.

Yen, J. T. 2001. Anatomy of the digestive system and nutritional physiology. In: Swine Nutrition, 2nd Ed. (Eds. A. J. Lewis and L. L. Southern). CRC press, Boca Raton, FL, USA. pp. 31-63.

18

Effectiveness of Phytogenic Feed Additive as Alternative to Bacitracin Methylene Disalicylate on Hematological Parameters, Intestinal Histomorphology and Microbial Population and Production Performance of Japanese Quails

M. Manafi*, M. Hedayati, and S. Khalaji

Department of Animal Science, Faculty of Agricultural Sciences, Malayer University, Malayer 65719-95863, Iran

ABSTRACT: This study was conducted to evaluate the effects of phytogenic additive and antibiotic growth promoter in laying Japanese quails. One hundred and sixty five quails were divided into three groups of 5 replicates and 11 quails (8 females and 3 males) in each replicate. Treatment 1 was fed control diet, treatment 2 was fed control diet supplemented with 0.05% bacitracin methylene disalicylate as antibiotic growth promoter and treatment 3 was fed control diet supplemented with 0.1% phytogenic feed additive (PFA) for two periods of 3 weeks each from 37 to 42 weeks of age. Results showed that egg production, eggshell strength, eggshell weight, villus height and villus height to crypt depth ratio were significantly ($p \leq 0.05$) increased and feed consumption, feed conversion ratio, albumen, Haugh unit, cholesterol, low-density lipoprotein, alanine transaminase, gamma glutamyltransferase, alkaline phosphatase, high-density lipoprotein, triglyceride, number of goblet cell, crypt depth and intestinal bacterial population of Coliforms, Salmonella and *E. coli* were significantly ($p \leq 0.05$) decreased in PFA fed group. It is concluded that addition of PFA containing phytomolecules and organic acids as main ingredients could significantly improve the production parameters and the general health of laying quails as an alternative to antibiotic growth promoters. (**Key Words:** Feed Antibiotic, Blood Constituents, Performance, Phytogenic Feed Additive, Japanese Quails)

INTRODUCTION

Due to the growing demand for the use of herbal ingredients in human daily food, a tendency to minimize the chemical feed additives in poultry diets is among the interests of the producers. Considering the ban of feed antibiotics in many countries including Iran, their removal from the diet may negatively affect profitability of the animals (Manafi, 2015).

The beneficial effects of different bioactive compounds are found to enhance poultry productivity. The term 'phytogenic compound' refers to the plant parts (*e.g.* garlic, oregano, thyme, rosemary, coriander, and cinnamon) as well as to their respective extracts in the form of essential oils. Many beneficial properties of phytogenic compounds originate from their bioactive molecules (*e.g.* carvacrol, thymol, cineole, linalool, anethole, allicin, capsaicin, allylisothiocynate, and piperine) (Grashorn, 2010). A large number of *in vitro* and *in vivo* studies have established a wide range of phytobiotic activities in poultry nutrition like stimulation of feed consumption (FC), antimicrobial, coccidiostatic, anthelmintic, immunestimulating (Manafi, 2015) antibacterial and antioxidant functions (Windisch et al., 2008), enhancement of digestive enzyme activities (Jamroz et al., 2006), positive effects on performance and feed conversion ratio (FCR), quality and carcass meat safety and lowering the levels of cholesterol, low-density lipoprotein (LDL) and high-density lipoprotein (HDL) (Stanaćev et al., 2011) in birds.

Recently, rearing of quail has gained importance in the poultry industry. The consumption of quail eggs and meat has significantly increased and there is scope for increasing

* Corresponding Author: M. Manafi.
E-mail: manafim@malayeru.ac.ir

healthy quail products in the market. Quail meat has extremely low skin fat and low cholesterol value. It is rich in micronutrients and a wide range of vitamins including the B complex, E and K and folate. The effects of individual essential oils have been studied earlier in quails; however, to the author's knowledge, there have been limited studies investigating the effects of bioactive components combination in laying quails. Therefore, this study was planned and performed to investigate the effects of phytomolecules bioactive compounds (allicin, peppermint, thymol, and carvacrol) and organic acids (propionic acid and fumaric acid) in comparison with antibiotic growth promoter in production performance, serum biochemistry, illeal bacterial status, immune response and intestinal morphology of Japanese laying quails.

MATERIALS AND METHODS

Experimental birds and diets

One hundred and sixty five layer Japanese quails (*Coturnixcoturnix japonica*) at 37 weeks of age with average initial body weight of 224±5 g were divided into 3 treatment groups, with 5 replicates and 11 quails (8 females and 3 males) in each replicate. Data were recorded from the quail when they were aged between 37 to 39 and 40 to 42 weeks, according to Malayer University approved animal care rules and protocols. Quails were housed in thermostatically controlled batteries (152.4×45.6×26.7 cm) with raised wire floors in an environmentally controlled building having forced air ventilation for 6 weeks duration. Treatment groups were fed a corn-soybean meal basal diet (control); control plus 0.05% bacitracin methylene disalicylate (BMD) and control plus 0.1% phytogenic feed additive (PFA). The PFA used in this study was Natusol, a combination of phytomolecules bioactive compounds (allicin, peppermint, thymol, and carvacrol) and organic acids (propionic acid and fumaric acid) from natural sources, provided by a commercial company (Zeus Biotech Limited, Mysore, India). The basal commercial quail breeder diet was fed as mash and prepared with the same batch of ingredients. Diet was formulated isonitrogenic and isoenergic. The ingredients and chemical composition of the coccidiostat-free basal diets are shown in Table 1. All quails were fed *ad libitum* feed and water throughout the study period and exposed to 17 h light. The PFA and BMD were procured from local market in powder form and added on top to the basal diet and were included into the basal diet according to experimental treatments. The feeding and collection protocols used in the present experiment were approved by the bioethical committee of Malayer University (protocol No. 84/5-1-246) under the guidelines of animal protection used for experimental and other scientific purposes. The quails were also raised and cared based on the recommendation of Iranian Council on Animal Care.

Table 1. Ingredients and chemical composition of the basal diet of layer Japanese quail (37 to 42 weeks of age)

Item	%
Maize	38.00
Soybean meal (44% CP)	25.00
Sunflower meal (34.7% CP)	10.00
Meat-bone meal (58.5% CP)	2.00
Barley	12.70
Vegetable oil	4.00
Dicalcium phosphate	0.4
Limestone	7.300
Sodium chloride	0.30
Vitamin premix[1]	0.15
Mineral premix[2]	0.15
Calculated composition	
ME (kcal/kg)	2819
CP (%)	20.5
Dig Lys (%)	1.08
Dig Met (%)	0.35
Dig Met+Cys (%)	0.75
Dig Thr (%)	0.79
Ca (%)	3
Available P (%)	0.40

CP, crude protein; ME, metabolizable energy.

[1] Each kg of vitamin premix contains 6,000,000 IU vit A; 600,000 IU vit D_3; 20,000 IU vit E; 2 g vit K; 1.2 g vit B_1; 2.4 g vit B_2; 2 g vit B_6; 12 mg vit B_{12}; 10 g niasin; 300 mg folic acid; 4 g calcium panthotenate; 50 mg D-Biotin.

[2] Each g of mineral premix contains 80 g Mn; 30 g Fe; 60 g Zn; 5 g Cu; 0.5 g Co; 2 g I; 235.68 g Ca.

Production parameters

Prior to start of the experiment, two weeks pre-experimental feed was given to quails for adaptation. The production parameters measured were hen day egg production (EP, %), egg weight (EW, g), FC (g/week) and FCR. FC was calculated as the average of each replicate and the FCR was expressed as grams of feed consumed per gram of egg produced (g feed/g egg). Daily EP and EW (g) were recorded during 6 weeks. Treatment means for these traits were not different at the start of experiment ($p>0.05$). Mass EP was calculated (EW×EP), accordingly.

Egg characteristics

At the end of the experiment (week 42), ten eggs per treatment (two eggs per replicate) were randomly chosen, examined and measured for eggshell thickness (mm), eggshell strength (kgf/cm^2), albumen (mm), Haugh unit scores and eggshell weight (g). Eggshell thickness (without inner and outer shell membranes) was measured at 3 different points (top, middle, and bottom) using an ultrasonic micrometer (Sanovo Technology A/S, Odense NV, Denmark) without cracking the eggshell. Likewise, eggshell thickness was calculated from the middle of the egg using a micrometer instrument (Norouzi et al., 2013). Albumen (height) was measured using standard micrometer (CE 300). The Haugh unit score was calculated using Roush's formula (Roush, 1981). Eggshell weight was measured after depleting internal contents and placing them at room temperature for 48 h by digital pan balance with 0.001 g accuracy.

Serum biochemical parameters

At the end of the experiment (week 42), ten quails per treatment (two quails per pen) were selected, and individual blood samples were collected separately in non-heparinized tubes. The serum was separated, stored at –20°C for the further use and later was analyzed for glucose, cholesterol, LDL, HDL, alanine transaminase (ALT), gamma glutamyltransferase (GGT), alkaline phosphatase (ALP), and triglyceride by enzyme-linked immunosorbent assay technique using commercial kits (Boehringer Mannheim Hitachi 704 automatic analyzer, Tokyo, Japan). The methodology and the set of reagents for each parameter were recommended by the manufacturer of the analyzer system. Treatment-wise means of values were computed and presented.

Intestinal morphology parameters

At the end of the trial, upon obtaining the permission of Ethical Committee of the University, ten birds from each treatment (two quails per replicate) were randomly selected, stunned and killed by cutting the jugular vein. The abdominal cavity was then opened and the digestive tract with contents was removed aseptically and each carcass was subjected to detailed postmortem examination. Ileum tissue samples were separated from the Meckel's diverticulum up to 1 cm proximal to the ileocecal junction and then dried with desiccant paper. A 2-cm section of ileum was taken from the middle of the ileum and gently flushed with phosphate-buffered saline (pH 7.2). Tissue sections were immediately fixed in 10% neutral buffered formalin and changed 3 times to complete the fixation process. A single 0.5 cm sample was cut from each ileal section, dehydrated with increasing concentrations (70%, 80%, 95%, and 100%) of ethanol, cleared with xylene and placed into polyfin embedding wax. Tissue sections (2 μm) were cut by microtome (Leitz-1512 Microtome, Leitz, Wetzlar, Germany), floated onto slides and stained with hematoxylin (Gill no. 2, Sigma, St. Louis, MO, USA) and eosin (Sigma). To measure villus height and crypt depth, images from samples were taken using a digital camera with light microscopy. Twelve images from 4 tissue sections of each ileal section were taken and 24 villus heights and crypt depths were measured by imaging software. Measurements for each villus length were taken from the tip of the villus to the valley and measurements for crypt depth were taken from the valley to the basolateral membrane (Xu et al., 2003). For determining the number of goblet cells in 1mm of villus length, all samples were dehydrated in ethanol and embedded in paraffin wax. Sections were stained with Alcian blue and periodic acid-Schiff to visualize goblet cells (Rohana and Thomas, 2009).

Selected intestinal bacterial population

To determine the changes in some selected microbial populations in the gastrointestinal tract of quails at week 42, cecal contents immediately after killing the quails were gently collected in sterile sampling tube and transferred on ice to the laboratory for microbial study. The cecal contents of each bird were pooled for serial dilution. Microbial populations were determined by serial dilution (10^{-4} to 10^{-6}) of cecal samples in anaerobic diluents before inoculation onto Petri dishes of sterile agar as described by Bryant and Burkey (1953). *E. coli* was grown on eosin methylene blue agar, Salmonella in *Salmonella Shigella* (Merck, Darmstadt, Germany) and Coliforms on McConkey agar (Darmstadt, Germany). *E. coli* was incubated aerobically at 37°C. Plates were counted between 24 and 48 h after inoculation. Colony forming units were defined as distinct colonies measuring at least 1mm in diameter. Then, 9 sterile test tubes with lids containing 9 mL of phosphate buffer solution as diluent were prepared. Approximately 1 g of the cecal contents taken by sterile swab and homogenized for 3 min before transferring to microbiology lab in cold condition and mixed (Bryant and Burkey, 1953). Then 1ml out of 10 mL of buffer plus cecal sample in the test tube was removed by 1,000 μL sampler and was transferred to the tubes and was mixed thoroughly. Similarly, it was transferred to other new tubes and this procedure was repeated until a dilution of 9 was completed. Later, 1ml of contents of each test tube was transferred to one of three selective media agar in petri plates, respectively, and each petri was plate incubated in 37°C for 24 h. Finally, the intestinal bacterial colony populations formed in each plate was counted and adjusted to $\times 10^9$ manually and then was reported.

Statistical analysis

Data were analyzed as a completely randomized design manner using the general linear model procedure of SAS (SAS Institute, 2007) for those traits which were measured only once during the experimental period. MIXED procedure of SAS (SAS Institute, 2007) was used for analysis of traits that were measured repeatedly throughout the experimental period with age considered as a new main factor and initial values considered as a covariate effect in the model. Differences between treatment means were tested using Duncan's multiple comparison tests for main effects and Tukey's test for interactions. Statistical significance was declared at p≤0.05.

RESULTS AND DISCUSSION

Production parameters

Mean EP, EW, FC, and FCR in control, BMD and PFA groups are shown in Table 2. In period I (weeks 37 to 39), EP and FCR values showed a significant (p≤0.05) improvement by addition of PFA, whereas, BMD group showed no significant changes compared with control. However, FCR was significantly similar in all studied treatment groups. In period II (weeks 40 to 42), the values for EP and FCR were significantly (p≤0.05) higher in BMD and PFA fed quails when compared with control group. In both periods, EW and FC values remained statistically non-significant (p≤0.05).

In current experiment, the PFA treated group consumed a similar amount of feed while producing more eggs due to utilization of feed more efficiently than control. In many previous studies—in agreement to our findings—addition of phytogenic/herbal compounds showed performance and FCR improvements (Sahin et al., 2010) and a better economic efficiency in Japanese quails (Christaki et al., 2011), commercial broilers (Tiihonen et al., 2010) and laying hens and broiler breeders (Bozkurt et al., 2012). However, these findings contradict some other reports (Bölükbaşi et al., 2008). In current experiment, EW did not vary among different treatments. These findings are in accordance with those of Bozkurt et al. (2012). In contrast, increased EW of layer hens by addition of essential oils has been reported by Bölükbaşi et al. (2008).

While exploring the current experiment data, it is worth mentioning that in period II, the EP drastically decreased compared with period I. This is also mainly due to the aging of the quails, though many factors like stock density, nutritional manipulation and environmental conditions can be the cause for these changes. González (1995) reported that the quail EP was significantly lowered by advancing quail age.

The PFA product used in this study contains phytomolecules (allicin, carvacrol, paprika, cinnamaldehyde, peppermint, and thymol) and organic acids (propionic acid and fumaric acid). The possible mechanism of action of herbal components might be due to the increased digestibility and improved utilization of feed thanks to their antioxidant properties and phenolic compounds. The intestinal availability of essential nutrients may result in better absorption through stimulating the secretion of pancreatic and endogenous digestive enzymes (Lovkova et al., 2001). In addition, the active substances of essential oils present in PFA had a diversifying power to synergistically improve nutrient utilization. As a result, energy and protein digestibilities are improved leading to optimized performance of laying quails (Sultan et al., 2015). Some other possible ways include increasing feed palatability due to their aromatic characteristics which promote FC, regulating of the gut microbial flora through changing the gastrointestinal bacterial load and modifying mucin biosynthesis (Windisch et al., 2008). On the other hand, organic acids present in PFA may favorably affect the host by lowering pH through inhibiting the pathogenic intestinal bacteria and decreasing the level of their toxic products. Modulation of the gut microbiota has a critical role in maintaining host health (Tollba et al., 2012). Organic acid in combination with essential oil showed beneficial impacts on digestive enzyme activities of both pancreas and intestinal mucosa, leading to increased growth performance of broilers (Jang et al., 2007). In turkeys, a mixture of essential oils with a blend of organic acids increased body weight and decreased FCR (Mikulski et al., 2008). The boosting effect of BMD could be attributed to the active ingredients of bacitracin which can inhibit the synthesis of bacterial cell membrane of gram positive bacteria. BMD can also enhance the antibacterial activity and increase the

Table 2. Effect of dietary feed additives on performance parameters of layer Japanese quail (37 to 42 weeks of age)

Treatments		Egg production* (%)		Egg weight (g)		Feed consumption (g/week)		Feed conversion ratio (g feed/g egg)	
BMD	PFA	37-39 week	40-42 week	37-39 week	40-42 week	37-39 week	40-42 week	37-39 week	40-42 week
0	0	72.73^{bc}	66.14^{c}	11.54	11.74	322.30	323.50	2.46^{a}	2.72^{a}
0.05%	0	77.20^{a}	73.76^{a}	11.93	11.83	321.70	320.33	2.22^{b}	2.30^{b}
0	0.1%	74.47^{b}	71.92^{b}	11.85	11.82	320.25	331.28	2.31^{ab}	2.44^{b}
ANOVA									
SEM		0.525	0.810	0.181	0.136	13.992	17.986	0.037	0.058
p value		≤0.0001	≤0.0001	0.6800	0.9654	0.0721	0.0625	0.0195	0.0040

BMD, bacitracin methylene disalicylate (antibiotic growth promoter); PFA, phytogenic feed additive; ANOVA, analysis of variance; SEM, pooled standard error of column wise means comparison.

* Hen day mass egg production (%).

Feed conversion ratio was calculated as kilograms of feed consumed per kilogram of egg produced.

Means with different letters within the same column are significantly different (p≤0.05).

product stability after combination with bacitracin (Attia et al., 2003).

Egg characteristics

The effect of BMD and PFA supplementation on the egg quality parameters of quails at 42 weeks of age is given in Table 3. Notably, eggshell thickness was not influenced by dietary treatments; whereas, eggshell strength and eggshell weight were increased and albumen and Haugh unit scores were reduced significantly ($p \leq 0.05$) by addition of PFA into the basal diet. Incorporation of BMD also revealed almost similar results with PFA, when compared with control group. PFA was found to be more effective than BMD at the end of trial (week 42).

Previous literature reveals contradictory results about the effects of herbal feed additives/phytogenic components on egg quality characteristics of laying birds. It was reported that Haugh unit scores, shell thickness, eggshell weight, yolk color, yolk weight and albumen and yolk indices were not altered by supplementing essential oils into the daily diet of laying quails (Bozkurt et al., 2012). The achievement on eggshell strength in current study is supported by the findings of Kaya et al. (2013), who reported that plant extract of *Origanum vulgare*, *Thymus vulgaris*, thyme oil, origanum oil, garlic oil, anise oil and fennel oil enhanced eggshell strength and eggshell thickness. However, Bozkurt et al. (2012) reported that essential oils supplementation had no effect on eggshell strength and eggshell thickness of laying hens.

There are several factors influencing the quality characteristics of quail eggs. The improvement of the eggshell strength found in current trial by consumption of PFA in quails could be attributed to the stimulatory role of the essential oils on enzymes secretion and amino acids production, which are required for the formation of eggshell (Nazligul et al., 2001). Digestibility of most amino acids in ileum may increase by addition of essential oils into the diet (Maenner et al., 2011). The other possible reason is the influence of essential oils on the metabolic activity of beneficial bacteria, within the intestine, which positively influences mineral absorption rate, especially those of Ca^{2+} and Mg^{2+} (Roberfroid, 2000).

On the other hand, organic acids may affect the integrity of microbial cell membranes and interfere with nutrient transport and energy metabolism (Hedayati et al., 2014). It is also believed that lactic acid may decrease pathogenic microorganisms in the crop leading to less competition for nutrients between quails and bacteria. In addition, there is a report stating the lactic acid is palatable for birds which increases FC leading to enhance egg quality characteristics (Laury et al., 2009). More so, it is presumed that the organic acid mixture contributes to the production of longer and wider quail eggs with thicker shells, larger surface areas and increased breaking strength in quails.

Serum biochemical parameters

The effect of BMD and PFA feed supplements on the blood biochemical parameters of laying quails are presented in Table 4. It was found that PFA addition significantly ($p \leq 0.05$) decreased cholesterol, LDL, HDL, ALP, and triglyceride. By contrast, increased GGT levels were observed by inclusion of PFA with no impact on ALT levels, compared with control group. Likewise, the BMD also showed less effectiveness than PFA on all above-mentioned parameters. The glucose content was significantly ($p \leq 0.05$) decreased in BMD group and remained unchanged among PFA and control treatments, but it was significantly higher in PFA than control and BMD fed groups.

In many studies, it is reported that essential oils from different plant origins have increased serum triglyceride, total cholesterol and glucose of Japanese quail (Soltan et al., 2008; Khaksar et al. 2012). In contrary, there is evidence of reduced serum cholesterol, plasma triglyceride and phospholipids in birds with the incorporation of plant derivatives into the diets (Babazadeh et al., 2011).

Previous reports have demonstrated the antioxidant and antimicrobial activity roles of PFA. Furthermore, other effects such as anti-inflammatory, anti-fungal, anti-infectious and anti-toxigenic have been reported in many studies (Swiatkiewicz and Arczewska-Wlosek, 2012). Blood metabolites indices of liver, renal functions and hematological contents may provide good instances of

Table 3. Effect of dietary feed additives on egg characteristics of layer Japanese quail at 42 weeks of age

Treatments		Eggshell thickness	Eggshell strength	Albumen height	Haugh unit	Eggshell weight (g)
BMD	PFA	(mm)	(kgf/cm^2)	(mm)		
0	0	0.20	2.10^b	4.91^a	91.33^a	1.08^b
0.05%	0	0.20	2.26^a	4.24^c	86.10^c	1.14^a
0	0.1%	0.22	2.28^a	4.46^b	89.33^b	1.13^a
ANOVA						
SEM		0.010	0.26	0.79	0.552	0.10
p value		0.4413	0.0132	≤0.0001	≤0.0001	0.0067

BMD, bacitracin methylene disalicylate (antibiotic growth promoter); PFA, phytogenic feed additive; ANOVA, analysis of variance; SEM, pooled standard error of column wise means comparison.

Means with different letters within the same column are significantly different ($p \leq 0.05$).

Table 4. Effect of dietary feed additives on serum biochemical of layer Japanese quail at 42 weeks of age

Treatments		Glucose	Cholesterol	LDL	HDL	ALT	GGT	ALP	Triglyceride
BMD	PFA	(mg/dL)	(mg/dL)	(mg/dL)	(mg/dL)	(Iu/L)	(Iu/L)	(U/L)	(mg/dL)
0	0	304.29[a]	356.04[a]	115.45[a]	99.33[a]	2.01[a]	264.13[b]	3,660.67[a]	382.32[a]
0.05%	0	293.50[b]	224.52[b]	98.20[b]	96.78[b]	2.01[a]	300.33[a]	2,997.04[b]	293.11[b]
0	0.1%	304.65[a]	218.24[c]	84.29[c]	90.81[c]	1.95[b]	253.66[c]	2,486.83[c]	281.67[c]
ANOVA									
SEM		1.26	6.25	3.09	0.97	0.01	3.20	116.81	3.42
p value		≤0.0001	≤0.0001	≤0.0001	≤0.0001	0.0032	≤0.0001	≤0.0001	≤0.0001

LDL, low-density lipoprotein; HDL, high-density lipoprotein; ALT, alanine aminotransferase; GGT, gamma-glutamyltransferase; ALP, alkaline phosphatase; BMD, bacitracin methylene disalicylate (antibiotic growth promoter); PFA, phytogenic feed additive; ANOVA, analysis of variance; SEM, pooled standard error of column wise means comparison.
Means with different letters within the same column are significantly different (p≤0.05).

animal health status and physiological condition (Toghyani et al., 2010). The lipophilic property and chemical structure of phenolic compounds present in PFA could play a role in manipulation of the enzyme activities inside the body of broilers (Attia et al., 2003). The lower serum cholesterol content may be the result of an increased lipid digestibility due to a higher secretion of bile and digestive enzymes indicating enhanced nutrient supply and transport (Manafi, 2015). The increased nutrient supply for growth is reflected in enhanced nutrient transport in the blood. Moreover, the PFA could cause a circulatory enzyme elevation. Generally GGT and ALT are considered as liver enzymes which are increased at the time of liver damage (heptato-cellular degeneration), so the decrease in these enzymes may provide proof for the occurrence of hepato-protective effect of essential oils present in PFA. Brenes and Roura (2010) have reported that serum total protein, albumin concentrations and AST activities significantly enhanced in laying hens receiving organic acids but lipid metabolism markers (cholesterol, HDL, LDL, triglyceride, and total lipid concentrations) remained unchanged. The glucose level in blood samples is a biochemical indicator of stress that decreases in quails fed PFA compared with BMD in current study and could be justified by the stress-lowering effect of this feed additive (Yesilbag and Coplan, 2006).

Morphology of intestine

This study revealed a significant (p≤0.05) increase in illeal villus height and villus height to crypt depth ratio (VH/CD), decrease in crypt depth and number of goblet cell in 1 mm of villus height in quails fed both PFA and BMD, compared to respected control groups (Table 5). The variation between the data of different treatments and studied parameters are more evident in PFA fed group.

The findings of present experiment in intestinal morphology parameters are in consistence with reports of Adibmoradi et al. (2006), who stated that jejunal villus height was increased and crypt depth was decreased leading to increased villus height to crypt depth ratio in birds fed graded levels of garlic in the diet. In contrast, some reports stating that the addition of essential oils did not affect villus height (Jamroz et al., 2006; Reisinger et al., 2011). Moreover, Garcia et al. (2007) reported an increase in crypt depth when essential oil was supplemented into the broiler diet. In contrast to our results, inclusion of essential oils in the diet of broilers increased goblet cell numbers reported by Garcia et al. (2007), Reisinger et al. (2011).

Villus height and crypt depth are considered as indicators of a well-functioning intestine. Increased villus height provides a bigger surface area for nutrients absorption and therefore performance enhancement (Awad et al., 2008). On the other hand, a drop in villus height can

Table 5. Effect of dietary feed additives on morphology of intestine of layer Japanese quail at 42 weeks of age

Treatments		Villus height	Number of goblet cells[1]	Crypt depth	Villus height to crypt
BMD	PFA	(μm)		(μm)	depth ratio
0	0	5.14[b]	10.19[a]	0.96[a]	5.34[c]
0.05%	0	6.48[a]	9.29[b]	0.76[b]	8.51[a]
0	0.1%	6.16[a]	9.38[b]	0.79[b]	8.81[b]
ANOVA					
SEM		0.155	0.104	0.024	0.034
p value		≤0.0001	≤0.0001	≤0.0001	≤0.0001

BMD, bacitracin methylene disalicylate (antibiotic growth promoter); PFA, phytogenic feed additive; ANOVA, analysis of variance; SEM, pooled standard error of column wise means comparison.
[1] Number of goblet cells in each 1 mm of villus length.
Means with different letters within the same column are significantly different (p≤0.05).

reduce nutrient absorption by decreasing the intestinal surface area for absorption. Hence, reduction in nutrient absorption will lead to decreased resistance to diseases and diminished performance. Escalation in secretion of digestive tract is the negative consequence of deeper crypt and shorter villi (Xu et al., 2003). Goblet cells are liable for the production of intestinal mucins that are the major component of the mucus layer coats of the broilers intestine. Many scientists believe that essential oils have the ability to reduce the growth of *E. coli* and *C. perfringens* and increase the numbers of *Lactobacillus* spp. when fed to birds (Brenes and Roura, 2010).

The PFA products stimulate gut hormones exerting diverse actions in the gastro intestinal tract leading to intestinal growth, enhancing mucosal blood flow and most importantly nutrient assimilation and thereby changing FC pattern by providing a feedback signal to the brain (Guban et al., 2006). On the subject of the intestinal morphology, normally, heavier broilers are linked with longer villi, greater villus width and higher villus surface area (Adibmoradi et al., 2006). It has been proposed that longer villi would result in an increased surface area and higher absorption of nutrients (Incharoen et al., 2010). This higher absorptive ability of the intestine in PFA fed birds is supported by higher blood nutrient concentrations. According to Yanishlieva et al. (1999), essential oils seem to have positive effects on blocking the radical chain process by interaction with peroxide radicals. Organic acids have direct effects on the poultry performance. In study of Antongiovanni et al. (2007) inclusion of organic acids in broiler diets increased the weight and length of small intestine. On the contrary, Vieira et al. (2008) reported that inclusion of some organic acids did not significantly influence the villus height at any age of broiler's life. Sultan et al. (2015) also demonstrated that organic acids improve the protein digestibility by decreasing endogenous nitrogen loss and by producing ammonia which ultimately lead to better intestinal morphological properties of birds.

Selected intestinal bacterial population

The bacterial population of intestine of quails fed BMD and PFA is shown in Table 6. Results showed that addition of BMD and PFA significantly ($p \leq 0.05$) lowered the cecal number of coliforms, *Salmonella*, and *E. coli*, with more reduction emphasis in PFA fed group. Addition of PFA decreased the population of cecal bacterial count of coliforms, *Salmonella*, and *E. coli* significantly when compared with control. At present, there are prominent studies indicating that essential oils could control the common intestinal pathogen growth of poultry (Dorman and Deans, 2000). Addition of thyme essential oil in Japanese quail diet significantly ($p \leq 0.05$) increased the number of Lactobacillus, decreased *E. coli* population in the

Table 6. Effect of dietary feed additives on intestinal bacterial count ($\times 10^9$) of layer Japanese quail at 42 weeks of age

Treatments		Coliforms	Salmonella	*E. coli*
BMD	PFA			
0	0	3.55^a	7.15^a	3.25^a
0.05%	0	2.31^b	4.25^c	2.12^b
0	0.1%	2.43^b	5.42^b	2.22^b
ANOVA				
SEM		0.155	0.292	0.126
p value		≤0.0001	≤0.0001	≤0.0001

BMD, bacitracin methylene disalicylate (antibiotic growth promoter); PFA, phytogenic feed additive; ANOVA, analysis of variance; SEM, pooled standard error of column wise means comparison.
Means with different letters within the same column are significantly different ($p \leq 0.05$).

ileum and inhibited the growth of *Salmonella typhimurium* due to its antimicrobial properties (Dorman and Deans, 2000). Siragusa et al. (2008) reported that addition of Humuluslupulus decreased *C. perfringens* population in the broiler intestine. A blend of essential oils containing carvacrol and thymol is reported to have the ability to decrease *E. coli* and *C. perfringens* in broilers (Jang et al., 2007). In study of Jamroz et al. (2006), a significant reduction in *E. coli* numbers was reported following addition of natural plant extract in broiler diet. However, there are a few studies conversely reporting that dietary essential oils had no effect on the intestinal microflora populations (Hermans et al., 2011). Organic acids including short chain fatty acids (SCFAs) such as acetate, propionate and butyrate have been found to control *Salmonella enteritidis*, showing growth promoting impact on the beneficial intestinal microflora (Hansen et al., 1997). Positive beneficial effects of organic acids on production performance and carcass traits of broilers were reported by Leeson et al. (2005) and Antongiovanni et al. (2007).

Poultry caeca have a range of microorganisms with strong impacts on the performance and the health of broilers. Therefore, the aim of PFA addition into the diet is to create a favorable gut microflora by decreasing pathogenic bacteria (Apajalahti et al., 2001). It is assumed that gut microflora reduces nutrients availability to host animal by enforcing the intestinal cell turnover and thereby increasing the intestinal requirement for nutrients. Moreover, intestinal microflora and epithelial cells have to compete for nutrients (Dibner and Richards, 2005). A higher absorptive capacity in the intestine of PFA fed animals is also supported by higher blood nutrient concentrations of those quails, as observed in the present study. However, many factors influence the effectiveness of dietary supplementation of phyto-additives such as dose of extracts in the mucus layer in the intestine (Hermans et al., 2011). More so, fatty acids cause changes in intracellular pH of intestine. Butyric acid is considered the prime enterocytes energy source necessary for the correct development of the gut associated lymphoid tissue (Antongiovanni et al., 2007). In conclusion, the outcomes of current study revealed that this novel eubiotic feed additive having phytomolecules and organic acids as main ingredients, at an inclusion level of 0.1%, did improve the performance, most of studied egg characteristics, serum biochemical parameters, immunity, morphology of intestine and reduced population of harmful bacteria in quails. It can be concluded that addition of combined bioactive components of herbal plants along with organic acid as PFA at the dosage of 0.1% can be used as an alternative to antibiotic growth promoters in laying Japanese quails.

ACKNOWLEDGMENTS

This study was financially supported by Director of Research, Malayer University, Malayer, Iran. The authors would like to thank Mr. Reza Soleymani, laboratory staff of Department of Animal Science, Malayer University Iran for his skillful technical assistance.

REFERENCES

Adibmoradi, M., B. Navidshad, J. Seifdavati, and M. Royan. 2006. Effect of dietary garlic meal on histological structure of small intestine in broiler chickens. J. Poult. Sci. 43:378-383.

Antongiovanni, M., A. Buccioni, F. Petacchi, S. Leeson, S. Minieri, A. Martini, and R. Cecchi. 2007. Butyric acid glycerides in the diet of broiler chickens: effects on gut histology and carcass composition. Ital. J. Anim. Sci. 6:19-25.

Apajalahti, J. H. A., H. Kettunen, M. Bedford, and W. E. Holben. 2001. Percent G+C profiling accurately reveals diet-related differences in the gastrointestinal microbial community of broiler chickens. Appl. Environ. Microbiol. 67:5656-5667.

Attia, Y. A., A. E. Tag El-Din, H. S. Zeweil, and M. A. Arafat. 2003. Nutritional evaluation of nigella seed meal and the effect of microbial phytase and amino acids supplementations on its feeding value for Japanese quail. Egypt. J. Nutr. Feeds 6:201-217.

Awad, W., K. Ghareeb, and J. Böhm. 2008. Intestinal structure and function of broiler chickens on diets supplemented with a synbiotic containing *Enterococcus faecium* and oligosaccharides. Int. J. Mol. Sci. 9:2205-2216.

Babazadeh, D., T. Vahdatpour, H., Nikpiran, M. A. Jafargholipour, and S. Vahdatpour. 2011. Effects of probiotic, prebiotic and synbiotic intake on blood enzymes and performance of Japanese quails (*Coturnix japonica*). Ind. J. Anim. Sci. 81:870-874.

Bölükbaşi, Ş. C., M. K. Erhan, and Ö. Kaynar. 2008. The effect of feeding thyme, sage and rosemary oil on laying hen performance, cholesterol and some proteins ratio of egg yolk and *Escherichia coli* count in feces. Arch Geflugelk. 72:231-237.

Bozkurt, M., K. Küçükyilmaz, A. U. Çatli, M. Çinar, E. Bintas, and F. Çöven. 2012. Performance, egg quality, and immune response of laying hens fed diets supplemented with mannan-oligosaccharide or an essential oil mixture under moderate and hot environmental conditions. Poult. Sci. 91:1379-1386.

Brenes, A. and E. Roura. 2010. Essential oils in poultry nutrition: Main effects and modes of action. Anim. Feed Sci. Technol. 158:1-14.

Bryant, M. P. and L. A. Burkey. 1953. Cultural methods and some characteristics of some of the more numerous groups of bacteria in the bovine rumen. J. Dairy Sci. 36:205-217.

Christaki, E. V., E. M. Bonos, and P. C. Florou-Paneri. 2011. Use of anise seed and/or α-tocopheryl acetate in laying Japanese quail diets. S. Afr. J. Anim. Sci. 41:126-133.

Dibner, J. J. and J. D. Richards. 2005. Antibiotic growth promoters in agriculture: history and mode of action. Poult. Sci. 84:634-643.

Dorman, H. J. and S. G. Deans. 2000. Antimicrobial agents from plants: Antibacterial activity of plant volatile oils. J. Appl. Microbiol. 88:308-316.

Garcia, V., P. Catala-Gregori, F. Hernandez, M. D. Megias, and J. Madrid. 2007. Effect of formic acid and plant extracts on growth, nutrient digestibility, intestine mucosa morphology and meat yield of broilers. J. Appl. Poult. Res. 16:555-562.

González, M. 1995. Influence of age on physical traits of Japanese quail (*Coturnixcoturnix japonica*) eggs. Ann. Zootec. 44:307-312.

Grashorn, M. A. 2010. Use of phytobiotics in broiler nutrition – an alternative to in feed antibiotics? J. Anim. Feed Sci. 19:338-347.

Guban, J., D. R. Korver, G. E. Allison, and G. W. Tannock. 2006. Relationship of dietary antimicrobial drug administration with broiler performance, decreased population levels of *Lactobacillus salivarius*, and reduced bile salt deconjugation in the ileum of broiler chickens. Poult. Sci. 85:2186-2194.

Hansen, L. L., A. E. Larsen, B. B. Jensen, and J. Hansen-Moller. 1997. Short time effect of zinc bacitracin and heavy fouling with faeces plus urine on boar taint. J. Anim. Sci. 64:351-363.

Hedayati, M., M. Manafi, M. Yari, and A. Avara. 2014. The influence of an Acidifier feed additive on biochemical parameters and immune response of broilers. Annu. Res. Rev. Biol. 4:1637-1645.

Hermans, D., A. Martel, K. van Deun, F. van Immerseel, M. Heyndrickx, F. Haesebrouck, and F. Pasmans. 2011. The cinnamon-oil ingredient trans-cinnamaldehyde fails to target *Campylobacter jejuni* strain KC 40 in the broiler chicken cecum despite marked *in vitro* activity. J. Food Prot. 74:1729-1734.

Incharoen, T., K. Yamauchi, T. Erikawa, and H. Gotoh. 2010. Histology of intestinal villi and epithelial cells in chickens fed low-protein or low-crude fat diets. Ital. J. Anim. Sci. 9:e82.

Jamroz, D., T. Wertelecki, M. Houszka, and C. Kamel. 2006. Influence of diet type on the inclusion of plant origin active substances on morphological and histochemical characteristics of the stomach and jejunum walls in chicken. J. Anim. Physiol. Anim. Nutr. 90:255-268.

Jang I. S., Y. H. Ko, S. Y. Kang, and C. Y. Lee. 2007. Effect of commercial essential oils on growth performance, digestive enzyme activity, and intestinal microflora population in broiler chickens. Anim. Feed Sci. Technol. 134:304-315.

Kaya, A., H. Kaya, M. Macit, S. Celebi, N. Esenbuga, M. A. Yoruk, and M. Karaoglu. 2013. Effects of dietary inclusion of plant extract mixture and copper into layer diets on egg yield and quality, yolk cholesterol and fatty acid composition. Kafkas Univ. Vet. Fak. Derg. 19:673-679.

Khaksar, V., M. van Krimpen, H. Hashemipour, and M. Pilevar. 2012. Effects of thyme essential oil on performance, some blood parameters and ilealmicroflora of Japanese quail. J. Poult. Sci. 49:106-110.

Laury, A. M., M. V. Alvarado, G. Nace, C. Z. Alvarado, J. C. Brooks, A. Echeverry, and M. M. Brashears. 2009. Validation of a lactic acid- and citric acid-based antimicrobial product for the reduction of *Escherichia coli* O157: H7 and *Salmonella* on beef tips and whole chicken carcasses. J. Food Prot. 72:2208-2211.

Leeson, S., H. Namkung, M. Antongiovanni, and E. H. Lee. 2005. Effect of butyric acid on the performance and carcass yield of broiler chickens. Poult. Sci. 84:1418-1422.

Lovkova, M. Y., G. N. Buzuk, S. M. Sokolova, and N. I. Kliment'eva. 2001. Chemical features of medicinal plants (Review). Appl. Biochem. Microbiol. 37:229-237.

Maenner, K., W. Vahjen, and O. Simon. 2011. Studies on the effects of essential-oil-based feed additives on performance, ileal nutrient digestibility, and selected bacterial groups in the gastrointestinal tract of piglets. J. Anim. Sci. 89:2106-2112.

Manafi, M. 2015. Comparison study of a natural non-antibiotic growth promoter and a commercial probiotic on growth performance, immune response and biochemical parameters of broiler chicks. J. Poult. Sci. 52:274-281.

Mikulski, D., Z. Zduńczyk, J. Jankowski, and J. Juśkiewicz. 2008. Effects of organic acids or natural plant extracts added to diets for turkeys on growth performance, gastrointestinal tract metabolism and carcass characteristics. J. Anim. Feed Sci. 17:233-246.

Nazligul, A., K. Turkyilmaz, and H. E. Bardakçioglu. 2001. A study on some production traits and egg quality characteristics of Japanese quail. Turk. J. Vet. Anim. Sci. 25:1007-1013.

Norouzi, S., A. Yaghobfar, M. Shokrpour, and A. Safamehr. 2013. Determination of nutritive value and effect of different levels of prosopis juliflora pods on performance of laying hens. Res. Anim. Prod. 4:62-77.

Reisinger, N., T. Steiner, S. Nitsch, G. Schatzmayr, and T. J. Applegate. 2011. Effects of a blend of essential oils on broiler performance and intestinal morphology during coccidial vaccine exposure. J. Appl. Poult. Res. 20:272-283.

Roberfroid, M. B. 2000. Prebiotics and probiotics: are they functional foods? Am. J. Clin. Nutr. 71:1682S-1687S.

Rohana, A. and F. Thomas. 2009. Thermotolerance-induced goblet cell activity confers protection in post-operative gut barrier dysfunction. Int. J. Surg. 7:237-242.

Roush, W. B. 1981. TI 159 calculator program for Haugh unit calculation. Poult. Sci. 60:1086-1088.

Sahin, K., C. Orhan, M. Tuzcu, S. Ali, N. Sahin, and A. Hayirli. 2010. Epigallocatechin-3-gallate prevents lipid peroxidation and enhances antioxidant defense system via modulating hepatic nuclear transcription factors in heat-stressed quails. Poult. Sci. 89:2251-2258.

SAS Institute. 2007. SAS/STAT User's Guide. Version 9.2 edn. SAS Institute Inc., Cary, NC. USA.

Siragusa, G. R., G. J. Haas, P. D. Matthews, R. J. Smith, R. J. Buhr, N. M. Dale, and M. G. Wise. 2008. Antimicrobial activity of lupulone against *Clostridium perfringens* in the chicken intestinal tract jejunum and caecum. J. Antimicrob. Chemother. 61:853-858.

Soltan, M. A., R. S. Shewita, and M. I. El-Katcha. 2008. Effect of dietary anise seeds supplementation on growth performance, immune response, carcass traits and some blood parameters of broiler chickens. Int. J. Poult. Sci. 7:1078-1088.

Stanaćev, V., D. Glamočić, N. Milošević, N. Puvača, V. Stanaćev, and N. Plavša. 2011. Effect of garlic (*Allium sativum* L.) in fattening chicks nutrition. Afr. J. Agric. Res. 6:943-948.

Sultan, A., T. Ullah, S. Khan, and R. U. Khan. 2015. Effect of organic acid supplementation on the performance and ileal microflora of broiler during finishing period. Pak. J. Zool. 47:635-639.

Swiatkiewicz, S. and A. Arczewska-Wlosek. 2012. Prebiotic fructans and organic acids as feed additives improving mineral availability. World's Poult. Sci. J. 68:269-279.

Tiihonen, K., H. Kettunen, M. H. Bento, M. Saarinen, S. Lahtinen, A. C. Ouwehand, H. Schulze, and N. Rautonen. 2010. The effect of feeding essential oils on broiler performance and gut microbiota. Br. Poult. Sci. 51:381-392.

Toghyani, M., M. Toghyani, A. Gheisari, G. Ghalamkari, and M. Mohammadrezaei. 2010. Growth performance, serum biochemistry and blood hematology of broiler chicks fed different levels of black seed *(Nigella sativa)* and peppermint (*Mentha piperita*). Livest. Sci. 129:173-178.

Tollba, A. A. H., S. A. M. Shabaan, and M. A. A. Abdel-Mageed. 2012. Effects of using aromatic herbal extract and blended with organic acids on productive and physiological performance of poultry. Egypt. Poult. Sci. J. 30:229-248.

Vieira, S. L., O. A. Oyarzabal, D. M. Freitas, J. Berres, J. E. M. Peña, C. A. Torres, and J. L. B. Coneglian. 2008. Performance of broilers fed diets supplemented with sanguinarine-like alkaloids and organic acids. J. Appl. Poult. Res. 17:128-133.

Windisch, W., K. Schedle, C. Plitzner, and A. Kroismayr. 2008. Use of phytogenic products as feed additives for swine and poultry. J. Anim. Sci. 86:140-148.

Xu, Z. R., C. H. Hu, M. S. Xia, X. A. Zhan, and M. Q. Wang. 2003. Effects of dietary fructooligosaccharide on digestive enzyme activities, intestinal microflora and morphology of male broilers. Poult. Sci. 82:1030-1036.

Yanishlieva, N. V, E. M. Marinova, M. H. Gordon, and V. G. Raneva. 1999. Antioxidant activity and mechanism of action of thymol and carvacrol in two lipid systems. Food Chem. 64:59-66.

Yesilbag, D. and I. Çolpan. 2006. Effects of organic acid supplemented diets on growth performance, egg production and quality and on serum parameters in laying hens. Rev. Med. Vet. 157:280-284.

19

Effects of Dietary Glucose on Serum Estrogen Levels and Onset of Puberty in Gilts

Fangfang Li, Yujing Zhu, Lan Ding, and Yong Zhang*
Department of Animal Nutrition, Faculty of Animal Sciences and Veterinary,
Shenyang Agricultural University, Shenyang 110866, China

ABSTRACT: Metabolic signals and the state of energy reserves have been shown to play a crucial role in the regulation of reproductive function. This study was carried out to investigate the effects of dietary glucose levels on puberty onset in gilts. Weight-matched, landrace gilts (n = 36) 162±3 days old, weighing about 71.05±4.53 kg, were randomly assigned to 3 dietary treatment groups of 12 gilts each. The trial lasted until the onset of puberty. Gilts in each group were supplied with diets containing different levels of glucose as follows: i) starch group (SG) was free of glucose, contained 64% corn derived starch; ii) low-dose group (LDG) contained 19.2% glucose and 44.8% corn derived starch; iii) high-dose group (HDG) contained 30% glucose and 30% corn derived starch. Results indicated: i) The growth performance of gilts were not affected by the addition of glucose, but the age of puberty onset was advanced significantly ($p<0.05$); ii) Compared with the SG, the concentration of insulin significantly increased before puberty in HDG ($p<0.05$); iii) There was no difference in serum progesterone (P) levels amongst the different feed groups, however, levels of estradiol (E_2), luteinizing hormone, and follicle-stimulating hormone were significantly higher at puberty onset in HDG ($p<0.05$). Overall, our findings indicate that glucose supplementation significantly advances puberty onset, which can have practical purposes for commercial breeding.
(**Key Words:** Glucose, Gilt, Growth Performance, Age of Puberty, Estrogen)

INTRODUCTION

In recent years, the advancement of pubertal onset in gilts due to diet composition has received increased attention. Advancing pubertal onset by 30 days also advances the age of breeding, which saves feed and resources. When animals reach a certain stage, the sensitivity of hypothalamus to gonadal steroid hormones is reduced; this is coupled with increased synthesis and secretion of gonadotropin-releasing hormone (GnRH) (Fauser, 2003; Plant et al., 2004). As the sensitivity of pituitary and gonads to GnRH is elevated, increased secretion of luteinizing hormone (LH) and follicle-stimulating hormone (FSH) then promotes the synthesis of progesterone (P) and estradiol (E_2), resulting in the start of estrus (Jin and Yang, 2014).

* Corresponding Author: Yong Zhang.
E-mail: syau2011@163.com

Insulin, a hormone involved in glucose signaling, can function on hypothalamus directly and transmit signals regarding energy metabolism, which thereby regulates breeding performance of the animals. *In vitro*, it has been identified that insulin can enhance hypothalamic GnRH activity and mRNA expression, through the early growth response-1 protein (DiVall et al., 2007). Additionally, in pituitary cells cultured *in vitro*, insulin can promote the release of LH and FSH. Lastly, insulin can act on the ovary to affect the development of follicle by promoting ovulation and reducing follicular atresia (Castellano et al., 2009). Carbohydrates, proteins, and fats are the major dietary energy sources (Zhou et al., 2010). Yet glucose (composing carbohydrates), above all, contributes to the secretion of serum insulin and is important for insulin signaling. In this way, the practice of using fats as energy, instead of carbohydrates, may decrease reproductive performance in lactating sows through the absence of the effect of insulin (Hansen et al., 2014).

Glucose molecules, by and large, are absorbed by

intestinal epithelial cells to provide energy for animals, while excess glucose is stored as hepatic glycogen or converted to other storage forms. Diet energy sources are mainly starch and fat, however, the role of glucose as the third energy source has still only occasionally been reported. In this study, feeds supplemented with different levels of glucose, starch, and fat were utilized to investigate whether glucose supplementation in diets affects growth, the age of puberty onset, and the production of serum estrogen in gilts.

MATERIALS AND METHODS

Diets and feed manufacture

Feeds used were commercial-type corn-soybean meal diet formulated to meet nutritional requirements of growing gilts as recommended by the NRC (1998) and were provided by Liaoning Wellhope Agri-Tech Co., Ltd. Each group was fed with feeds consisting of the same energy level, but from different energy sources: i) starch group (SG) was free of glucose, contained 64% corn derived starch; ii) low-dose group (LDG) contained 19.2% glucose and 44.8% corn derived starch; iii) high-dose group (HDG) contained 30% glucose and 30% corn derived starch. The diet composition and nutrients level are shown in Table 1. Gilts were allowed access to feed and water *ad libitum*. The trial lasted 60 days until the onset of puberty.

Gilt experiment

The animal care and use protocol was reviewed and approved by the Animal Care and Use Committee, Shenyang Agricultural University. Thirty six Landrace gilts at age of 162±3 days, weighing about 71.05±4.53 kg, were randomly divided into four groups, with 12 animals in each group. After 3 days of feed adaptation, gilts were weighed. The animals were weighed again at the end of the trial. Average daily gain (ADG), average daily feed intake (ADFI), and feed/gain ratio (F/G) were recorded daily. As an endogenous indicator, AIA was used for determining nutrient digestibility. Five gilts of similar condition were selected from each group and about 200 g of fecal sample from gilt was collected at 06:00 to 08:00 am on day 57, 58, and 59 respectively, and 20 mL 10% tartaric acid solution was added to fecal samples to avoid ammonia volatilization. Each fecal sample was sealed and frozen for nutrient digestibility analysis. Blood samples were collected from the superior vena cava on the day prior to the feeding trial, and on day 8, day 20 (prepubescent) after feeding, and the day of puberty onset. After incubating at room temperature for 15 min, blood samples were centrifuged at 2,000 rpm for 20 min to obtain serum.

Table 1. Composition and nutrient level of the test diets (air dry basis)

Items	SG[1]	LDG[2]	HDG[3]
Ingredients			
Corn (%)	64.41	42.63	28.11
Soybean meal (%)	11.24	15.27	17.95
Wheat middlings (%)	15.00	15.00	15.00
DDGS (%)	5.00	5.00	5.00
Limestone (%)	1.17	1.42	1.36
Premix[4]	0.40	0.40	0.40
Salt (%)	0.60	0.60	0.50
Choline chloride 50%	0.10	0.10	0.10
Calcium hydrogen phosphate (%)	2.08	1.58	1.58
Glucose	0.00	18.00	30.00
Total (%)	100	100	100
Nutrient levels			
CP (%)	14.00	14.00	14.00
Degesitable energy (MJ/kg)[5]	14.89	14.91	14.92
EE (%)	4.12	3.20	2.59
Crude ash (%)	6.07	5.92	5.79
C-fiber (%)	3.79	3.59	3.46
Ca (%)	1.00	1.00	0.98
TP (%)	0.76	0.65	0.62
Lys	0.97	0.97	0.97
Met+Cys	0.43	0.43	0.43

DDGS, distillers dried grains with soluble elements; CP, crude protein; EE, ether extract; TP, total phosphorus.

[1] The starch group (SG) was free of glucose, but containing 64% corn derived starch.

[2] The low-dose group (LDG) containing 19.2% glucose and 44.8% corn derived starch.

[3] The high-dose group (HDG) containing 30% glucose and 30% corn derived starch.

[4] Provided per kg of diet: vitamin A, 240,000 IU; vitamin B_1, 30 mg; vitamin B_2, 120 mg; vitamin B_6, 60 mg; vitamin B_{12}, 360 mg; vitamin D_3, 60,000 IU; vitamin E, 720 IU; vitamin K_3, 30 mg; biotin, 0.1 mg; folic acid, 6 mg; D-pantothenic acid, 300 mg; nicotinic acid, 600 mg; Cu (copper sulfate), 4 g; Fe (as ferrous sulfate), 4 g; Mn (manganese sulfate), 1.0 g; Zn (zinc sulfate), 2.5 g; I (potassium iodide), 2 mg, Ca (calcium), 150 g, Se (sodium selenite), 6 mg; P (phosphorus), 300 g; sodium chloride 48 g.

[5] DE was a calculated value.

Other nutrient levels were measured values.

Genital observation was used to identify puberty onset in gilts during the trial period. The gilts were observed from 8:00 to 10:00 am and 15:00 to 17:00 pm daily. Onset of puberty and time of standing heat were determined by an experienced stockperson based on behavioral and vulval characteristics. Behaviorally, standing still under applied back pressure was used as criterion to establish onset of estrus.

ADG, ADFI, and F/G were determined and used as parameters of growth performance. Blood glucose was tested using a kit purchased from Beijing Beihuan Kangtai Clinical Reagents Co., Ltd. Insulin, glucagon, E_2, and P were detected by the Beijing Biotechnology Institute. Nutrient utilization was determined using the acid insoluble ash (AIA) endogenous indicator. The content of AIA in fecal samples was determined and the apparent digestibility of crude protein and crude fat was calculated.

Statistics analyses

The SPSS 14.0 software was used to determine the significance of the experimental data and the least significant difference method was used for multiple comparisons. Difference significance was taken at $p<0.05$. Results are expressed as mean±standard deviation (SD).

RESULTS

As shown in Table 2, the ADG in glucose supplemented groups has no different with other groups ($p>0.05$). In LDG, the ADG was 12.7% higher than the SG ($p<0.05$). The difference of feed conversion between groups was not significant ($p>0.05$).

There was no significant difference in apparent protein digestibility between groups ($p>0.05$), the digestibility of nitrogen free extract (NFE) and the apparent digestibility of crude fat were not significant different amongst the groups ($p>0.05$) (Table 3).

Puberty onset of gilts in LDG was significantly advanced compared to glucose free groups ($p<0.05$). Notably, puberty onset in the HDG was 26 days earlier compared to the SG, and 19 days earlier compared to the LDG (Table 4) ($p<0.01$). Significant changes in the body weight of pubertal gilts was not observed ($p>0.05$).

Table 5 shows that serum glucose concentration in glucose supplemented groups did not increase. The insulin level of gilts in HDG was extremely elevated compared to the other groups before puberty ($p<0.05$). However, no difference was observed among groups at the onset of puberty ($p>0.05$). Serum P and glucagon concentrations of gilts were not affected ($p>0.05$). The E_2, LH, and FSH concentration in HDG was significantly higher than in other groups at the onset of puberty ($p<0.05$), but no differences were discovered before puberty ($p>0.05$).

Table 2. Effects of glucose on growth of gilts

Items	SG	LDG	HDG	p value
Initial weight (kg)	70.26±2.93	67.22±3.29	71.27±2.53	0.44
Final weight (kg)	83.22±2.53	79.00±3.63	82.99±2.04	0.25
ADG (kg)	0.57±0.10	0.51±0.03	0.53±0.05	0.71
ADFI (kg)	1.96±0.21^{a}	2.21±0.14^{c}	2.05±0.31^{b}	0.00
F/G	3.82±0.43	4.47±0.24	4.21±0.39	0.54

SG, starch group; LDG, low-dose group; HDG, high-dose group; ADG, average daily gain; ADFI, average daily feed intake; F/G, feed/gain ratio. In the same row, values with different superscripts differ ($p<0.05$).

DISCUSSION

Palatability is a major factor affecting feed intake and it has been suggested that glucose addition can improve feed palatability. However, feed intake levels of the HDG (containing 30% glucose) declined compared to other groups, indicating that an optimal range of glucose addition may exist. Additionally, body weight gain of animals in the glucose supplemented groups was less than the SG In contrast, the feed conversion ratio acted oppositely, the feed conversion in SG was the lowest, followed by HDG, and LDG. The apparent ability of gilts to digest their diet was not significantly different. This is likely due to the fact that the digestive system of gilts is fully developed at the age of 160 days, resulting in complete digestion of the NFE and fat (Gerrits et al., 2001; Li et al., 2012).

The mammalian reproductive cycle is controlled by the complex interactions of hormones secreted by hypothalamus, pituitary, and gonads. In this study, puberty onset in animals receiving high glucose feed was significantly advanced compared with the other feed composition groups. It has been previously shown that glucose-induced insulin secretion leads to the GnRH secretion from the hypothalamus, increasing LH and FSH secretion and follicular development, thereby affecting pituitary and ovarian function (Jin and Yang, 2014). In a

Table 3. Effects of glucose on digestion

Items	SG	LDG	HDG	p value
AIA (%)	11.05±0.43	10.94±0.35	10.64±0.76	0.67
CP apparent digestibility (%)	74.08±0.14	75.81±0.93	78.50±0.21	0.96
NFE digestibility (%)	88.6±0.68	88.2±0.42	87.9±0.59	0.58
EE apparent digestibility (%)	91.2±0.94	90.9±0.28	87.8±0.84	0.51

SG, starch group; LDG, low-dose group; HDG, high-dose group; AIA, acid insoluble ash; CP, crude protein; NFE, nitrogen free extract; EE, ether extract. In the same row, values with different superscripts differ ($p<0.05$).

Table 4. Effects on the advent of puberty and body weight of gilts after glucose supplementation

Items	SG	LDG	HDG	p value
Age of puberty (d)	222.51±4.89^{b}	215.80±3.52^{b}	196.42±1.84^{a}	0.00
Body weight at onset of puberty (kg)	98.84±2.65	95.41±3.56	102.44±4.63	0.54

SG, starch group; LDG, low-dose group; HDG, high-dose group.
In the same row, values with different superscripts differ ($p<0.05$).

Table 5. Serum parameters of gilts of different ages after glucose supplementation

Items	Age	SG	LDG	HDG	p value
Glucose (mmol/L)	Beforre puberty	5.93±0.27	5.42±0.22	5.38±0.34	0.57
	Puberty	5.77±0.20	5.53±0.39	5.51±0.31	0.06
Insulin (μIU/mL)	Before puberty	6.10±1.99[a]	10.33±2.02[b]	28.66±1.67[c]	0.01
	Puberty	13.78±1.55	12.15±1.91	14.59±1.49	0.59
Glucagon (pg/mL)	Before puberty	155.25±5.74	132.75±10.22	146.48±16.72	0.53
	Puberty	161.89±13.8	155.23±10.2	164.30±10.80	0.83
E_2 (pg/mL)	Before puberty	13.55±1.83	16.91±1.56	18.59±1.15	0.56
	Puberty	25.66±1.52[a]	27.78±2.31[a]	33.62±2.43[b]	0.02
P (pg/mL)	Before puberty	10.34±0.17	9.95±0.15	10.79±0.80	0.71
	Puberty	13.31±0.12	24.04±0.27	13.24±0.56	0.08
LH (ng/mL)	Before puberty	0.21±0.03	0.23±0.02	0.23±0.04	0.91
	Puberty	1.55±0.32[a]	1.61±0.30[a]	1.83±0.26[b]	0.04
FSH (ng/mL)	Before puberty	1.86±0.25	1.83±0.14	1.93±0.30	0.87
	Puberty	2.10±0.41[a]	2.20±0.36[a]	2.61±0.39[b]	0.02

SG, starch group; LDG, low-dose group; HDG, high-dose group; E_2, estradiol; P, progesterone; LH, luteinizing hormone; FSH, follicle-stimulating hormone.
In the same row, values with different superscripts differ ($p<0.05$)

previous study conducted in SD rats, diet supplementation with multivitamin-glucose was added to diets, vaginal opening (an external marker of puberty onset in rodents) occurred earlier, indicating that glucose addition advanced puberty onset in rats (Fan et al., 1997; Matsui et al., 2004), which is similar to our findings in the pig.

The addition of dietary glucose did not affect the levels of serum glucose. This is likely due to the fact that glucose can induce the secretion of insulin and pancreatic glucagon to regulate blood glucose levels within a stable range. This is similar to findings in rats, where it was shown that high-glucose and high-fat diets have no effect on the blood glucose levels (Steger and Rabe, 1997). Glucose and insulin play an important role in the reproductive function of prepubertal gilts (Kemp et al., 1995). Kapelanski et al. (2008) suggested that providing an insulin-promoting diet, via glucose supplementation for sows, was a means of stimulating sexual maturity and inducing insulin cycling. The reports mentioned above are very consistent with our current findings in gilts. Insulin as peripheral nutrition signal, not only affects the hypothalamic - pituitary - gonadal axis, but also signals into the reproductive axis via insulin-like growth factor (IGF), growth hormone (Kemp et al., 1995). At present, there are many theories on the relationship between insulin, and other insulin-like factors in regulating reproduction, but much of the data is contradictory and further studies will need to be performed in order to tease apart these relationships. Additionally, studies investigating the optimal levels and timing of glucose supplementation for estrus induction will also be important to conduct.

Our results indicated that serum E_2, LH, and FSH concentrations were significantly higher upon with glucose supplementation. Previous studies have indicated that IGF-1 can act on ovarian granulosa cells and endometrial cells to produce androgen, which is converted into estrogen via aromatase (Jin and Yang, 2014). A high concentration of E_2 in pubertal gilts is essential for the start of estrus. When gilts got puberty onset, GnRH secretion increased, followed by the secretion of LH and FSH for the start of estrus (Jin and Yang, 2014). P concentrations were lower in adolescent animals, even though the concentration of E_2 increased before the start of estrus, and P content was maintained at low levels (Irwig et al., 2005; Zhou et al., 2014). After uterus, E_2 levels decreased, while concentration of P began to rise.

IMPLICATIONS

In conclusion, current study demonstrated that growth performance of gilts supplemented with glucose in feed was unaffected, but their ADFI improved. Glucose supplementation also significantly advanced puberty onset, significantly increased serum insulin concentrations, slightly affected serum E_2 concentrations, but had no significant effects on nutrient digestibility, and body weight gain. Appropriate supplementation of glucose in diet can promote the reproductive performance of gilts and have practical uses for breeding.

ACKNOWLEDGMENTS

The work was supported by the funding of Chinese NSFC, grant Nos. 31440082 and 31101253.

REFERENCES

Castellano, J. M., V. M. Navarro, J. Roa, R. Pineda, M. A. Sanchez-Garrido, D. Garcia-Galiano, E. Vigo, C. Dieguez, E. Aguilar, L. Pinilla, and M. Tena-Sempere. 2009. Alterations in hypothalamic KiSS-1 system in experimental diabetes: Early changes and functional consequences. Endocrinology 150:784-794.

DiVall, S. A., S. Radovick, and A. Wolfe. 2007. Egr-1 binds the GnRH promoter to mediate the increase in gene expression by insulin. Mol. Cell Endocrinol. 270:64-72.

Fan, B., X. Sun, J. Wang, and X. Zhang. 1997. Effect of dietary energy restriction on reproduction in rats. Wei Sheng Yan Jiu. 26:327-329 (Article in Chinese).

Gerrits, W. J. J., M. J. W. Heetkamp, T. Zandstra, and J. W. Schrama. 2001. Effect of synchronizing dietary protein and glucose supply on nitrogen retention in growing pigs. J. Anim. Sci. 79:S3-S6.

Hansen, A. V., A. B. Strathe, P. K. Theil, and E. Kebreab. 2014. Energy and nutrient deposition and excretion in the reproducing sow: model development and evaluation. J. Anim. Sci. 92:2458-2472.

Irwig, M. S., G. S. Fraley, J. T. Smith, B. V. Acohido, S. M. Popa, M. J. Cunningham, M. L. Gottsch, D. K. Clifton, and R. A. Steiner. 2005. Kisspeptin activation of gonadotropin releasing hormone neurons and regulation of KiSS-1 mRNA in the male rat. Neuroendocrinology 80:264-272.

Jin, J. M. and W. X. Yang. 2014. Molecular regulation of hypothalamus-pituitary-gonads axis in males. Gene 551:15-25.

Kapelanski, W., M. Biegniewska, and H. Jankowiak. 2008. Results of sow reproductive performance due to application of insulingenic diet and natural stimulation of estrus. Res. Pig Breed. 2:19-21.

Kemp, B., N. M. Soede, F. A. Helmond, and M. W. Bosch. 1995. Effects of energy source in the diet on reproductive hormones and insulin during lactation and subsequent estrus in multiparous sows. J. Anim. Sci. 73:3022-3029.

Li, X. F., Y. S. Lin, J. S. Kinsey-Jones, and K. T. O'Byrne. 2012. High-fat diet increases LH pulse frequency and kisspeptin-neurokinin B expression in puberty-advanced female rats. Endocrinology 153:4422-4431.

Matsui, H., Y. Takatsu, S. Kumano, H. Matsumoto, and T. Ohtaki. 2004. Peripheral administration of metastin induces marked gonadotropin release and ovulation in the rat. Biochem. Biophys. Res. Commun. 320:383-388.

NRC (National Research Council). 1998. Nutrient Requirements for Swine, 10th edn. National Academy Press, Washington DC, USA.

Plant, T. M. and M. L. Barker-Gibb. 2004. Neurobiological mechanisms of puberty in higher primates. Hum. Reprod. Update 10:67-77.

Steger, R. W. and M. B. Rabe. 1997. The effect of diabetes mellitus on endocrine and reproductive function. Proc. Soc. Exp. Biol. Med. 214:1-11.

Fauser, B. C. J. M. 2003. Reproductive medicine: molecular, cellular and genetic fundamentals. J. Am. Med. Assoc. (JAMA). 290:3005-3005.

Zhou, D., Y. Zhuo, L. Che, Y. Lin, Z. Fang, and D. Wu. 2014. Nutrient restriction induces failure of reproductive function and molecular changes in hypothalamus-pituitary-gonadal axis in postpubertal gilts. Mol. Biol. Rep. 41:4733-4742.

Zhou, D. S., Z. F. Fang, D. Wu, Y. Zhuo, S. Y. Xu, Y. Z. Wang, P. Zhou, and Y. Lin. 2010. Dietary energy source and feeding levels during the rearing period affect ovarian follicular development and oocyte maturation in gilts. Theriogenology 74:202-211.

20

Standardized Ileal Amino Acid Digestibility of Commonly used Feed Ingredients in Growing Broilers

Zafar Ullah[1,2,*], Gulraiz Ahmed[3], Mehr un Nisa[4], and Muhammad Sarwar[1]

[1] Institute of Animal Sciences, University of Agriculture, Faisalabad 38000, Pakistan

ABSTRACT: This experiment was conducted to determine standardized ileal amino acid digestibility (SIAAD) of commonly used feed ingredients in poultry diets in Pakistan. These feed ingredients included corn, rice broken (RB), rice polishings (RP), wheat bran (WB), sunflower meal (SFM), cottonseed meal (CSM), guar meal (GM), soybean meal (SBM) from India and Argentine and fish meal (FM). The SIAAD of each ingredient was determined in triplicate using 21-days-old broilers. Day-old male broiler chicks (Hubbard× Hubbard) were reared on corn-SBM based diet from 1 to 13 days and thereafter birds were fed experimental diets from day 14 to 21. Each diet was fed to 36 birds kept in six replicate cages, each cage had six birds. In cereals, the SIAAD of corn's amino acid (AA) (90.1%) was similar ($p>0.05$) to RB (89.0%). Isoleucine (97.8%) and lysine (96.9%) were highly digestible AA in corn and RB, respectively. Among cereal-by products, WB's SIAAD (76.9%) was same ($p>0.05$) as RP (71.9%). Arginine from WB (82.5%) and RP (83.2%) was highly digestible. However, threonine in WB (72.7%) and leucine in RP (69.6%) were the lowest digestible AAs. In plant protein meals, AAs from Argentine-SBM (85.1%) and Indian-SBM (83.4%) had higher ($p<0.5$) SIAAD than other protein meals. However, SIAAD of SFM (77.1%) and CSM (71.7%) was intermediate while GM (60.3%) exhibited the lowest ($p<0.05$) SIAAD among all ingredients. Arginine from GM (76.9%), CSM (85.8%), SBM-India (89.5%) and SBM-Argentine (91.5%) was highly digestible from indispensable AAs. In SFM, methionine (91.4%) SIAAD was the greatest. The average SIAAD of FM was 77.6%. Alanine from FM had the highest (84.0%) but cysteine (62.8%) had the lowest SIAAD. In conclusion, cereals i.e. corn and RB had higher ($p<0.05$) SIAAD of the cereals by-products. The SIAAD of RP and WB was same ($p>0.05$). The SBM from plant protein meals had higher ($p<0.05$) SIAAD than other studied feed ingredients. However, the GM had the lowest ($p<0.05$) SIAAD among protein meals. (**Key Words:** Broilers, Cereals, Cereal By-products, Digestible Amino Acids)

INTRODUCTION

Amino acids (AA) are critical dietary components regulating the animal's physiological, metabolic and structural functions. To achieve optimum growth performance, the supply of these dietary AAs according to bird's requirement is mandatory. Provision of required dietary AA to birds results in their efficient utilization because any AA's excess or deficiency adversely affects bird's growth performance.

Interest in determining the AA digestibility has been increased since some scientists (Sibbald, 1987) developed rapid bioassay to estimate AA digestibility. Bioavailability of AAs to birds is key aspect in assessing the protein quality. The AA bioavailability is the portion of AA that is digested, absorbed and utilized by the animal. However, Ravindran et al. (2005) documented that under certain situations, AAs are absorbed in a form not suitable for animal utilization, making no contribution in animal's maintenance and production requirement. Thus, digestible AA (DAA) are more authentic in describing the available AA than total AA. Bioavailability values can be estimated through slope-ratio technique but it underestimates the AA digestibility (Stein et al., 2007) so DAAs assays are more valid approach (Stein

* Corresponding Author: Zafar Ullah.
E-mail: zu2005@yahoo.com

[2] Department of Animal Sciences, University of Sargodha, Sargodha 40100, Pakistan.

[3] Sadiq Feeds (Pvt.) Ltd., Rawalpindi 46300, Pakistan.

[4] Department of Food Science, Nutrition and Home Economics, Government College University, Faisalabad 38000, Pakistan.

Table 1. Nutrient composition[1] of cereals and cereal by-products, used in digestibility assays (as-fed basis)

Nutrient (%)	Ingredients			
	Corn	Rice broken	Rice polishings	Wheat bran
Moisture	8.67±0.44	9.87±0.40	6.84±0.62	8.20±1.28
Crude protein	8.81±0.54	10.07±0.21	12.38±0.90	13.50±0.71
Ether extract	3.61±0.07	1.21±0.05	10.74±1.44	3.33±0.47
Crude fiber	1.95±0.12	0.68±0.02	17.38±4.06	10.21±1.63
Ash	1.25±0.17	1.01±0.14	11.55±0.63	5.09±0.49
Acid insoluble ash	0.08±0.06	0.17±0.04	7.92±0.72	1.20±0.43

[1] Data were average of three samples of each feed ingredient.

et al., 2007). Moreover, some workers claimed that digestibility values determined in mature birds could not be used to formulate growing bird's diet because AA requirements change with bird's age (Adedokun et al., 2008); the ileal digestibility enables AA digestibility assay possible in growing birds and is more reliable tool to represent AA digestibility (Ravindran et al., 1999) than total tract digestibility. Standardized ileal AA digestibility (SIAAD) is widely used technique to present AA digestibility (Adedokun et al., 2008).

Reliable DAA values permitted more efficient broiler production (Lemme et al., 2004). Currently, nutritionists are formulating feeds on DAA basis (Huang et al., 2005). The low dietary protein was maximally utilized by birds for their maintenance and production requirements and this decreases both feed cost and nitrogen excretion into environment, decreasing limiting AA requirements (Dari et al., 2005). Most of the advanced countries have developed ileal-based DAA database of feedstuffs produced in their respective countries (Bryden et al., 2009). But unfortunately, the same database of locally produced ingredients in Pakistan is limited. Thus, the present experiment was conducted with the objective to determine SIAAD of various feed ingredients used in poultry diet in Pakistan.

MATERIALS AND METHODS

This experiment was conducted at Research and Development Center, Sadiq Feeds (Pvt.) Ltd. Rawalpindi, Pakistan with collaboration Institute of Animal Sciences, University of Agriculture Faisalabad, Pakistan.

Test ingredients

Feed ingredients used in this study were two cereals; corn (*Zea mays* L.) and rice broken (RB) (*Oryza sativa* L.); two cereal by-products; wheat (*Triticum* spp.) bran and rice (*Oryza sativa* L.) polishing; five oil seed meals; sunflower (*Helianthus annus* L.), cottonseed (*Gossypoum* spp.), guar (*Cyamopsis tetragonolobus* L.) and soybean (*Glycine max* L.) from Argentine and India and one animal protein meal, the fish meal (FM). The 3 samples of each feed ingredient were assayed for SIAAD. These ingredients were analyzed (Tables 1 and 2) for dry matter, crude protein (N×6.25) by LECO nitrogen analyzer (model FP-528, Leco Corporation, St. Joseph, MI, USA), ether extract, crude fiber, ash and acid insoluble ash content (AOAC, 2000).

Experimental diets

The thirty diets (10 ingredients×3 samples) were formulated such that all AA were provided from test ingredient in its respective diet. Cereals and cereal by-products were 91.8% of diet. The inclusion level of protein meal in diets was adjusted on crude protein (CP) basis so that dietary CP remained about 20% (Ravindran et al., 2005). Dextrose was used as energy source in these diets. Acid insoluble ash, an external digestibility marker, was added at 2% to each diet (Ravindran et al., 2005). Calcium and phosphorus supplementation was identical in diets formulated using sunflower meal (SFM), cottonseed meal (CSM), guar meal (GM), soybean meal (SBM)-Argentine

Table 2. Nutrient composition[1] of protein meals, used in digestibility assays (as-fed basis)

Nutrient (%)	Ingredients					
	Sunflower meal	Guar meal	Cotton seed meal	Fish meal	Soybean meal (India)	Soybean meal (Argentine)
Moisture	6.85±1.75	5.05±0.33	6.83±0.26	11.06±2.41	8.14±0.52	7.86±0.27
Crude protein	27.07±1.91	40.46±0.39	37.85±4.83	44.92±2.35	50.70±0.16	46.38±0.30
Ether extract	0.75±0.26	5.35±0.23	1.88±1.68	17.76±4.17	0.53±0.15	1.18±0.15
Crude fiber	24.81±3.55	13.06±0.46	13.87±3.65	-	5.08±1.02	3.66±0.40
Ash	6.51±0.74	5.75±0.29	6.71±0.46	23.58±3.95	7.39±0.14	6.57±0.15
Acid insoluble ash	1.93±0.86	1.26±0.32	0.48±0.12	10.97±1.74	1.15±0.20	1.02±0.08

[1] Data were average of three samples of each feed ingredient.

Table 3. Ingredient composition of experimental diets used in digestibility assays-selected examples (as-fed basis)

Ingredients (%)	Ingredients						
	Cereal and cereal by-products	Cotton seed meal	Sunflower meal	Guar meal	Fish meal	Soybean meal (India)	Soybean meal (Argentine)
Test ingredient[1]	91.8	52.84	72.53	47.53	44.52	39.44	43.29
Dextrose	-	35.06	15.37	40.37	47.28	48.46	44.61
Sunflower oil	2.0	6.0	6.0	6.0	2.0	6.0	6.0
Arbocel[2] (cellulose)	-	-	-	-	3.0	-	-
Celite	2.0	2.0	2.0	2.0	2.0	2.0	2.0
Dicalcium phosphate	1.7	1.9	1.9	1.9	-	1.9	1.9
Limestone	1.3	1.0	1.0	1.0	-	1.0	1.0
Vit/min premix[3]	0.7	0.7	0.7	0.7	0.7	0.7	0.7
Choline chloride	0.3	0.3	0.3	0.3	0.3	0.3	0.3
Salt	0.2	0.2	0.2	0.2	0.2	0.2	0.2

[1] Test ingredient served as sole source of amino acid. Except cereals and cereal by-products, test ingredients were included to supply approximately 20% dietary crude protein.

[2] Arbocel, Insoluble raw fiber concentrate, Holzmuhle, Rosenberg, Germany.

[3] Provided per kg of diet: retinyl acetate, 4,400 IU; cholecalciferol, 118 µg; DL-α-tocopheryl acetate, 12 IU; menadione sodium bisulphite, 2.40 mg; thiamine, 2.5 mg; riboflavin, 4.8 mg; niacin, 30 mg; D-pentothenic acid, 10 mg; pyridoxine, 5 mg; biotin, 130 µg; folic acid, 2.5 mg; cyanocobalamin, 19 µg; manganese, 85 mg ($MnSO_4 \cdot H_2O$); iron, 80 mg ($FeSO_4 \cdot H_2O$); zinc, 75 mg (ZnO); copper, 6 mg ($CuSO_4 \cdot 5H_2O$); iodine, 1 mg (ethylene diamine dihydroiodide); selenium, 130 µg (Na_2SeO_3).

and SMB-India. However, calcium and phosphorus supplementation was not added in 3 diets containing FM but these diets contained 3% Arbocel as fiber source (Ravindran et al., 2005). Vitamin and mineral supplementation was similar across all diets (Table 3). A nitrogen free diet (NFD) was formulated (Table 4) to determine endogenous AA losses (EAA) (Table 7) to calculate SIAAD (Adedokun et al., 2007).

Table 4. Ingredient composition of nitrogen free diet (as-fed basis)

Ingredients	Percentage
Corn starch	16.9
Dextrose	64.0
Sunflower oil	5.0
Arbocel (cellulose)[1]	5.0
Celite	2.0
Dicalcium phosphate	1.9
Limestone	1.3
Vit/min premix[2]	0.7
Choline chloride	0.3
Salt	0.2
Sodium bicarbonate	1.5
Potassium chloride	1.2

[1] Arbocel, Insoluble raw fiber concentrate, Holzmuhle, Rosenberg, Germany.

[2] Provided per kg of diet: retinyl acetate, 4,400 IU; cholecalciferol, 118 µg; DL-α-tocopheryl acetate, 12 IU; menadione sodium bisulphite, 2.40 mg; thiamine, 2.5 mg; riboflavin, 4.8 mg; niacin, 30 mg; D-pentothenic acid, 10 mg; pyridoxine, 5 mg; biotin, 130 µg; folic acid, 2.5 mg; cyanocobalamin, 19 µg; manganese, 85 mg ($MnSO_4 \cdot H_2O$); iron, 80 mg ($FeSO_4 \cdot H_2O$); zinc, 75 mg (ZnO); copper, 6 mg ($CuSO_4 \cdot 5H_2O$); iodine, 1 mg (ethylene diamine dihydroiodide); selenium, 130 µg (Na_2SeO_3).

Bird's management

A total of 1,116 day-old male broiler chicks (Hubbard× Hubbard) were arranged from commercial hatchery (SB Hatchery, Rawalpindi, Pakistan) and kept in cages. All chicks were reared under identical managemental conditions. Room temperature was maintained at 32°C±1°C during 1st week and gradually decreased to 24°C by the end of 3rd week. Birds received continuous fluorescent light throughout the experimental period. Chicks were vaccinated against Newcastle Disease (ND), Infectious Bronchitis (IB) and Infectious Bursal Disease (IBD). Supply of fresh and clean water was made available round the clock. Chicks were fed *ad-libitum* corn-SBM starter diet in crumble form to fulfill their nutritional requirements (NRC, 1994) from day 1 to 13. On day 14, all the chicks were fasted overnight, individually weighed and randomly distributed to 186 replicate cages (6 birds in each cage) in similar cumulative body weight manner among all cages. The variation in mean body weight among replicates was ±10 g. Experimental diets were offered to birds from day 14 to 21 of age. Each experimental diet was offered *ad-libitum* to 36 birds kept in 6 replicates; each replicate had 6 birds. So, 3 diets based on 3 samples of same feed ingredient were fed to birds of 18 replicate cages. The NFD diet was offered *ad-libitum* to 36 birds placed in 6 replicate cages.

Ileal digesta collection

On day 21, all birds were euthanized by intravenous injection, Ketamax (Ketamine hydrochloride). Contents of ileum from vitelline diverticulum (formally named as Meckel's diverticulum) to 40 mm proximal to ileo-caecal junction (Bandegan et al., 2009) were collected in plastic zip bags by gently flushing through long tip syringe containing distilled water and air pressure. The ileal digesta of all birds in a replicate was pooled, immediately stored at –20°C and subsequently freeze-dried (Kong and Adeola, 2014). Dried ileal digesta then ground in coffee grinder (MC3001 coffee grinder; Moulinex Ltd. Weston, ON, Canada) to pass through 0.5 mm sieve and stored in plastic tubes at –4°C for further analyses (Bandegan et al., 2009).

Chemical analyses

Raw ingredients, test diets and ileal digesta samples were analyzed for dry matter (DM: AOAC, 2000) and CP (N×6.25) by LECO nitrogen analyzer (model FP-528, Leco Corporation, USA). Acid insoluble ash of both diets and ileal digesta samples were determined (AOAC, 2000). The AA profile of test ingredients (Tables 5 and 6) and ileal digesta was determined by the procedure used by Palliyeguru et al. (2010) using AA analyzer (Biochrom 30 plus, Biochrom Ltd. Cambridge, UK). Samples were oxidized with hydrogen peroxide-formic acid-phenol solution and sodium disulfite was used to decompose excess oxidation reagent. After oxidation, samples were hydrolyzed using 6*M* HCl for 24 hours. The pH of hydrolysate was adjusted at 2.20, centrifuged, filtered and AA profile was determined.

Calculations

The EAA concentration was calculated as milligrams of AA flow per kg DM intake as described by Moughan et al. (1992).

leal AA flow, mg/kg DMI

$$= \left[\text{AA in ileal digesta, mg/kg} \times \left(\frac{\text{Diet marker, mg/kg}}{\text{Ileal marker, mg/kg}}\right)\right]$$

Apparent ileal AA digestibility (AIAAD), %

$$= \left[1 - \left(\frac{\text{Marker in diet}}{\text{Marker in ileal digesta}}\right) \times \left(\frac{\text{AA in ileal digesta}}{\text{AA in diet}}\right)\right] \times 100$$

The endogenous ileal AA losses (mg/kg of DM intake) were used (Table 7) to calculate SIAAD by using following equation.

Standardized ileal AA digestibility (SIAAD), %

$$= \text{AIAAD, \%} + \left[\left(\frac{\text{Ileal AA flow, g/kg of DMI}}{\text{AA in raw material, g/kg of DMI}}\right) \times 100\right]$$

Statistical analyses

The mean and standard deviation of ingredient's nutrient composition, ingredient's AA profile and SIAAD of test ingredients were calculated to provide information on variability. The average SIAAD values between cereal and cereal by products as well as protein meals were analyzed using Analysis of variance techniques using SAS (2009). Means were compared by Tukey's test.

RESULTS AND DISCUSSION

Total amino acid concentration

Cereals and cereal by-products: The average AA concentration of three samples in corn, RB, rice polishings (RP), and wheat bran (WB) ranged from 0.19% to 1.42%, 0.20% to 1.59%, 0.25% to 1.67%, and 0.23% to 2.81%, respectively. In all cereals and cereal by-products, glutamic acid concentration was highest. Its value in corn, RB, RP, and WB was 1.42%, 1.59%, 1.67%, and 2.81%, respectively. Methionine was the lowermost in corn (0.19%), RP (0.25%) and WB (0.23%); however, in RB, cysteine concentration (0.20%) was the lowest (Table 5). The AA concentration of feed ingredients in this study was commensurate with other literature (Heartland Lysine, 1996; Evonik, 2010). The minor variation in WB's AA concentration in present study compared to those reported by Evonik (2010) might be because of difference in CP content. Generally, AA concentration increases with increasing protein level (Ravindran et al., 2005). Varying plant breeding program,

Table 5. Total amino acid composition[1] of cereals and cereal by-products, used in digestibility assays (as-fed basis)

Amino acid (%)	Ingredients			
	Corn	Rice broken	Rice polishings	Wheat bran
Dry matter	91.33±0.44	90.13±0.40	93.16±0.62	91.80±1.28
Indispensable amino acid				
Arginine	0.42±0.06	0.71±0.05	0.88±0.04	1.00±0.08
Histidine	0.27±0.05	0.21±0.01	0.33±0.03	0.42±0.02
Isoleucine	0.27±0.04	0.37±0.02	0.41±0.04	0.45±0.08
Leucine	0.83±0.09	0.74±0.05	0.81±0.08	0.88±0.09
Lysine	0.28±0.05	0.33±0.02	0.54±0.03	0.61±0.07
Methionine	0.19±0.03	0.25±0.02	0.25±0.04	0.23±0.04
Phenylalanine	0.39±0.08	0.48±0.03	0.56±0.08	0.58±0.12
Threonine	0.28±0.04	0.32±0.02	0.45±0.05	0.50±0.04
Valine	0.41±0.07	0.53±0.03	0.65±0.08	0.69±0.13
Dispensable amino acid				
Alanine	0.53±0.06	0.52±0.03	0.71±0.06	0.72±0.06
Aspartic acid	0.51±0.08	0.80±0.04	1.02±0.04	1.06±0.05
Cysteine	0.21±0.03	0.20±0.02	0.26±0.02	0.30±0.03
Glycine	0.33±0.05	0.40±0.02	0.65±0.06	0.79±0.04
Glutamic acid	1.42±0.27	1.59±0.10	1.67±0.18	2.81±0.29
Serine	0.35±0.03	0.44±0.03	0.51±0.02	0.61±0.03

[1] Data were average of three samples of each feed ingredient.

agronomic practices, cultivation season and soil conditions are reported to influence chemical composition and AA contents of feed ingredients (Ravindran et al., 2014).

Protein meals: The AA contents of SFM, GM, CSM, FM, SBM-India and SBM-Argentine ranged from 0.49% to 5.64%, 0.40% to 7.01%, 0.64% to 8.60%, 0.28% to 4.39%, 0.64% to 9.24%, and 0.57% to 8.20% (Table 6), respectively. Glutamic acid concentration was the highest in all ingredients. Methionine was the lowest in GM (0.40%), CSM (0.64%) and SBM from both regions. However, in SFM (0.49%) and FM (0.28%), cysteine concentration was the lowest. Similar results were reported by other workers (Heartland Lysine, 1996; Ravindran et al., 2005).

Table 7. Concentration of endogenous amino acid losses used to standardize the amino acid digestibility

Amino acid	Endogenous amino acid concentration (mg/kg DMI)
Indispensable amino acid	
Arginine	179
Histidine	189
Isoleucine	349
Leucine	341
Lysine	225
Methionine	49
Phenylalanine	202
Threonine	412
Valine	396
Dispensable amino acid	
Alanine	108
Aspartic acid	168
Cysteine	141
Glycine	120
Glutamic acid	237
Serine	136

DMI, dry matter intake.

Standardized ileal amino acid digestibility

Cereals and cereal by-products: The mean SIAAD of corn, RB, RP, and WB ranged from 79.4% to 97.8%, 82.1% to 96.9%, 64.9% to 83.2%, and 71.1% to 82.8%, respectively (Table 8). The overall mean AA digestibility of corn (90.1%) was same ($p>0.05$) compared to RB (89.0%). Likewise, the WB exhibited equal ($p>0.05$) SIAAD (76.9%) compared to RP (71.9%). The numerically lower RP's SIAAD than WB, supported the findings of Warren and Farrell (1991). Ravindran et al. (2005) reported higher apparent digestibility of wheat middlings than RP. Corn's SIAAD (90.1%) was concordant with those reported by Sauvant et al. (2004), Rostagno et al. (2005) and Evonik (2010). However, it was higher than those reported by Ravindran et al. (2005). The SIAAD of WB (76.9%) was also concordant with the values reported by Evonik (2010), Rostagno et al. (2005) and Sauvant et al. (2004). The SIAAD of RB (89.0%) of our study supported the findings reported by Sauvant et al. (2004). However, some variations were also reported by Evonik (2010). The SIAAD of RP (71.9%) in present study was supported by the findings of Evonik (2010). However, contrasting results were reported by Sauvant et al. (2004), Rostagno et al. (2005) and

Table 6. Total amino acid composition[1] of protein meals, used in digestibility assays (as-fed basis)

Amino acid (%)	Ingredients					
	Sunflower meal	Guar meal	Cotton seed meal	Fish meal	Soybean meal (India)	Soybean meal (Argentine)
Dry matter	93.15±1.75	94.95±0.33	93.17±0.26	88.94±2.41	91.86±0.52	92.14±0.27
Indispensable amino acid						
Arginine	2.29±0.38	4.50±0.20	4.38±0.39	1.57±0.13	3.62±0.09	3.17±0.16
Histidine	0.73±0.14	0.97±0.04	1.20±0.05	0.72±0.08	1.35±0.04	1.23±0.03
Isoleucine	1.19±0.23	1.10±0.02	1.39±0.06	1.45±0.30	2.31±0.08	1.86±0.25
Leucine	1.80±0.32	2.11±0.02	2.51±0.05	2.42±0.38	3.81±0.08	3.31±0.17
Lysine	0.99±0.18	1.51±0.12	1.56±0.17	1.99±0.32	3.01±0.06	2.69±0.14
Methionine	0.66±0.17	0.40±0.02	0.64±0.01	0.76±0.12	0.64±0.06	0.57±0.03
Phenylalanine	1.32±0.25	1.45±0.09	2.37±0.02	1.38±0.16	2.58±0.12	2.19±0.19
Threonine	1.05±0.18	1.14±0.06	1.41±0.06	1.22±0.15	1.91±0.05	1.78±0.04
Valine	1.43±0.27	1.29±0.02	1.77±0.07	1.78±0.32	2.39±0.11	1.90±0.30
Dispensable amino acid						
Alanine	1.21±0.18	1.52±0.03	1.80±0.03	2.43±0.30	2.12±0.02	1.97±0.03
Aspartic acid	2.52±0.35	3.66±0.04	3.95±0.12	2.85±0.31	5.60±0.08	5.11±0.04
Cysteine	0.49±0.08	0.44±0.04	0.69±0.02	0.28±0.07	0.71±0.02	0.63±0.07
Glycine	1.70±0.28	2.00±0.01	1.87±0.02	2.60±0.23	2.80±0.02	1.91±0.03
Glutamic acid	5.64±1.26	7.01±0.16	8.60±0.12	4.39±0.40	9.24±0.54	8.20±0.06
Serine	1.18±0.15	1.63±0.12	1.82±0.15	1.15±0.09	2.46±0.10	2.32±0.05

[1] Data were average of three samples of each feed ingredient.

Ravindran et al. (2005). Reason of this varying AA digestibility might be difference in age of birds and apparent digestibility values without correcting with EAA losses. The apparent digestibility is generally low due to higher EAA proportion in terminal ileum. The AA intake reduced in birds fed diets containing low protein/AA content (Kim et al., 2012b). The decreased AA intake resulted in increased EAA. Moter and Stein (2004) also reported wider variance in apparent and standardized digestibility for low protein ingredients. The method used to estimate EAA losses also influenced SIAAD values (Bryden et al., 2009). Adedokun et al. (2008) reported varying corn SIAAD when EAA were determined by feeding either NFD or 10% casein diet. In literature, several terminologies and methods have been used to present AA digestibility results (Ravindran et al., 2005; Bryden et al., 2009; Kim et al., 2012a,b). These terms included apparent or true digestibilities using ileal content (intact or cecectomized) or excreta (Ravindran et al., 1999) of varying age roosters, cockerels and broilers (Huang et al., 2005). Therefore, direct comparison of SIAAD values, determined in this study using 21-days old broilers, with reported literature seems really valid.

Some endogenous and exogenous factors have been reported to influence digestibility. The endogenous factors included the specie, strain, age (Kim and Corzo, 2012) and sex (Huang et al., 2005) while exogenous aspects were the presence of antinutritional substances, dietary AA balance and stress level on animals. These factors affect physical, chemical and physiological state of digestive environment (Huang et al., 2005) which may influence AA digestibility. The ingredient's natural variability and their processing also affect AA digestibility (Bell, 1993; Adedokun et al., 2007). More variation in RP's SIAAD in this study may be because of varying both processing techniques and chemical composition of samples. Not only the ingredient's chemical composition influences its digestibility but linkages of protein to other nutrients such as carbohydrates, fats and proteins are also important in this regard (Bryden et al., 2009). The SIAAD of all cereals and cereal by-products differed from each other. The reason may be that each ingredient was a mixture of various proteins and each of them was digested at different rate.

Among cereal and cereal by-product; Isoleucine, lysine, arginine and glutamic acid were highly digestible in corn (97.8%), RB (96.9%), RP (83.2%), and WB (82.8%), respectively. In indispensable AAs, threonine was the lowest digestible in corn (86.6%) and WB (72.7%), methionine in RB (82.1%) and leucine in RP (69.6%). From dispensable AAs, glycine had the lowest digestibility in corn (79.4%), RP (64.9%) and WB (71.1%). The cysteine (84.0%) from RB had also the lowest digestibility. The highest glutamic acid digestibility among dispensable AAs (Table 8) supported the findings of other workers (Bryden et al., 2009).

Protein meals: The SIAAD of SBM-Argentine, SBM-India, SFM, CSM, and GM ranged from 77.2% to 91.5%, 77.1% to 89.5%, 60.0% to 91.4%, 56.8% to 85.8%, and

Table 8. Standardized ileal amino acid digestibility percentage of cereals and cereal by-products in broiler chicks[1]

Amino acid (%)	Ingredients				p-value
	Corn	Rice broken	Rice polishings	Wheat bran	
Indispensable amino acid					
Arginine	$93.3^{a}\pm1.27$	$94.4^{a}\pm1.87$	$83.2^{b}\pm3.81$	$82.5^{b}\pm1.39$	***
Histidine	$91.5^{a}\pm2.29$	$92.6^{a}\pm2.76$	$75.2^{b}\pm1.92$	$77.5^{b}\pm0.86$	***
Isoleucine	$97.8^{a}\pm1.15$	$93.2^{a}\pm3.68$	$75.0^{b}\pm7.12$	$81.0^{b}\pm0.87$	***
Leucine	$94.7^{a}\pm1.31$	$89.5^{a}\pm3.28$	$69.6^{b}\pm6.16$	$78.6^{b}\pm1.87$	***
Lysine	$91.0^{a}\pm2.73$	$96.9^{a}\pm2.03$	$80.6^{b}\pm4.01$	$78.6^{b}\pm2.91$	***
Methionine	$94.9^{a}\pm0.99$	$82.1^{ab}\pm4.62$	$72.1^{b}\pm8.82$	$81.7^{b}\pm1.66$	**
Phenylalanine	$94.2^{a}\pm0.47$	$88.6^{a}\pm3.28$	$70.1^{b}\pm7.13$	$72.8^{b}\pm3.11$	***
Threonine	$86.6^{a}\pm3.29$	$91.1^{a}\pm4.87$	$70.7^{b}\pm5.13$	$72.7^{b}\pm1.59$	***
Valine	$92.6^{a}\pm0.76$	$91.6^{a}\pm3.23$	$71.4^{b}\pm5.60$	$76.1^{b}\pm1.57$	***
Dispensable amino acid					
Alanine	$90.1^{a}\pm1.70$	$87.0^{a}\pm3.13$	$70.6^{b}\pm3.36$	$73.5^{b}\pm2.31$	***
Aspartic acid	$83.4^{a}\pm2.02$	$88.4^{a}\pm2.55$	$70.6^{b}\pm4.84$	$74.1^{b}\pm1.29$	***
Cysteine	$88.6^{a}\pm3.05$	$84.0^{ab}\pm4.90$	$66.2^{c}\pm5.73$	$76.3^{bc}\pm0.78$	***
Glycine	$79.4^{a}\pm1.54$	$86.5^{a}\pm2.96$	$64.9^{b}\pm4.95$	$71.1^{b}\pm0.96$	***
Glutamic acid	$91.3^{a}\pm1.91$	$85.0^{a}\pm3.78$	$71.7^{b}\pm4.86$	$82.8^{a}\pm1.49$	***
Serine	$82.3^{ab}\pm3.43$	$84.7^{a}\pm3.21$	$66.8^{c}\pm6.55$	$73.8^{bc}\pm1.48$	**
Average	$90.1^{a}\pm5.18$	$89.0^{a}\pm4.23$	$71.9^{b}\pm4.97$	$76.9^{b}\pm3.87$	***

[1] Data were average of 3 samples of each feed ingredient; diet based on each sample was fed to six replicates each containing six birds.
[a,b,c,d] Values sharing different superscripts within rows are statistically significant ($p<0.05$).
* $p<0.05$; ** $p<0.01$; *** $p<0.001$.

Table 9. Standardized ileal amino acid digestibility percentage of protein meals in broiler chicks[1]

Amino acid (%)	Ingredients						p-value
	Sunflower meal	Guar meal	Cotton seed meal	Fish meal	Soybean meal (India)	Soybean meal (Argentine)	
Indispensable amino acid							
Arginine	90.9^{a}±1.08	76.9^{b}±9.15	85.8^{ab}±1.63	82.6^{ab}±1.55	89.5^{a}±1.16	91.5^{a}±2.65	**
Histidine	77.0^{b}±3.20	65.1^{c}±3.89	75.0^{b}±2.72	75.5^{b}±3.90	87.1^{a}±1.64	87.9^{a}±2.96	***
Isoleucine	83.9^{ab}±1.60	58.5^{d}±3.92	70.5^{c}±2.78	78.5^{bc}±5.84	86.1^{ab}±2.20	87.8^{a}±2.17	***
Leucine	82.1^{a}±1.51	60.3^{c}±2.85	73.9^{b}±1.97	82.6^{a}±1.90	84.6^{a}±2.18	87.1^{a}±2.45	***
Lysine	78.3^{bc}±2.93	57.3^{d}±3.74	56.8^{d}±4.31	72.7^{c}±4.34	86.7^{ab}±2.08	89.6^{a}±3.55	***
Methionine	91.4^{a}±0.62	63.8^{c}±4.14	75.0^{b}±2.02	79.1^{b}±1.18	87.9^{a}±3.37	89.8^{a}±2.13	***
Phenylalanine	78.9^{ab}±7.39	60.4^{b}±4.55	71.2^{ab}±14.15	81.1^{a}±1.32	84.7^{a}±0.31	85.4^{a}±2.78	**
Threonine	73.1^{bc}±3.09	52.6^{d}±4.23	67.7^{c}±1.90	77.3^{ab}±3.23	82.0^{a}±1.91	83.7^{a}±3.66	***
Valine	78.7^{bc}±1.96	53.6^{d}±5.73	68.9^{c}±2.50	77.1^{ab}±4.34	84.5^{a}±2.45	86.5^{a}±2.53	***
Dispensable amino acid							
Alanine	79.5^{ab}±2.68	51.4^{c}±6.61	72.3^{b}±1.50	84.0^{a}±2.67	83.2^{a}±2.35	84.8^{a}±0.94	***
Aspartic acid	78.7^{ab}±2.43	63.7^{c}±1.41	74.7^{b}±3.00	76.1^{b}±4.17	84.5^{ab}±4.25	82.1^{a}±1.24	***
Cysteine	69.4^{ab}±2.66	51.0^{c}±4.42	60.9^{bc}±4.37	62.8^{bc}±4.70	77.1^{a}±3.26	77.2^{a}±7.44	***
Glycine	60.0^{cd}±6.31	56.6^{d}±1.23	68.3^{bc}±2.51	77.9^{ab}±4.01	85.1^{a}±0.49	87.6^{a}±3.67	***
Glutamic acid	87.6^{a}±1.58	71.0^{c}±1.86	81.6^{ab}±2.43	80.7^{b}±2.36	80.0^{ab}±1.42	82.4^{a}±3.64	***
Serine	68.3^{bc}±4.0	62.4^{c}±4.23	73.0^{b}±2.18	75.8^{ab}±5.13	84.6^{a}±1.95	83.0^{a}±1.46	***
Average	77.1^{b}±8.05	60.3^{d}±7.24	71.7^{c}±7.12	77.6^{b}±5.14	83.4^{a}±3.32	85.1^{a}±3.84	***

[1] Data were average of 3 samples of each feed ingredient; diet based on each sample was fed to six replicates each containing six birds.
[a,b,c,d] Values sharing different superscripts within rows are statistically significant ($p<0.05$).
* $p<0.05$; ** $p<0.01$; *** $p<0.001$.

51.0% to 76.9%, respectively (Table 9). The SBM-Argentine and SBM-India had the highest SIAAD among vegetable protein meals. Most of AAs from SBM-Argentine and SBM-India had higher SIAAD than other vegetable protein meals. However, SIAAD of SFM was more ($p<0.05$) than CSM. Guar meal had the lowest SIAAD ($p<0.05$).

The low SIAAD of SFM (77.1%) in our study was due to higher fiber content. The fiber fraction was >24% in the samples studied and fiber content was negatively correlated with AA digestibility (Senkoylu and Dale, 1999). Bell (1993) reported reduced digestibility due to high hull portion while studying canola meal. The SIAAD of CSM (71.7%) had almost similar values reported by Rostagno et al. (2005), Sauvant et al. (2004) and Evonik (2010). However, contrasting values were reported by Ravindran et al. (2005). The higher fiber, gossypol contents (Phelps, 1966) and condensed tannins (Yu et al., 1993) were the cause of reduced AA digestibility in CSM (Ravindran et al., 2005). These anti-nutritional substances not only increase digesta viscosity but interact with digestive enzymes making them unavailable and consequently reducing the AA digestibility. The GM's SIAAD (60.3%) was the lowest among all ingredients in this study. This low digestibility when compared to findings of Nadeem et al. (2005) might be due to different animal used, sample sources (Wang and Parsons, 1998) or chemical composition. Difference in agronomic practices, environmental, soil conditions (Ravindran et al., 2014) and processing may influence feed ingredient's digestibility. The substances like β-mannans and gums in GM may form complex with digestive enzymes and increase digesta viscosity and thus reduce AA digestibility. Data regarding AA digestibility in GM were limited in different database systems reported by some workers (Evonik, 2010).

The SIAAD of SBM from Argentine and Indian regions was similar to those reported by other workers (Evonik, 2010; Ravindran et al., 2014). Similarly, SIAAD of SBM-Argentine and SBM-India did not differ between them (Evonik, 2010). However, in this trial, numerically higher SIAAD of SBM-Argentine (85.1%) than that of SBM-Indian (83.4%) was supported by the findings of Ravindran et al. (2014). The higher fiber in Indian-SBM might have reduced AA digestibility. The reason of increased SBM-Argentine digestibility might be the varying nutrient composition, processing conditions (Ravindran et al., 2014) and temperature of cultivar conditions (Piper and Boote, 1999). Generally, processing enhances AA availability. However, under extreme conditions like high temperature either it may lead to Maillard reactions making AAs unavailable to birds (Ahmad et al., 2007) or damage AAs particularly the basic AAs (Fenwick and Curtis, 1980). Other than applying heat and pressure during processing, even just grinding of ingredient is not without any effect. For example, grinding may change shape and size of particle without affecting chemical composition. Enhanced particle's surface area during grinding allows more enzymatic action, enhancing the digestion (Bryden et al., 2009). In present study, the SIAAD of FM was 77.6% which supported the findings of Evonik (2010) but were lower than those reported by Rostagno et al. (2005) and Sauvant et al. (2004). Possible reason of varying FM's SIAAD might be the difference in fish variety because FM is prepared from a variety of fish and each variety varies in its nutrient composition and AA digestibility.

Arginine from GM, CSM, SMB-Argentine, and SBM-India had the highest SIAAD. Cysteine from GM, FM, and SBM from both regions while glycine (60.0%) from SFM and Lysine (56.8%) from CSM had the lowest SIAAD. However, SIAAD of methionine from SFM (91.4%) and alanine from FM (84%) was high (Table 9). It is well recognized that AAs are heat susceptible during processing especially the SBM (Parsons et al., 1992), canola meal (Newkirk et al., 2003) and SFM (Zhang and Parsons, 1994). The pressurized steam or heat during processing may destroy or alter AA, making them unavailable to animal's requirements (Wang and Parsons, 1998). The lower lysine digestibility from CSM in present study supported the findings of other scientists (Ravindran et al., 2005). Ravindran et al. (2005) documented that the reason of reduced lysine digestibility in CSM might be the formation of indigestible complex between lysine and gossypol during processing. The heat supply in presence of free gossypol accelerated the process of complex formation especially with free AAs (Baliga and Lyman, 1957). The other reason of lysine's reduced digestibility is its free α-amino group which is highly susceptible to damage (Ravindran et al., 2005). The reduced lysine availability is not only specific for CSM but was evident in several other oilseed meals (Zhang and Parsons, 1994). Generally, in animal protein meals, cysteine was the lowest digestible (Heartland Lysine, 1996; Evonik, 2010) because it is the most affected AA by high temperature and pressure during processing (Wang and Parsons, 1998). The exact phenomenon by which processing adversely affects the cysteine is not clear (Wang and Parsons, 1998) but might be due to lysinoalanine or lanthionine formation (Robbins et al., 1980).

CONCLUSION

The corn and RB had same ($p>0.05$) SIAAD in cereals. Similarly in cereal byproducts, WB and RP had equal SIAAD value ($p>0.05$). Among protein meals, the SBM-Argentine, SBM-India had higher SIAAD values ($p<0.05$) than those of SFM, FM, and CSM. However, the GM had the lowest ($p<0.05$) SIAAD value among protein meals studied. Formulating broiler diets using SIAAD values of various feed ingredients may not only make broiler production a cost effective enterprise but will also reduce the environmental pollution.

ACKNOWLEDGMENTS

This study was supported by Higher Education Commission of Pakistan. The authors are grateful to Sadiq Feeds (Pvt.) Ltd. Rawalpindi, Pakistan, for providing chicks, feeds and laboratory facilities.

REFERENCES

Adedokun, S. A., C. M. Parsons, M. S. Lilburn, O. Adeola, and T. J. Applegate. 2007. Endogenous amino acid flow in broiler chicks is affected by the age of birds and method of estimation. Poult. Sci. 86:2590-2597.

Adedokun, S. A., O. Adeola, C. M. Parsons, M. S. Lilburn, and T. J. Applegate. 2008. Standardized ileal amino acid digestibility of plant feedstuffs in broiler chickens and turkey poults using a nitrogen-free or casein diet. Poult. Sci. 87:2535-2548.

Ahmad, G., T. Mushtaq, M. A. Mirza, and Z. Ahmad. 2007. Comparative bio-efficacy of lysine from L-lysine hydrochloride or L-lysine sulfate in basal diets containing graded levels of canola meal for female broiler chickens. Poult. Sci. 86:525-530.

AOAC. 2000. Official Methods of Analysis. 17th ed. AOAC Int., Gaithersburg, MD, USA.

Baliga, B. P. and C. M. Lyman. 1957. Preliminary report on the nutritional significance of bound gossypol in cottonseed meal. J. Am. Oil Chem. Soc. 34:21-24.

Bandegan, A., W. Guenter, D. Hoehler, G. H. Crow, and C. M. Nyachoti. 2009. Standardized ileal amino acid digestibility in wheat distillers dried grains with solubles for broilers. Poult. Sci. 88:2592-2599.

Bell, J. M. 1993. Factors affecting the nutritional value of canola meal: A Review. Can. J. Anim. Sci. 73:679-697.

Bryden, W. L., X. Li, G. Ravindran, L. I. Hew, and V. Ravindran. 2009. Ileal Digestible Amino Acid Values in Feedstuffs for Poultry. Rural Industrial Research and Development Corporation. RIRDC Publication no 09/071. Gatton, Australia.

Dari, R. L., A. M. Penz Jr., A. M. Kessler, and H. C. Jost. 2005. Use of digestible amino acids and the concept of ideal protein in feed formulation for broilers. J. Appl. Poult. Res. 14:195-203.

Evonik. 2010. AMINODat 4.0. Evonik Industries, Evonik Degussa GmbH, Hanau-Wolfgang, Germany.

Fenwick, G. R. and R. F. Curtis. 1980. Rapeseed meal and its usage in poultry diets: A review. Anim. Feed Sci. Technol. 5:255-298.

Heartland Lysine. 1996. Digestibility of Essential Amino Acids for Poultry and Swine, version 3.3. Heartland Lysine, Inc., Chicago, IL, USA.

Huang, K. H., V. Ravindran, X. Li, and W. L. Bryden. 2005. Influence of age on the apparent ileal amino acid digestibility of feed ingredient for broiler chickens. Br. Poult. Sci. 46:236-245.

Kim, E. J. and A. Corzo. 2012. Interactive effect of age, sex, and strain on apparent ileal amino acid digestibility of soybean meal and an animal by-product blend in broilers. Poult. Sci. 91:908-917.

Kim, E. J., P. L. Utterback, and C. M. Parsons. 2012a. Comparison of amino acid digestibility coefficients for soybean meal, canola meal, fish meal and meat and bone meal among 3 different bioassays. Poult. Sci. 91:1350-1355.

Kim, E. J., P. L. Utterback, and C. M. Parsons. 2012b. Comparison of amino acid digestibility coefficients for corn, corn gluten meal and corn distillers dried grains with solubles among 3 different bioassays. Poult. Sci. 91:3141-3147.

Kong, C. and O. Adeola. 2014. Invited review: Evaluation of amino acid and energy utilization in feedstuff for swine and poultry diets. Asian Australas. J. Anim. Sci. 27:917-925.

Lemme, A., V. Ravindran, and W. L. Bryden. 2004. Ileal digestibility of amino acid in feed ingredients for broilers. World's Poult. Sci. J. 60:423-437.

Moter, V. and H. H. Stein. 2004. Effect of feed intake on endogenous losses and amino acid and energy digestibility by growing pigs. J. Anim. Sci. 82:3518-3525.

Moughan, P. J., G. Schuttert, and M. Leenaars. 1992. Endogenous amino acid flow in the stomach and small intestine of the young growing rat. J. Sci. Food Agric. 60:437-442.

Nadeem, M. A., A. H. Gillani, A. G. Khan, and M. Nisa. 2005. True metabolizable energy values of poultry feedstuffs in Pakistan. Int. J. Agric. Biol. 6:990-994.

Newkirk, R. W., H. L. Classen, and M. J. Edney. 2003. Effects of prepress-solvent extraction on the nutritional value of canola meal for broiler chickens. Anim. Feed Sci. Technol. 104:111-119.

NRC. 1994. Nutrient Requirements of Poultry. 9th edn. National Academy Press, Washington DC, USA.

Palliyeguru, M. W. C. D., S. P. Rose, and A. M. Mackenzie. 2010. Effect of dietary protein concentration on the incidence of subclinical necrotic enteritis and growth performance of broiler chicken. Poult. Sci. 89:34-43.

Parsons, C. M., K. Hashimoto, K. J. Wedekind, Y. Han, and D. H. Baker. 1992. Effect of over processing on availability of amino acids and energy in soybean meal. Poult. Sci. 71:133-140.

Phelps, R. A. 1966. Cottonseed meal for poultry: from research to practical application. World's Poult. Sci. J. 22:86-112.

Piper, E. L. and K. J. Boote. 1999. Temperature and cultivar effects on soybean seed oil and protein concentrations. J. Am. Oil Chem. Soc. 76:1233-1241.

Ravindran, V., L. I. Hew, and W. L. Bryden. 2005. Apparent ileal digestibility of amino acids in dietary ingredients for broiler chickens. J. Anim. Sci. 81:85-97.

Ravindran, V., L. I. Hew, G. Ravindran, and W. L. Bryden. 1999. A comparison of ileal digesta and excreta analysis for the determination of amino acid digestibility in food ingredients for poultry. Br. Poult. Sci. 40:266-274.

Ravindran, V., M. R. Abdollahi, and S. M. Bootwalla. 2014. Nutrient analysis, metabolizable energy and digestible amino acids of soybean meals of different regions for broilers. Poult. Sci. 93:2567-2577.

Robbins, K. R., D. H. Baker, and J. W. Finley. 1980. Studies on the utilization of lysinoalanine and lanthionine. J. Nutr. 110:907-915.

Rostagno, H. S., L. F. T. Albino, J. L. Donzele, P. C. Gomes, R. F. Oliveira, D. C. Lopes, A. S. Ferreira, and S. L. T. Barreto. 2005. Brazilian Tables for Poultry and Swine: Composition of Feedstuffs and Nutritional Requirements, 2 ed. Universidade Federal de Viçosa, Viçosa, Brazil.

SAS (Statistical Analysis System), 2009: User's Guide: Statistics, SAS Institute, Inc., Cary, NC, USA.

Sauvant, D., J. M. Perez, and G. Tran. 2004. Tables of composition and nutritive value of feed materials: Pigs, poultry, cattle, sheep, goats, rabbits, horses, fish (Eds. D. Sauvant, J. M. Perez, and G. Tran). Paris, France.

Senkoylu, N. and N. Dale. 1999. Sunflower meal in poultry diets: a review. World's Poult. Sci. J. 55:153-174.

Sibbald, I. R. 1987. Estimation of bioavailable amino acids in feeding stuffs for poultry and pigs: A review with emphasis on balance experiments. Can. J. Anim. Sci. 67:291-300.

Stein, H. H., B. Seve, M. F. Fuller, P. J. Moughan, and C. F. M. de Lange. 2007. Invited review: Amino acid bioavailability and digestibility in pig feed ingredients: Terminology and application. J. Anim. Sci. 85:172-180.

Wang, X. and C. M. Parsons. 1998. Effect of raw material source, processing system, and processing temperatures on amino acid digestibility of meat and bone meals. Poult. Sci. 77:834-841.

Warren, B. E. and D. J. Farrell. 1991. The nutritive value of full-fat and defatted Australian rice bran. V. The apparent retention of minerals and apparent digestibility of amino acids in chicken and adult cockerels fitted with cannulae. Anim. Feed Sci. Technol. 27:232-242.

Yu, F., T. N. Barry, P. J. Moughan, and G. F. Wilson. 1993. Condensed tannin and gossypol concentrations in cotton seeds and in processed cottonseed meal. J. Sci. Food Agric. 63:7-15.

Zhang, Y. and C. M. Parsons. 1994. Effects of over processing on the nutritional quality of sunflower meal. Poult. Sci. 73:436-442.

21

Substitution of Wheat for Corn in Beef Cattle Diets: Digestibility, Digestive Enzyme Activities, Serum Metabolite Contents and Ruminal Fermentation

Y. F. Liu[1,2,a], H. B. Zhao[1,2,a], X. M. Liu[1,2], W. You[1,2], H. J. Cheng[1,2], F. C. Wan[1,2,*],
G. F. Liu[1,2], X. W. Tan[1,2], E. L. Song[1,2], and X. L. Zhang[1,2]
[1] Institute of Animal Science and Veterinary Medicine, Shandong Academy of Agricultural Sciences,
Ji'nan, Shandong 250100, China

ABSTRACT: The objective of this study was to evaluate the effect of diets containing different amounts of wheat, as a partial or whole substitute for corn, on digestibility, digestive enzyme activities, serum metabolite contents and ruminal fermentation in beef cattle. Four Limousin×LuXi crossbred cattle with a body weight (400±10 kg), fitted with permanent ruminal, proximal duodenal and terminal ileal cannulas, were used in a 4×4 Latin square design with four treatments: Control (100% corn), 33% wheat (33% substitution for corn), 67% wheat (67% substitution for corn), and 100% wheat (100% substitution for corn) on a dry matter basis. The results showed that replacing corn with increasing amounts of wheat increased the apparent digestibility values of dry matter, organic matter, and crude protein ($p<0.05$). While the apparent digestibility of acid detergent fiber and neutral detergent fiber were lower with increasing amounts of wheat. Digestive enzyme activities of lipase, protease and amylase in the duodenum were higher with increasing wheat amounts ($p<0.05$), and showed similar results to those for the enzymes in the ileum except for amylase. Increased substitution of wheat for corn increased the serum alanine aminotransferase concentration ($p<0.05$). Ruminal pH was not different between those given only corn and those given 33% wheat. Increasing the substitution of wheat for corn increased the molar proportion of acetate and tended to increase the acetate-to-propionate ratio. Cattle fed 100% wheat tended to have the lowest ruminal NH_3-N concentration compared with control ($p<0.05$), whereas no differences were observed among the cattle fed 33% and 67% wheat. These findings indicate that wheat can be effectively used to replace corn in moderate amounts to meet the energy and fiber requirements of beef cattle. (**Key Words:** Beef Cattle, Digestibility, Digestive Enzyme Activities, Serum Metabolite Contents, Ruminal Fermentation, Wheat)

INTRODUCTION

Corn is a commonly grown crop throughout China and is extensively fed to beef cattle. However, the increasing price of corn, plus the additional subsidy the government is offering farmers to encourage them to grow more corn for the production of ethanol mean that there is growing interest in the use of other crops as an alternative feed source for cattle (Akbar, 2013). Wheat is also a prevalent crop in China, but its nutritive value for ruminants is different from that of corn. Being intermediate in soluble protein, higher in protein and cellulose, and similar in digestible energy and price, wheat is a suitable replacement for corn (Doepel et al., 2009). Studies in beef cattle suggest superior or similar weight gain and feed efficiency when fed wheat vs corn and alfalfa silage. The growth response to wheat silage appears comparable to that with corn silage. Adequate effective fiber with sufficient but not very high fermentable starch, and modest amounts of degradable proteins, make wheat a unique ingredient for sustainable modern ruminant production (Martin et al., 1999). Firkins et al. (2001) reported that total tract dietary of dry matter

* Corresponding Author: F. C. Wan.
E-mail: wanfc@sina.com
[2] Shandong Key Lab of Animal Disease Control and Breeding, Ji'nan, Shandong 250100, China.
[a] These authors contributed equally to this work.

(DM), organic matter (OM), and neutral detergent fiber (NDF) were not affected by dietary source of carbohydrate; however, total tract acid detergent fiber (ADF) digestibility tended to be lower ($p = 0.10$) in cows fed the wheat-based total mixed ration (TMR) compared with those fed the corn-based TMR. Nitrogen tended to be lower in cows fed the wheat-based diets compared with cows fed the corn-based diets. Urinary purine derivative excretion was similar in cows fed the corn and wheat-based diets (Gozho et al., 2008).

To explain the responses of beef cattle offered corn and wheat, a more comprehensive understanding of the effects of the diets on ruminal metabolism and digestion is required, but currently only limited data are available. The objectives of this study were to compare cattle fed wheat vs. corn, at three different levels of substitution, on digestibility, digestive enzyme activities, serum metabolite concentrations and ruminal fermentation of beef cattle.

MATERIALS AND METHODS

Study animals

Animals were cared for and handled in accordance with the guidelines for the care and use of laboratory animals of the Animal Nutrition Research Institute of Shandong Agricultural University and the Ministry of Agriculture of China. Four ruminal, proximal duodenum and the terminal ileum cannulated LiLu (Limousin×LuXi) beef cattle, with average live weight of 400±10 kg, were used in this study.

Study design

The four cattle were used in a 4×4 Latin square design experiment and housed in individual pens with free access to fresh water at the farm of Shandong Academy of Agricultural Sciences (Ji'nan, Shandong, China). There were 4 experimental periods with 13 d for dietary adaptation and 3 d for sample collection in each period. During the experimental period, the temperature of the ambient temperature was maintained between 20°C and 28°C.

The treatments comprised a control group (100% corn), 33% wheat group (33% substitution for corn), 67% wheat group (67% substitution for corn), and 100% wheat group (100% substitution for corn) on a DM basis. The experimental diets were designed to achieve the same amounts of crude protein (CP) and metabolizable energy. During the collection period, total fecal and urine collections were carried out for 3 consecutive days. The diets were balanced to be equal in net energy (MJ/kg) and CP (Table 1) and the cattle were fed twice daily, at 0700 and 1700, in equal portions. Supplementary concentrate was offered at 1% per kg body weight (BW), and the concentrate-to-silage ratio was 4:6. The experimental diets met the requirements of all animals for growth and BW gain according to the cattle feeding standard (NY/T815-2004). The wheat and corn were ground using a hammer mill; the geometric mean particle size of the ground wheat and corn was 1.5 mm. The ingredients and chemical composition (mean±standard deviation) of the basal diets (air-dry basis) used to supplement the cattle are shown in Table 1.

Table 1. Ingredients and chemical composition of the concentrate used to supplement the cattle

Items	Diet			
	Control	33% wheat	67% wheat	100% wheat
Ingredients (% DM)				
Corn	66.00	44.22	21.78	-
Wheat	-	21.78	44.22	66.00
Cottonseed meal	12.00	3.00	1.70	-
DDGS	14.00	21.00	13.00	6.00
Bran	2.50	4.50	12.80	20.80
Soybean oil	-	-	1.00	1.70
Sodium bicarbonate	1.50	1.50	1.50	1.50
NaCl	1.00	1.00	1.00	1.00
$CaCO_3$	0.80	0.80	0.80	0.80
Stone dust	0.16	0.16	0.16	0.16
Trace mineral premix[1]	0.04	0.04	0.04	0.04
Vitamin premix[2]	0.20	0.20	0.20	0.20
Total	100	100	100	100
Chemical composition				
CP (% of DM)	15.42	15.43	15.41	15.42
EE (% of DM)	3.68	4.09	3.27	2.53
ADF (% of DM)	7.00	6.08	5.95	5.80
NDF (% of DM)	12.87	15.00	16.63	18.34
Starch (% of DM)	43.38	45.09	47.76	50.35
NE[3] (MJ/kg)	8.32	8.28	8.31	8.27

DM, dry matter; DDGS, distiller's dried grains with solubles; CP, crude protein; EE, ether extract; ADF, acid detergent fiber; NDF, neutral detergent fiber; NE, net energy.

[1] The trace mineral premix provided the following, per kg of concentrate: Mn, 48 mg; Zn, 36 mg; Fe, 60 mg; Cu, 10 mg; I, 0.30 mg; Se, 0.24 mg; Co, 0.12 mg.

[2] The vitamin premix provided the following, per kg of concentrate: vitamin A (VA), 2,640 IU; vitamin D (VD), 302 IU; vitamin E (VE), 17 mg.

[3] The value of NE was calculated and other nutrient levels were measured.

Determination of digestibility

The feed intake (kg/d) for each animal was calculated as the difference between the feed offered and that refused during the last 3 d of each period. Total collection of feces and urine was conducted during 3 consecutive days for each period. The excreted feces were collected using pans placed behind the animal and weighed daily. The feces were then thoroughly mixed and one portion was taken to determine the fresh feces nitrogen; a second portion was dried at 55°C for at least 96 h, then sifted through a 40 mesh sieve, and stored at 4°C for determination of DM, OM, ADF, and NDF. The DM was determined by oven drying at 135°C for 2 h (AOAC, 1995; method 930.15). The OM content was calculated as the difference between 100% and the percentage of ash (AOAC, 1995; method 942.05). The NDF was determined as described by Van Soest et al. (1991) using heat-stable α-amylase without sodium sulfite. The ADF was determined according to AOAC (1995; method 973.18). The NDF and ADF values are expressed inclusive of residual ash. For the measurement of CP (N×6.25), samples were ground using a ball mill to a fine powder. Total N was determined by flash combustion and thermal conductivity detection.

Total urine was collected from the bladder of each individual using an indwelling balloon catheter. The urine samples from 3 consecutive days for each period were thoroughly mixed, and 10% 6 mol/L hydrochloric acid at a ratio of 1:50 (acid:urine) was added. About 5% of the volume of this mixture was taken and stored at –20°C for analysis of urinary nitrogen.

Assessment of digestive enzyme activities

The duodenal and ileal digesta were collected through the duodenal and ileal fistulas of each animal on the last day before the morning feed and stored at –80°C for measurement of trypsin, lipase and amylase activity. The digesta were thawed at 4°C, and 8 g portions were added to physiological saline according to a weight/volume ratio of 1:1 at 4°C. The mixture was homogenized for 15 s at low temperatures and then centrifuged at 4°C for 10 min at 1,500×*g*. The supernatant was collected for measurement of trypsin, lipase and amylase activity by colorimetry (Zhang et al., 2015). The kits were purchased from Nanjing Jiancheng bio Limited (Nanjing, China).

Analysis of blood variables

Blood samples were obtained from the jugular vein into ethylenediaminetetraacetic acid (EDTA, 3 mL) and non-EDTA (10 mL) containing vacutainer tubes before feed delivery on d 0 and at the end of the study, and the complete blood counts recorded. Plasma was then harvested by centrifugation at room temperature for 5 min at 1,500×*g* and frozen at –20°C for analysis of blood serum metabolites. Serum concentrations of triglycerides, blood urea nitrogen, total cholesterol, high-density lipoprotein, low-density lipoprotein, total protein, blood albumin, aspartate aminotransferase (AST), alanine aminotransferase (ALT) and alkaline phosphatase (ALP) were determined with an Automatic Biochemical Analyzer (7170S, HITACHI, Tokyo, Japan). The concentrations of white blood cells (WBC), lymphocytes, red blood cells, and hemoglobin, and the hematocrit, mean corpuscular volume, mean corpuscular hemoglobin (MCH), mean corpuscular hemoglobin concentration (MCHC), platelets, and platecrits were determined using and Automatic Blood Analyzer (KX-21, SYSMEX, Kobe, Japan).

Measurement of ruminal pH and fermentation

Ruminal pH was monitored continuously for 72 h during d 14 to 16 of each experimental period using a Portable PHB-10 pH meter as described by Penner et al. (2006). The daily ruminal pH data were averaged for each 2-h period and used to calculate the mean pH. Ruminal contents were collected on d 15 of each period. Samples were taken at –1, 1, 3, 5, 7, and 9 h after feeding. Approximately 0.5 L of ruminal contents was obtained from multiple sites within the rumen and strained through nylon mesh (pore size 355 μm). Subsamples (5 mL) of filtrate were preserved with 1 mL of 25% (wt/vol) HPO_3 and 1 mL of 1% H_2SO_4 for determination of volatile fatty acid (VFA) and ammonia nitrogen (NH_3-N) concentrations, respectively. The samples were stored frozen at –20°C until analyzed. Ruminal VFA were separated and quantified using a gas chromatograph (model 7890A, Agilent Technotogies, Santa Clara, CA, USA) with a capillary column (30 m×0.32 mm i.d., 1 μm phase thickness, Zebron ZB-FAAP, Phenomenex, Torrance, CA, USA). The concentration of NH_3-N in the ruminal contents was determined as described by Rhine et al. (1998).

Statistical analyses

The data were analyzed using the MIXED procedures (SAS, Raleigh, NC, USA) for a 4×4 Latin square design. The mixed model included the fixed effects of treatment and the random effects of square, heifer within square, and period within square. Day (or time within day) was considered a repeated measure for variables measured over time. For the repeated measures, various covariance structures were tried, with the final choice depending on low values for the Duncan information criteria. The effect of increasing amounts of wheat in the diet was examined through linear and quadratic orthogonal contrasts using the contrast statement of SAS. Differences were declared significant at $p<0.05$. Trends were discussed at $0.05 \le P < 0.10$ unless otherwise stated.

RESULTS

Digestibility

The apparent digestibility of DM, OM, and CP was not different among the control, 33% wheat and 67% wheat groups, but greater ($p<0.05$) for 100% wheat than for control (Table 3). In contrast, the apparent digestibility of NDF and ADF was lower ($p<0.05$) for 100% wheat than for

Table 2. Chemical composition of forage used to supplement the cattle

Items	
Mixed concentrate (% of DM)	30.95
Leymus chinensis hay (% of DM)	69.05
Chemical composition	
DM (%)	92.39
CP (%)	6.98
EE (%)	2.28
Ash (%)	5.31
ADF (%)	38.74
NDF (%)	64.61
NEmf (MJ/kg)	17.39
Ca (% of DM)	0.48
P (% of DM)	0.07

DM, dry matter; CP, crude protein; EE, ether extract; ADF, acid detergent fiber; NDF, neutral detergent fiber; NEmf, combined net energy.

control. Increased substitution of wheat for corn tended to increase the digestibility of DM and OM as well as that of CP. The digestibility of NDF and ADF decreased as more corn was replaced with wheat.

Digestive enzyme activities

Table 4 showed that increased substitution of wheat for corn significantly increased the activities of trypsin and lipase in the duodenal digesta as well as in the ileal digesta ($p<0.05$). The activity of amylase in the duodenal digesta increased as more corn was replaced with wheat ($p<0.05$), but the activity in the ileal digesta did not differ among the groups ($p>0.05$).

Serum metabolites and whole blood indicators

The hemogram and blood serum chemistry of cattle consuming different amounts of wheat are presented in Tables 5 and 6, respectively. Increased substitution of wheat for corn significantly increased the ALT concentration ($p<0.05$). There were no significant differences among the groups for the other metabolites ($p>0.05$). The whole blood indicators did not differ among the groups ($p>0.05$).

Ruminal pH and fermentation

Ruminal pH over 24 h for all the treatments was greatest just before the 0700 h feed; ruminal pH then declined, then showed another peak at 1,000 h, followed by

Table 3. Effect of diets containing different amounts of wheat as a substitute for corn on nutrient digestibility in beef cattle

	Control	33% wheat	67% wheat	100% wheat	SEM	p
DM (%)	59.75^{a}	60.82^{ab}	64.28^{ab}	67.38^{b}	0.305	0.035
OM (%)	61.72^{a}	62.59^{ab}	68.58^{ab}	70.95^{b}	0.362	0.032
CP (%)	54.21^{a}	55.32^{a}	57.85^{ab}	59.35^{b}	0.681	0.05
NDF (%)	58.57^{a}	55.67^{b}	53.03^{bc}	49.55^{c}	0.543	0.039
ADF (%)	53.70^{a}	49.84^{b}	51.23^{bc}	46.74^{c}	0.693	0.027

SEM, standard error of the mean; DM, dry matter; OM, organic matter; CP, crude protein; NDF, neutral detergent fiber; ADF, acid detergent fiber.
[ab] Means within a column with different letters differ significantly ($p<0.05$).

Table 5. Effect of diets containing different amounts of wheat as a substitute for corn on serum metabolite concentrations in beef cattle

	Control	33% wheat	67% wheat	100% wheat	SEM	p
ALT (U/L)	11.00^{c}	12.75^{bc}	16.75^{ab}	18.00^{a}	1.335	0.0092
AST (U/L)	47.75	46.75	40.75	46.25	4.626	0.7137
ALP (U/L)	59.5	55.25	53.25	58.75	4.948	0.7867
TP (g/L)	80.21	75.53	78.18	74.93	2.187	0.359
ALB (g/L)	28.53	26.93	26.92	29.1	1.392	0.6027
GLOB (g/L)	51.5	48.6	51.23	45.83	2.245	0.2917
BUN (mmol/L)	3.56	3.78	3.72	3.63	0.178	0.8253
TG (mmol/L)	0.29	0.29	0.29	0.26	0.043	0.9424
TC (mmol/L)	1.82	1.79	1.99	1.90	0.261	0.9465
HDL (μmmol/L)	0.73	0.65	0.7	0.73	0.072	0.8305
LDL (μmmol/L)	0.3	0.34	0.37	0.29	0.043	0.5801
LDH (U/L)	916.8	999.5	960.5	754.5	136.062	0.6119

SEM, standard error of the mean; ALT, alanine aminotransferase; AST, aspartate aminotransferase; ALP, alkaline phosphatase; TP, total protein; ALB, blood albumin; GLOB, globulin; BUN, blood urea nitrogen; TG, triglycerides; TC, total cholesterol; HDL, high-density lipoprotein; LDL, low-density lipoprotein; LDH, lactic dehydrogenase.
[ab] Means within a column with different letters differ significantly ($p<0.05$).

Table 4. Effect of diets containing different amounts of wheat as a substitute for corn on digestive enzyme activities in beef cattle

		Control	33% wheat	67% wheat	100% wheat	SEM	p
Duodenal	Lipase	187.67^{a}	194.20^{a}	222.64^{b}	264.47^{c}	9.057	0.0002
	Trypsin	50.93^{a}	56.32^{a}	73.70^{b}	78.13^{b}	4.623	0.0032
	Amylase	0.22^{a}	0.30^{ab}	0.41^{bc}	0.49^{c}	0.051	0.0136
Ileal	Lipase	148.28^{a}	155.47^{ab}	157.38^{ab}	168.17^{b}	4.257	0.0419
	Trypsin	68.25^{a}	77.75^{b}	78.61^{b}	81.28^{b}	2.501	0.016
	Amylase	0.34	0.37	0.37	0.44	0.036	0.2978

SEM, standard error of the mean.
[ab] Means within a column with different letters differ significantly ($p<0.05$).

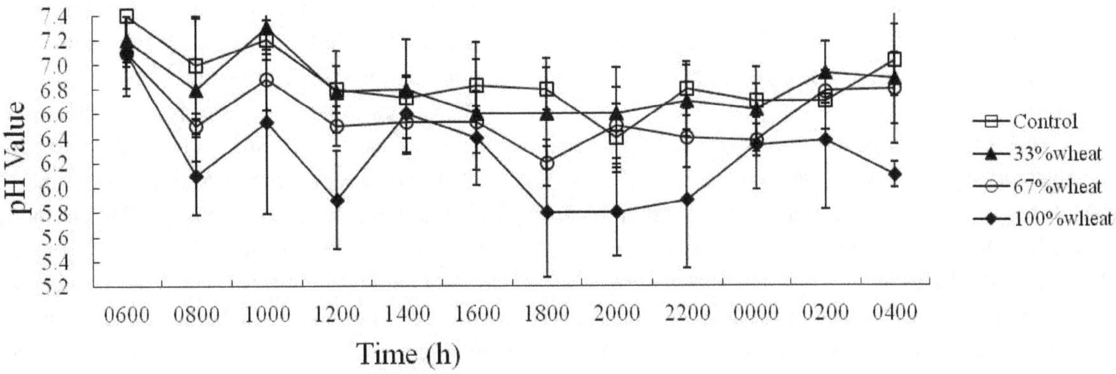

Figure 1. Effect of different diets on ruminal pH. Cattle were fed diets containing different amounts of wheat as a substitute for corn.

a decline until 2,000 h (Figure 1). This pattern was generally similar among treatments, but the pH in cattle fed 100% wheat diets declined more rapidly than that in cattle fed corn, resulting in a lower ruminal pH between 8,000 and 2,200 h. The ruminal pH was not different between the control and 33%, 67% wheat groups. However, increasing replacement of corn with wheat tended to decrease the mean pH (Table 7).

Table 6. Effect of diets containing different amounts of wheat as a substitute for corn on complete blood counts in beef cattle

	Control	33% wheat	67% wheat	100% wheat	SEM	p
WBC (10^9/L)	10.35	10.55	12.78	11.25	2.447	0.8932
LYS (%)	6.5	6.48	8.88	6.91	1.954	0.7963
RBC (10^{12}/L)	6.19	6.81	7.13	6.67	0.549	0.6839
HGB (g/L)	97.75	98.75	104.25	98.33	6.575	0.8874
HCT (%)	29.35	32.83	34.73	32.38	1.640	0.1935
MCV (fL)	48.53	48.8	49.55	49.34	3.002	0.9944
MCH (pg)	16.03	14.48	14.68	14.85	0.605	0.3118
MCHC (g/L)	333.5	300.25	299.5	303.5	12.335	0.2108
PLT (10^9/L)	488	345	505.8	452.5	99.005	0.6706
PCT (%)	0.36	0.29	0.48	0.36	0.105	0.6549

SEM, standard error of the mean; WBC, white blood cells; LYS, lymphocytes; RBC, red blood cells; HGB, hemoglobin; HCT, hematocrit; MCV, mean corpuscular volume; MCH, mean corpuscular hemoglobin; MCHC, mean corpuscular hemoglobin concentration; PLT, platelets; PCT, platecrits.

The contents of individual VFA, such as propionate and butyrate, did not differ between the control and other groups. However, increasing the substitution of wheat for corn increased the molar proportion of acetate and tended to increase the acetate-to-propionate ratio. Cattle fed 100% wheat tended to have the lowest ruminal NH_3-N concentration compared with control ($p<0.05$), whereas no differences were observed among the cattle fed 33% and 67% wheat, compared with control.

DISCUSSION

The attractiveness of wheat is due to its potential for increased yield, increased intake, and ease of preservation compared with other grain sources (GS) (Keady, 2005). Philippeau et al. (1999) reported that the GS (wheat or corn) had little or no effect on DM, OM, or starch intakes. The DM and OM digestibilities in the total tract were greater ($p<0.01$) for the wheat-based diet than for the corn-based diets. In this study, increasing the content of wheat in the diet resulted in a progressive increase in apparent digestibility of DM, OM, and CP, in agreement with the report by Philippeau et al. (1999). This suggests that increasing the proportion of wheat in the diet may positively affect digestibility, largely reflecting variations in ruminal starch digestion. Starch digestibility was almost complete for all wheat treatments, indicating excellent utilization of the wheat grain, as reported by Sinclair et al.

Table 7. Effect of diets containing different amounts of wheat as a substitute for corn on ruminal fermentation in beef cattle

	Control	33% wheat	67% wheat	100% wheat	SEM	p
Mean pH	6.90	6.90	6.60	6.20	0.206	0.188
Total VFA (mmoL/L)	84.04[a]	86.47[ab]	86.86[ab]	88.28[b]	0.035	0.021
Acetic acid (mmoL/L)	55.73[a]	57.06[ab]	57.14[ab]	58.81[b]	0.451	0.034
Propionic acid (mmoL/L)	13.36	13.50	13.73	13.08	0.064	0.143
Butyric acid (mmoL/L)	10.75	11.59	11.65	11.98	0.142	0.064
Acetic acid ratio propionic acid	4.26[a]	4.28[a]	4.19[a]	4.60[b]	0.039	0.019
Ammonia nitrogen (mg/100 mL)	9.26[a]	9.09[ab]	9.02[ab]	8.84[b]	0.024	0.031

SEM, standard error of the mean; VFA, volatile fatty acid.
[ab] Means within a column with different letters differ significantly ($p<0.05$).

(2003) for wheat of relatively similar chemical composition when using a fecal collection method. We have studied the effects of different methods of processing wheat on nutrient digestibility, digestive enzyme activities and serum metabolite concentrations of beef cattle (Jiang et al., 2015) and found that the digestibility of DM, OM, CP, NDF, and ADF in a puffed wheat diet was significantly higher than that in the other three diets ($p<0.05$). The digestibility of DM and OM in both the ground wheat diet and the crushed wheat diet was significantly higher than that in squashed wheat diet ($p<0.05$), and the digestibility of NDF and ADF in the ground wheat diet was significantly lower than that in the other diets ($p<0.05$). Mean particle size decreases markedly from the cereal to the duodenal contents (i.e., 2,209 vs 416 μm for wheat and 3,182 vs 728 μm for corn), as reported by Mc Geough et al. (2010). Moreover, the starch content in particles decreases markedly for wheat by 45%, for particles >2,000 μm, to 98%, for particles <500 μm, of the initial starch content. In contrast, for corn, the starch content remained almost constant in coarse particles and decreased markedly only for particles <1,000 μm. In this study, the wheat and corn were ground using a hammer mill and the geometric mean particle size of the ground wheat and corn was 1,500 μm. This may explain the lower digestibility of OM, DM, and CP for cattle fed corn diets in the present study.

The lower NDF digestibility of wheat compared with control may be explained by the large fraction of digestible NDF in corn. Nuez-Ortin and Yu (2009) reported greater *in situ* 48-h disappearance of NDF from corn (45%) vs wheat (51%). Whether the increase in total gastrointestinal tract NDF digestibility that occurs with dietary inclusion of wheat results from increased ruminal or intestinal digestion, or both, is not known. However, the observed ruminal acetate proportion supports the notion of greater ruminal NDF digestion with a wheat diet. The results indicate that substitution of wheat for corn may improve the digestibility of DM, OM and CP, although the digestibility of NDF and ADF may be decreased.

Little information is available on the distribution of starch digestion in the lower digestive tract (small intestine vs hindgut). In contrast to the study of Philippeau et al. (1999), where postruminal digestion of starch was greater in steers fed corn than those fed wheat (34.1% vs 8.8% of starch intake, respectively), our study showed that the activities of trypsin and lipase in the duodenal digesta in steers fed wheat were greater than in those fed corn. Also, the activity of amylase in the duodenal digesta increased linearly ($p<0.05$) as more corn was replaced with wheat, but the activity in the ileal digesta was not different among groups ($p>0.05$). The results indicate that, with diets comprising a large amount of starch that escapes ruminal fermentation, a considerable amount of starch could be digested in the hindgut because of limitations on starch digestion in the small intestine. These limitations may involve factors related to the limited time and surface exposure in the small intestine, or additional factors such as a limited capacity of the gut tissue to produce starch-hydrolyzing enzymes (Owens et al., 1986; Harmon, 1993).

Serum concentrations of α-glutamyl transpeptidase, ALT, AST, ALP, and cholesterol are conventionally used to diagnose hepatic damage in humans and domestic animals (Silanikove and Tiomkin, 1992). The α-glutamyl transpeptidase has proved to be a sensitive indicator of minor bovine hepatic damage, whereas ALP and cholesterol are used to detect bile obstruction and mild liver damage (Silanikove and Tiomkin, 1992). In the present study, there were no differences ($p>0.10$) in the blood serum metabolites or hemogram of the cattle, except for ALT, which increased as wheat was increased in the diet; however, all values were within the normal range for cattle. In addition, there was no difference ($p>0.10$) in red and WBC counts, indicating that the wheat diet may not have a negative impact on the immune responses of cattle. Lebedeva et al. (2007) found that the activity of the enzyme ALT, which is involved in the integration of protein and carbohydrate metabolism, and its ratio with the activity of AST in the blood of heavy milking Black Pied cows in the dry period characterized the metabolic state of the animals and were associated with their further reproductive capacity. It was shown in the work of Van Kengsel et al. (2007) that the preceding lactation can affect the character and intensity of metabolic processes in the following reproductive cycle in cattle.

Ruminal fermentation characteristics are presented in Table 6 and Figure 1. As expected, the mean pH was less for the wheat-based diet than for the corn-based diets, except for the 33% wheat diet. The pH fell to 5.90 about 5 h after feeding in cattle fed wheat, whereas the minimum pH was 6.70 in cattle fed corn (Figure 1). The pH remained below 6.2 for twice as long in cattle fed wheat compared with cattle fed corn (Philippeau et al., 1999). The fall in pH below 6.2 in relation to the rapid microbial fermentation of starch in wheat may be also involved in the lower fibrolytic activity when wheat replaces corn in the diet (Huhtanen and Khalili, 1992; Martin and Michalet-Doreau, 1995). The increase in the ratio of acetate to propionate as corn was replaced completely resulted primarily from an increase in the molar proportion of acetate and a change in the molar proportion of propionate. Increasing the acetate-to-propionate ration should affect the feed conversion efficiency of growing cattle because less propionate will be provided for gluconeogenesis. The increased molar proportion of acetate suggests that ruminal NDF digestion was increased, possibly due to the reduced ruminal pH.

The ruminal NH_3 and plasma urea concentrations observed indicate that adequate amounts of N were available in the experimental diets, with all plasma urea values falling within the range (3.4 to 7.3 mmol/L) defined by Castejon and Leaver (1994) as being normal. The decrease in rumen NH_3 observed in response to increasing the content of wheat may reflect an increase in microbial N synthesis, facilitated by the increase in starch intake (Hristov and Ropp, 2003). These results indicate that partial substitution of wheat for corn has a minimal impact on ruminal pH and ruminal fermentability of feed.

IMPLICATIONS

In conclusion, substituting the corn content in feedlot finishing diets with different amounts of wheat improved the digestibility of DM, OM, and CP and increased the activities of digestive enzymes, but reduced the digestibility of NDF and ADF. Serum metabolites and whole blood indicators were not affected by replacement of corn by wheat, except for ALT. These results suggest that the feeding value of a wheat-based feedlot finishing diet is similar to or better than that of a corn-based diet. However, ruminal pH status, the ratio of acetate to propionate and ruminal NH_3 were not improved in finishing diets containing 100% wheat as a replacement for corn.

ACKNOWLEDGMENTS

Financial support was provided in part by funds from China Agriculture Research System (CARS-38), the Sustentative Research Project in the Rural Field of the China Ministry of Science and Technology (2012BAD39B05-02), and the Innovation team of cattle industry in the technological system of the Shandong modern agriculture industry (SDAIT-12-011-07).

REFERENCES

AOAC (Association of Official Analytical Chemists). 1995. Official Methods of Analysis. 16th edn. Association of Official Analytical Chemists, Arlington, VA, USA.

Akbar, N. 2013. Barley forages for modern global ruminant agriculture: A review. Russ. Agric. Sci. 39:206-213.

Castejon, M. and J. D. Leaver. 1994. Intake and digestibility of urea-treated whole-crop wheat and liveweight gain by dairy heifers. Anim. Feed Sci. Technol. 46:119-130.

Doepel, L., A. Cox, and A. Hayirli. 2009. Effects of increasing amounts of dietary wheat on performance and ruminal fermentation of Holstein cows. J. Dairy Sci. 92:3825-3832.

Firkins, J. L., M. L. Eastridge, N. R. St-Pierre, and S. M. Noftsger. 2001. Effects of grain variability and processing on starch utilization by lactating dairy cattle. J. Amin. Sci. 79:E218-E238.

Gozho, G. N. and T. Mutsvangwa. 2008. Influence of carbohydrate source on ruminal fermentation characteristics, performance, and microbial protein synthesis in dairy cows. J. Dairy Sci. 91:2726-2735.

Harmon, D. L. 1993. Nutritional regulation of postruminal digestive enzymes in ruminants. J. Dairy Sci. 76:2102-2111.

Hristov, A. N. and J. K. Ropp. 2003. Effect of dietary carbohydrate composition and availability on utilization of ruminal ammonia nitrogen for milk protein synthesis in dairy cows. J. Dairy Sci. 86:2416-2427.

Huhtanen, P. and H. Khalili. 1992. The effect of sucrose supplements on particle-associated carboxymethylcellulase (EC3.2.1.4) and xylanase (EC 3.2.1.8) acticities in cattle given grass silage based diets. Br. J. Nutr. 67:245-255.

Jiang, S. Z., Z. B. Yang., W. R. Yang, Z. Li, C. Y. Zhang, X. M. Liu, and F. C. Wan. 2015. Diets of differentially processed wheat alter ruminal fermentation parameters and microbial populations in beef cattle. J. Anim. Sci. 93:5378-5385.

Keady, T. W. J. 2005. Ensiled maize and whole-crop wheat forages for beef and dairy cattle: Effects on animal performance. In: Silage Production and Utilization. Proceedings of the XIVth International Silage Conference, a Satellite Workshop of the XXth International Grassland Congress, Belfast, N. Ireland. (Eds. R. S. Park and M. D. Stronge). Wageningen Academic Publishers, Wageningen, The Netherlands. pp. 65-82.

Lebedeva, I. Yu., V. B. Leibova, and L. K. Ernst. 2012. Activity of protein and carbohydrate metabolism enzymes in black pied heifer blood in relation to subsequent reproductive intensity. Russ. Agric. Sci. 38:247-250.

Martin, C. and B. Michalet-Doreau. 1995. Variations in mass and enzyme activity of rumen microorganisms: effect of barley and buffer supplements. J. Sci. Food Agric. 67:407-413.

Martin, C., C. Philippeau, and B. Michalet-Doreau. 1999. Effect of wheat and corn variety on fiber digestion in beef steers fed high-grain diets. J. Anim. Sci. 77:2269-2278.

Mc Geough, E. J., P. O'Kiely, K. J. Hart, A. P. Moloney, T. M. Boland, and D. A. Kenny. 2010. Methane emissions, feed intake, performance, digestibility, and rumen fermentation of finishing beef cattle offered whole-crop wheat silages differing in grain content. J. Anim. Sci. 88:2703-2716.

Nuez-Qrtin, W. G. and P. Yu. 2009. Nutrient variation and availability of wheat DDGS, corn DDGS, and blend DDGS from bioethanol plants. J. Sci. Food Agric. 89:1754-1761.

Penner, G. B., K. A. Beauchemin, and T. Mutsvangwa. 2006. An evaluation of the accuracy and precision of a stand-alone submersible continuous ruminal pH measurement system. J. Dairy Sci. 89:2132-2140.

Philippeau, C., C. Martin, and B. Michalet-Doreau. 1999. Influence of grain source on ruminal characteristics and rate, site, and extent of digestion in beef steers. J. Anim. Sci.

77:1587-1596.

Owens, F. N., R. A. Zinn, and Y. K. Kim. 1986. Limits to starch digestion in the ruminant small intestine. J. Anim. Sci. 63:1634-1648.

Rhine, E. D., R. L. Mulvaney, E. J. Pratt, and G. K. Sims. 1998. Improving the Berthelot reaction for determining ammonium in soil extracts and water. Soil Sci. Soc. Am. J. 62:473-480.

Silanikove, N. and D. Tiomkin, 1992. Toxicity induced by poultry litter consumption: effect on measurements reflecting liver function in beef cows. Anim. Prod. 54:203-209.

Sinclair, L. A., R. G. Wilkinson, and D. M. R. Ferguson. 2003. Effects of crop maturity and cutting height on the nutritive value of fermented whole crop wheat and milk production in dairy cows. Livest. Prod. Sci. 81:257-269.

Van Knegsel, A. T. M., H. van den Brand, J. Dijkstra, W. M. van Straalen, R. Jorritsna, S. Tamminga, and B. Kemp. 2007. Effect of glucogenic vs. lipogenic diets on energy balance, blood metabolites, and reproduction in primiparous and multiparous dairy cows in early lactation. J. Dairy Sci. 90:3397-3409.

Van Soest, P. J., J. B. Roberston, and B. A. Lewis. 1991. Methods for dietary fiber, neutral detergent fiber and nonstarch polysaccharides in relation to animal nutrition. J. Dairy Sci. 74:3583-3597.

Zhang, Y., D. W. Chen, B. Yu, J. He, J. Yu, X. B. Mao, J. X. Wang, J. Q. Luo, Z. Q. Huang, G. X. Cheng, and P. Zheng. 2015. Spray-dried chicken plasma improves intestinal digestive function and regulates intestinal selected microflora in weaning piglets. J. Anim. Sci. 93:2967-2976.

22

Effect of KiFAY on Performance, Insulin-like Growth Factor-1, and Thyroid Hormones in Broilers

Amit Kini*, CustanFernandes, and Dayaram Suryawanshi[1]

Vinayak Ingredients (India) Private Limited, Lower Parel (West), Mumbai - 400 013, India

ABSTRACT: A comparative study was performed to investigate the efficacy of KiFAY as a feed additive on performance parameters, thyroid, and pancreatic hormone levels in broilers. Ninety birds (Vencobb 400) were randomly divided into three groups viz., Control (no DL-methionine supplementation), Treatment1 (containing added DL-methionine) and Treatment 2 (containing KiFAY and without DL-methionine supplementation). The performance parameters (weekly body weight, body weight gain, feed intake, and feed consumption ratio) were recorded and calculated during the whole study of 4 weeks. Analyses of insulin and insulin-like growth factor (IGF 1), triiodothyronine (T3), thyroxine (T4) and thyroid stimulating hormone (TSH) were performed at the end of the study. The results show that birds on supplementation of KiFAY performed significantly ($p<0.001$) better than other treatments. The weekly body weight, body weight gain, feed in-take and feed consumption ratio improved in KiFAY treated birds. The study found an increase in insulin and IGF1 levels ($p<0.001$) in KiFAY compared with the other treatments. Serum T3, T4, and TSH levels in the Treatment 2 were higher than other treatments ($p<0.001$). The KiFAY supplementation was able to improve performance with associated responses at a hormonal level in broilers. (**Key Words:** KiFAY, DL-methionine, Growth Performance, Thyroid Hormone, Insulin-like Growth Factor 1, Broilers)

INTRODUCTION

Amino acid supplementation has been the backbone of poultry nutrition. This well-known fact is demonstrated with methionine being the first limiting amino acid for broilers raised on a corn-soy based diet (Burley, 2012). The utilization of synthetic methionine and improving its bioavailability in feed has played a key role in constantly improving poultry performance. There are two main synthetic methionine varieties produced chemically to be used in poultry viz., DL-methionine powder form (DLM, 99%), and 2-hydroxy-4-(methylthio) butyric acid liquid form (HMB, 88%). Recently a third variety derived from biomass has been introduced. The relative bio-efficacy of DLM and HMB differs as shown in laying hens. A study by Danner and Bessei (2002) demonstrated the efficacy of HMB relative to DLM, which found to be 69% (feed conversion ratio, FCR) and 67% (daily egg mass) lower than DLM. Another study by Meirelles et al. (2003) compared DL-methionine hydroxy analog-free acid (MHA-FA) to DLM supplementation and demonstrated the latter to be a better performing feed additive for methionine supplementation. Recently, a direct relation in the relative bioavailability of MHA-FA, a poly herbal ingredient (PHI) with DLM was proposed as a feed additive The PHI claims to contain a blend of dipeptides and oligopeptides of methionine as well as certain intermediates of the same amino acid. The study states a 58% and 4.5% relative bioavailability for MHA-FA and PHI to that of DLM on a product basis respectively (Sangali et al., 2014).

Poultry nutritionists formulate dietary levels as per ileal amino acid digestibility of poultry feed ingredients. The digestibility rates among ingredients show intra-ingredient interactions to alter digestibility of the complete feed formulation. Digestibility of protein relates to particle size and digestive rates of ingredients (Pacheho et al., 2013). Slowly digestible starches (SDS) have positive effects on

* Corresponding Author: Amit Kini.
E-mail: amit@vinayakcorporation.com
[1] Omega Laboratories, Satara, Maharashtra, 415521, India.

feed conversion rates in low protein diets (Weurding et al., 2003). Another study suggests that the interaction between protein (glutamine and casein) and SDS benefits FCR in broilers due to protein-sparing effects (Entinget al., 2005) and therefore enhances the growth efficiency of broiler chickens. It also suggests that starch digestion rate might alter the metabolic responses of insulin. The addition of synthetic amino acids at high levels also stimulates insulin secretion from pancreas (Murray et al., 1998).

The effects of amino acid availability on the avian endocrine system have been least studied. Reports indicate that plasma level of triiodothyronine (T3) were elevated in broilers with methionine, arginine, lysine and isoleucine deficient dietary levels compared to pair-fed controls (50/50 mixture of broiler feed and purified amino acid diet)(Carew et al., 1997). Later Carew et al. (2003) further reported Met deficiency alters the normal thyroid hormone metabolism along with depressed levels of insulin-like growth factor 2 (IGF-2), but the effect is dependent on the degree of deficiency. In the same experiment, IGF-1 did not show any variation even in a Met deficient diet. However, decrease in plasma IGF-1 levels were exhibited in the broilers with a low protein diet, which was brought to normal by feeding dietary protein (Rosebrough and McMurtry, 1993).

Onion (*Allium cepa*) and Garlic *(Allium sativum)* belong to the family Liliaceae (Ebesunun et al., 2007). These phytogenic derivatives show the presence of bioactive component, dialkylpolysulfides, which is a sulfur-containing organic compound. A study led by Aji et al. (2011) used onion and garlic to improve the productive performance of broiler chickens. Garlic and onion consumption also have shown insulin like effect on the body (Keusgen, 2002). Arginine also shows properties of being a metabolic regulator, stimulating protein synthesis and reducing protein stress induced catabolism by stimulating the secretion of insulin, growth hormone and glucagon (Frank et al., 2007).

KiFAY is a tan colored amorphous powder, composed of phytoconstituent fractions obtained from onion, garlic in diatomaceous earth blended with algal cell wall components. It claims to be a natural amino acid optimizer that improves utilization of methionine in the body. Thus, in this study, we investigated the effect of KiFAY in diets deficient in methionine in contrast to the DLM supplemented diet. The comparison is based on commercial productive performance parameters and hormonal correlations.

MATERIALS AND METHODS

Birds and management

The experimental protocol was verified as per Institutional Animal Ethics Committee (IAEC) guidelines. The birds were raised on a deep litter system, using rice husk as bedding. The chicks were housed in environmentally controlled sheds with a stocking density of one bird per square foot approximated for the bird size at the end of 4 weeks. The Feed in a mash form and water were provided *ad libitum*. The basal feed diet was formulated as per National Research Council (1994) recommendations. The ingredients and chemical compositions of the diets were analyzed and are shown in Table 1. Brooding was carried out for the first 5 days providing 1 watt per bird. The temperature and humidity in the house was maintained in the range of 25°C to 28°C at 70% humidity. The light and dark period was maintained with 23 h light:1 h dark program.

Experimental design

Ninety day-old broiler chicks (Vencobb 400) were divided in three equal groups with equal number of male and females as confirmed by wing sexing. The birds were individually wing banded to determine the flock uniformity during performance analysis. The treatments comprised three replicates of ten birds each. The Treatments composed of control (C), basal Diet deficient in methionine; Treatment 1 (T1), basal Diet with DL-methionine (starter phase-2.7 g/kg, grower phase-2.35 g/kg, and finisher phase-2.06 g/kg) supplementation and Treatment 2 (T2), basal Diet with KiFAY (starter phase-2.7 g/kg, grower-2.35 g/kg, and finisher-2.06 g/kg) supplementation. At the end of each week, body weight gain and feed intake were recorded. Weekly FCR was calculated and Mortality was recorded on a daily basis.

Blood collection from sample space

At 28 d, two birds from each replicate were sacrificed by a cervical bleeding method. Blood was collected in a Vacuette tube for hormone estimation.

Estimation of T3, T4, thyroid stimulating hormone, insulin and insulin-like growth factor 1

After the collection of whole blood, the blood was allowed to clot for 1 h. The tubes were centrifuged at 1,200 g to separate the clot from serum, for 20 minutes at 4°C. The serum was collected in a fresh tube and sent for testing at Suburban diagnostics laboratory, Satara, Pune, 411009. The IGF 1 was estimated using a sandwich enzyme-linked immunosorbent assay kit (Elabscience, Wuhan, Hubei, China) as per the manufacturer's instructions.

Statistical analysis

Statistical analyses were performed by GraphPad Instat version 5.0 from Microsoft Windows 7. The means were compared using one-way analysis of variance with the

Table 1. Dietary formulations and nutrient contents of experimental diets

Items	Starter (1-7 d)	Grower (8-14 d)	Finisher (15-28 d)
Ingredients (g/kg)			
Corn	625	659.36	672.16
Soybean meal	294	260	240
Wheat bran	10	10	10
Degummed soybean oil	30	30	40
Deflourinated phosphate	18	18	18
Ground limestone	10	10	10
Trace minerals[1]	1	1	1
Vitamin premix[2]	0.25	0.25	0.25
Enzyme mix[3]	0.45	0.45	0.45
Liver supplements	0.25	0.25	0.25
Toxin binder	1	1	1
coccidiostat	-	-	0.5
CTC	1	1	-
Antioxidant	0.1	0.1	0.1
Acidifier	1	1	-
Choline chloride	0.167	0.167	0.167
Sodium bicarbonate	0.167	0.167	0.167
Emulsifier	0.5	0.5	0.5
Common Salt	0.781	0.781	0.781
L-tryptophan	0.11	0.05	-
L-threonine	0.575	0.525	0.47
L-lysine	3.45	3.05	2.15
Methionine source[4]	2.7	2.35	2.06
Total	1,000	1,000	1,000
Chemical composition[5]			
Total crude protein	20.07	19	19.2
AME (Kcal/kg)	3,195.79	3,275.65	3,400
Non phytate P (%)	0.45	0.45	0.45
Calcium (%)	1	1	1
Lysine (%)	1.2	1.2	1.2
Threonine (%)	0.8	0.8	0.8

	Control	T1	T2	Control	T1	T2	Control	T1	T2
Methionine (%)[6]	0.25	0.55	0.3	0.22	0.465	0.26	0.2	0.44	0.25
Met+Cyst (%)	0.62	0.771	0.6	0.459	0.71	0.55	0.432	0.675	0.5

CTC; chlortetracycline; AME; apparent metabolizable energy.

[1] Vitamin B_{12}, 15 µg; vitamin B_2, 5 µg; vitamin D_3, 60 µg; D-Biotin, 0.2 mg; folate, 0.75 mg; vitamin B_3, 40 mg; vitamin B_5, 10 mg; vitamin A acetate, 3.5 mg; vitamin E, 15 mg; vitamin B_1, 1.63 mg; vitamin B_6, 3.26 mg.

[2] mg per 100 g of mineral mix: Iron, 20; manganese, 23.1; zinc, 7.4; copper, 1.6; potassium iodate, 0.12; calcium carbonate, 47.8; selenium, 0.2.

[3] Enzyme activity of each enzyme (IU/g); Phytase, 5,000; β-gluconase, 18,000; α-amylase, 35,000; β-xylanase, 1,700.

[4] The methionine source in Treatment 1 is DL-Methionine and Treatment 2 is Kifay.

[5] The total crude protein and amino acids were analyzed by Omega Laboratories, India.

[6] As per methionine type used in different treatments.

Tukey-Kramer multiple comparisons test. Less than 0.05 p value was considered statistically significant.

RESULTS AND DISCUSSION

The use of KiFAY in the current study focuses on the evaluation of performance and hormones used as a markers of methionine accretion in broilers. Table 2 summarizes the effect of different dietary supplementation on performance parameters. In the present study, we were able to achieve 10.06% better final weight with 7% better FCR at 28 days. The weight of the chicks on day zero differed non-significantly ($p>0.05$) among treatments. During the whole study, the feed intake differed significantly every week in each treatment ($p<0.001$). The highest feed intake was observed with control ($p<0.001$), whereas, the highest body weight gain was recorded with T2 ($p<0.001$). The FCR was found to be significantly better in T2 than the other

Table 2. Effect of dietary treatments on cumulative body weight, cumulative body weight gain, feed consumption and feed conversion ratio in broiler chickens

Week	Control	Treatment 1	Treatment 2	p value
Cumulative body weight (g)				
0 d	46.23±0.68[a]	44.43±0.75[a]	44.90±0.77[a]	NS
1 wk	148.33±0.59[a]	135.73±0.65[b]	142.76±0.43[c]	***
2 wk	331.63±0.72[a]	350.33±0.96[b]	371.13±0.59[c]	***
3 wk	582.16±0.83[a]	652.40±1.03[b]	709.96±1.02[c]	***
4 wk	976.23±4.24[a]	1,037.62±3.69[b]	1,153.58±4.12[c]	***
Cumulative body weight gain (g)				
1 wk	102.10±0.88[a]	91.30±0.98[b]	97.86±0.85[c]	***
2 wk	183.30±0.89[a]	214.60±1.17[b]	228.36±0.79[c]	***
3 wk	250.53±0.99[a]	302.06±1.25[b]	338.83±1.22[c]	***
4 wk	394.06±4.18[a]	385.13±3.76[ab]	443.62±4.31[c]	***
Feed consumption (g/bird)				
1 wk	171.16±1.75[a]	163.10±1.52[b]	170.93±1.23[ac]	***
2 wk	311.90±2.4[a]	303.03±6.36[ab]	321.40±4.28[ac]	*
3 wk	429.80±7.52[a]	493.70±1.26[b]	420.73±6.48[ac]	***
4 wk	852.67±5.64[a]	736.72±5.76[b]	815.51±13.66[c]	***
Feed conversion ratio (feed/gain)				
1 wk	1.67±0.018[a]	1.79±0.028[b]	1.74±0.018[ab]	**
2 wk	1.70±0.016[a]	1.41±0.03[b]	1.40±0.018[bc]	***
3 wk	1.71±0.03[a]	1.63±0.008[b]	1.24±0.02[c]	***
4 wk	2.17±0.02[a]	1.91±0.02[b]	1.84±0.03[bc]	***

NS, nonsignificant.

Results are provided as means of treatments along with standard mean error.

[a,b,c] Means in the same row with different superscript differ significantly, * $p<0.05$; ** $p<0.01$; *** $p<0.001$.

treatments ($p<0.001$). The feed consumption increased in control and T2. The increase in feed intake with consistent FCR in T2 could be due to the positive effect on palatability attributed to the presence of garlic extracts in the composition of KiFAY (Abouelfetouh and Moussa, 2012). The increase in feed intake was also suspected due to deficiency of vital nutrients in feed. The study led by Berhe and Gous (2008) reveals that broilers treated with low protein feed tend to have an increase feed intake in Cobb strain but decrease in Ross strain. On the other hand, a methionine deficient diet specifically resulted in reduced feed consumption in Hubbard strain (Carew et al., 2003). However, broadening of the FCR due to reduced methionine supplementation evident in the control group of the present study was not seen in KiFAY treated birds. Mortality during whole study was within permissible limit (less than 1%) in all groups.

Thyroid hormones play important role in growth regulation in chickens as demonstrated by thyroidectomy, which results in reduced growth rate (King and King, 1973). Thyroid hormone analysis revealed a comparative increase in plasma levels of T3, T4 and thyroid stimulating hormone (TSH) in Treatment 2 as shown in Figure 1 and Table 3. Serum thyroid hormones (T3 and T4) in control and treatment1 did not significantly differ ($p>0.05$), however, supplementation of KiFAY in Treatment 2 improved thyroid hormones levels than other treatments ($p<0.001$). The T3 levels were increased in T2 ($p<0.001$) (Figure 1a). These results correlated with findings of Saki et al. (2011) who reported a decreased level T3 in methionine deficient layers. However, these results contradict with that of Buchansak et al. (2006) and Carew et al. (2003) who explained the increased T3 levels results due to methionine deficient feed diet. This shows that thyroid hormone activity can be potential indicator to the metabolic activity in a commercial poultry system (Melesse et al., 2011). Moravej et al. (2006) who demonstrated increased T4 levels in high protein fed Lohmann strain broilers with improved body weight gain support the data. Thus, increase in serum T3 and T4 levels on KiFAY supplementation can be explained by a preliminary increase in TSH levels, which resulted in improving the metabolic activity in broilers. In the present study, there was a significant increase ($p<0.001$) in serum TSH levels in broilers supplemented with KiFAY as compared to other treatments (Figure 1c). This alteration could be attributed to the increase in serum insulin level where the thyroid axes connects the somatotropic axes.

Serum insulin and IGF1 level in birds supplemented with KiFAY significantly differ than other treatments ($p<0.001$) (Figure 2). Insulin and IGF-1 is directly related to the protein turnover rate in broilers. McGuinness and Cogburn (1990) report that IGF1 level peak in initial three

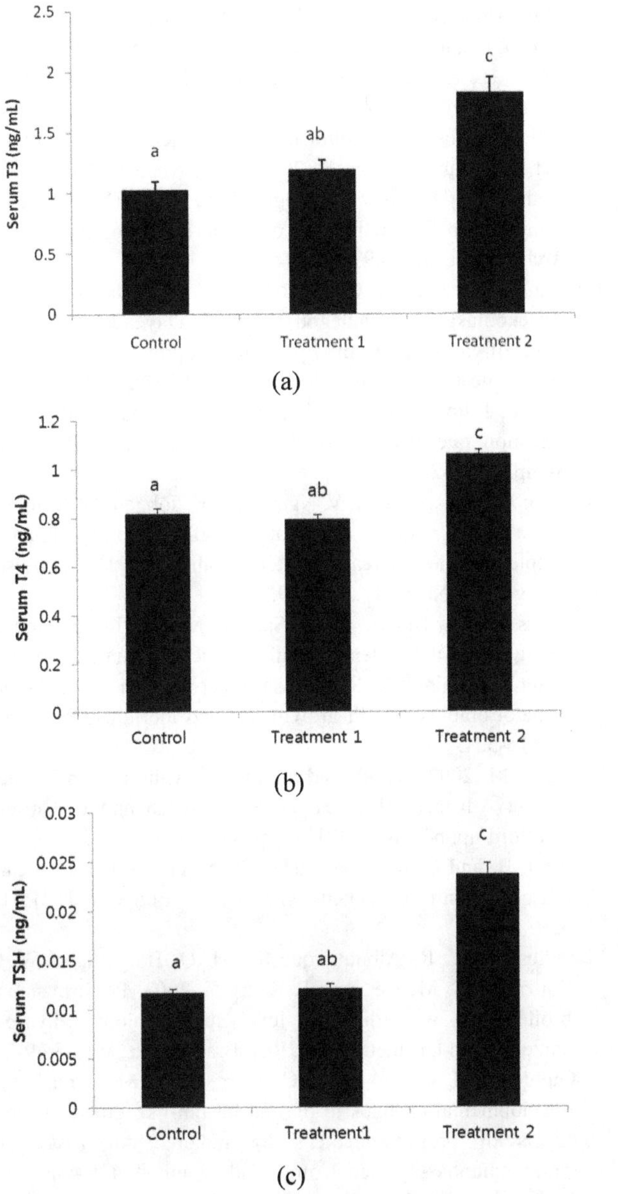

Figure 1. Effect of dietary treatments on (a) T3, (b) T4, and (c) thyroid stimulating hormone (TSH) levels of broilers at 28 days of age. Error bars indicate standard error mean of each treatment. Bars headed by a different letter are significantly different from all others ($p<0.05$).

weeks and T3 level in the fourth week with significant impact on broiler growth performance. This reveals a direct

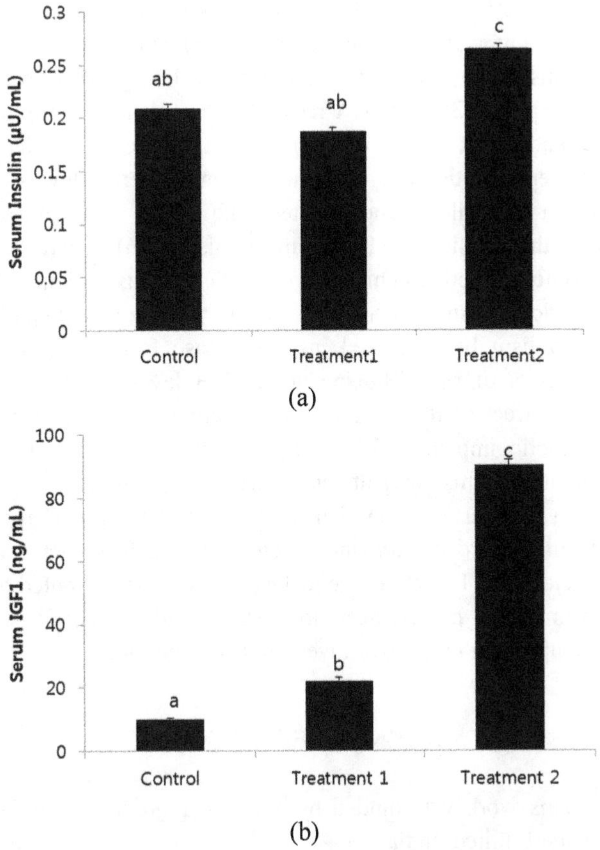

Figure 2. Effect of dietary treatments on (a) Insulin and (b) insulin-like growth factor 1 (IGF1) levels of broilers at 28 days of age. Error bars indicate standard error mean of each treatment. Bars headed by a different letter are significantly different from all others ($p<0.05$).

relation between T3 and IGF1 in broilers. This hormonal increase also overlaps with the period for which the maximum growth occurs in broilers and requires maximum protein accretion. The IGF-1 levels have a direct positive correlation with absolute weight gain and carcass cuts in broilers (Zhai et al., 2016). The improved muscle protein fractional synthesis rate was also shown to be maximum between 2 to 4 weeks of age for fast growing birds and decreased as age progressed as described by Tesseraud et al. (2000). Gonzales et al. (1999) attributed same trend with IGF-1 levels in broilers and showed them independent from the effect of thyroid axes. However, goitrogen

Table 3. Effect of dietary treatments on thyroid hormones (T3, T4, and TSH) and pancreatic hormones (insulin and IGF 1) in broilers

Hormonal parameters	Control	Treatment 1	Treatment 2
Serum T3 levels (ng/mL)	1.0271±0.065	1.1924±0.079	1.8227±0.129
Serum T4 levels (ng/mL)	0.8201±0.0186	0.7956±0.0176	1.06151±0.0188
Serum TSH Levels (ng/mL)	0.0117±0.00035	0.0121±0.00044	0.0237±0.00011
Insulin (µU/mL)	0.2091±0.0115	0.1868±0.004	0.265±0.004
IGF1 (ng/mL)	10±0.5	22±1.2	90±1.9

TSH, thyroid stimulating hormone; IGF1, insulin-like growth factor 1.
The values are provided as means of treatments along with standard mean error.

administration had reduced the growth rate in chicken (Rosebrough and McMurtry, 2003) and anti-thyroid axes administration resulted in decreased circulating IGF-I levels and hepatic IGF-I mRNA restorable by T4 administration (Tsukadaet al., 1998). This explains an indirect relationship between both the axes. The increase in the serum levels of insulin in broilers supplemented with KiFAY is in unison with the studies of Bnouham et al. (2006) where the presence of active compound of garlic such as Allicin and S-allylcysteinsulfoxide have shown to increase hepatic metabolism by increased insulin release from pancreatic beta cells of rats. The increased TSH levels could be a direct effect of insulin secretion affecting thyroid axes for metabolic impetus. The increase in IGF-1 could not be explained at this moment and deserves further study.

In conclusion, supplementation of KiFAY in feed optimized bird performance even when the feed diet was deficient in DL-methionine levels. This can be attributed to the improved protein accretion, which could be confirmed by employing protein turnover studies in the future.

ACKNOWLEDGMENTS

This work was funded by Vinayak Ingredients (India) Private Limited, India.

REFERENCES

Abouelfetouh, A. Y. and N. K. Moussa. 2012. Enhancement of antimicrobial activity of four classes of antibiotics combined with garlic. Asian J. Plant Sci. 11:148-152.

Aji, S. B., K. Ignatius, A. Y. Ado, J. B. Nahu, A. Abdulkarim, U. Aliyu, M. B. Gambo, M. A. Ibrahim, H. Abubakar, M. M. Bukar, H. M. Imam, and P. T. Numan. 2011. Effects of feeding onion (*Allium cepa*) and garlic (*Allium sativum*) on some performance characteristics of broiler chickens. Res. J. Poult. Sci. 4:22-27.

Berhe, E. T. and R. M. Gous. 2008. Effect of dietary protein content on growth, uniformity, and mortality of two commercial broiler strains. S. Afr. J. Anim. Sci. 38:293-302.

Bnouham, M., A. Ziyyat, H. Mekhfi, A. Tahri, and A. Legssyer. 2006. Medicinal plants with potential antidiabetic activity - A review of ten years of herbal medicine research (1990-2000). Int. J. Diabetes Metab. 14:1-25.

Bunchasak, C., T. Sooksridang, and R. Chaiyapit. 2006. Effects of adding methionine hydroxy analogue as methionine source at the commercial requirement recommendation on production performance and evidence of ascites syndrome of male broiler chicks fed corn-soybean based. Int. J. Poult. Sci. 5:744-752.

Burley, H. K. 2012. Enrichment of methionine from naturally concentrated feedstuffs for use in organic poultry diets. Ph.D. Thesis. The Pennsylvania State University, University Park, PA, USA.

Carew, L. B., J. P. McMurtry, and F. A. Alster. 2003. Effects of methionine deficiencies on plasma levels of thyroid hormones, insulin-like growth factors-I and -II, liver and body weights, and feed intake in growing chickens. Poult. Sci. 82:1932-1938.

Carew, L. B., K. G. Evarts, and F. A. Alster. 1997. Growth and plasma thyroid hormone concentrations of chicks fed diets deficient in essential amino acids. Poult. Sci. 76:1398-1404.

Danner, E. E. and W. Bessei. 2002. Effectiveness of liquid DL-methionine hydroxy analogue-free acid (DL-MHA-FA) compared to DL-methionine on performance of laying hens. Arch. Geflugelk. 66:97-101.

Ebesunun, M. O., O. O. Popoola, E. O. Agbedana, J. M. Olisekodiaka, J. A. Onuegbu, and A. A. Onyeagala. 2007. The effect of garlic on plasma lipids and lipoproteins in rats fed on high cholesterol enriched diet. Biokemistri 19:53-58.

Enting, H., J. Pos, R. E. Weurding, and A. Veldman. 2005. Starch digestion rate affects broiler performance. Aust. Poult. Sci. Symp. 17:17-20.

Frank, J. W., J. Escobar, H. V. Nguyen, S. C. Jobgen, W. S. Jobgen, T. A. Davis, and G. Wu. 2007. Oral N-carbamylglutamate supplementation increases protein synthesis in skeletal muscle of piglets. J. Nutr. 137:315-319.

Gonzales, E., J. Buyse, J. R. Sartori, M. M. Loddi, and E. Decuypere. 1999. Metabolic disturbances in male broilers of different strains. 2. Relationship between the thyroid and somatotropic axes with growth rate and mortality. Poult. Sci. 78:516-521.

Keusgen, M. 2002. Health and allium. In: Allium Crop Science: Recent Advances (Eds. H. D. Rabinowitch and L. Currah). Stratford-upon-Avon, CT, UK. pp. 357-373.

King, D. B. and C. R. King. 1973. Thyroidal influence on early muscle growth of chickens. Gen. Comp. Endocrinol. 21:517-529.

Meirelles, H. T., R. Albuquerque, L. M. O. Borgatti, L. W. O. Souza, N. C. Meister, and F. R. Lima. 2003. Performance of broilers fed with different levels of methionine hydroxy analogue and DL-methionine. Rev. Bras. Cienc. Avic. 5:69-74.

McGuinness, M. C. and L. A. Cogburn. 1990. Measurement of developmental changes in plasma insulin-like growth factor-I levels of broiler chickens by radioreceptor assay and radioimmunoassay. Gen. Comp. Endocrinol. 79:446-458.

Melesse, A., S. Maak, R. Schmidt, and G. von Lengerken. 2011. Effect of long-term heat stress on key enzyme activities and T3 levels in commercial layer hens. Int. J. Livest. Prod. 2:107-116.

Moravej, H., K. Homayoun, S. Mahmod, and M. Y. Hassan. 2006. Plasma concentrations of thyroid hormone and growth hormone in lohmann male broilers fed on different dietary energy and protein levels. Int. J. Poult. Sci. 5:457-462.

Murray, R. K., D. K. Granner, P. A. Mayes, and V. W. Rodwell, 1998. Harper's Illustrated Biochemistry. Appleton Lana, Norwalk, CT, USA. pp. 610-617.

NRC (National Research Council). 1994. Nutrient Requirements of Poultry. 9th rev. ed. National Academy Press, Washington, DC, USA.

Pacheco, W. J., C. R. Stark, P. R. Ferket, and J. Brake. 2013. Evaluation of soybean meal source and particle size on broiler performance, nutrient digestibility, and gizzard development. Poult. Sci. 92:2914-2922.

Rosebrough, R. W. and J. P. McMurtry. 2003. Methimzole and thyroid hormone replacement in broilers. Domest. Anim. Endocrinol. 24:231-242.

Rosebrough, R. W. and J. P. McMurtry. 1993. Protein and energy relationships in the broiler chicken. Effects of protein quantity and quality on metabolism. Br. J. Nutr. 70:667-678.

Saki, A. A., R. N. Harsini, M. M. Tabatabaei, P. Zamani, M. Haghighat, and H. R. H. Matin. 2011. Thyroid function and egg characteristics of laying hens in response to dietary methionine levels. Afr. J. Agric. Res. 6:4693-4698.

Sangali, C. P., L. D. G. Bruno, R. V. Nunes, A. Rodrigues de Oliveira Neto, P. C. Pozza, T. M. Moraes de Oliveira, R. Frank, and R. André Schöne. 2014. Bioavailability of different methionine sources for growing broilers. R. Bras. Zootec. 43:140-145.

Tesseraud, S., A. M. Chagnau, and J. Grizard. 2000. Muscle protein turnover during early development in chickens divergently selected for growth rate. Poult. Sci. 79:1465-1471.

Tsukada, A., T. Ohkubo, K. Sakaguchi, M. Tanaka, K. Nakashima, Y. Hayashida, M. Wakita, and S. Hoshino. 1998. Thyroid hormones are involved in insulin-like growth factor-I (IGF-I) production by stimulating hepatic growth hormone receptor (*GHR*) gene expression in the chicken. Growth Horm. IGF Res. 8:235-242.

Weurding, R. E., H. Enting, and M. W. Verstegen. 2003. The relation between starch digestion rate and amino acid level for broiler chickens. Poult. Sci. 82:279-284.

Zhai, W., E. D. Peebles, X. Wang, P. D. Gerard, H. A. Olanrewaju, and Y. Mercier. 2016. Effects of dietary lysine and methionine supplementation on Ross 708 male broilers from 21 to 42 d of age (III): serum metabolites, hormones, and their relationship with growth performance. J. Appl. Poult. Res. 25:223-231.

23

Effect of Dietary Supplementation of Red Ginseng By-product on Laying Performance, Blood Biochemistry, Serum Immunoglobulin and Microbial Population in Laying Hens

H. K. Kang, S.-B. Park, and C. H. Kim*

Poultry Science Division, National Institute of Animal Science, RDA, Cheonan 331-801, Korea

ABSTRACT: This study was carried out to investigate the effect of dietary supplementation of red ginseng by-product (RGB) on the laying performance, blood biochemistry, and microbial population in laying hens. A total of 120 Hy-Line Brown laying hens (75 weeks old) were randomly allotted to 1 of 3 dietary treatments with 4 replicates per treatment. A commercial-type basal diet was prepared, and 2 additional diets were prepared by supplementing 5.0 or 10.0 g/kg of RGB to the basal diet at the expense of corn. The diets were fed to hens on an *ad libitum* basis for 4 weeks. There were no differences in feed intake, egg weight, and feed conversion ratio during 4 weeks of the feeding trial. However, hen-day egg production was significantly greater ($p<0.05$) for the RGB treatment groups than that for the basal treatment group. There were no differences in triglyceride, aspartate aminotransferase, and alanine aminotransferase during the 4-week feeding trial. However, RGB supplementation increased ($p<0.05$) the serum immunoglobulin G (IgG) and IgM content compared with basal treatment group. The total cholesterol was lower ($p<0.05$) in the RGB treatments groups than that in the basal treatment group. The intestinal *Lactobacillus* population was greater ($p<0.05$) for the RGB treatments groups than that for the basal treatment group. However, the numbers of *Salmonella* and *Escherichia coli* were not different among dietary treatments. During the entire experiment, there was no significant difference in egg quality among all the treatments. In conclusion, in addition to improving hen-day production, there were positive effects of dietary RGB supplementation on serum immunoglobulin and cholesterol levels in laying hens.
(**Key Words:** Red Ginseng By-product, Laying Hen, Laying Performance, Serum Cholesterol)

INTRODUCTION

Ginseng (*Panax ginseng* Meyer) is a traditional medicinal plant, and has been widely used over 2,000 years in Korea, Japan, and China (Szeto et al., 2010). Ginseng contains saponin, ginsenosides, essential oils, polyacetylenic alcohol, peptides, vitamins, and polysaccharides (Cheung et al., 2007; Kang et al., 2007; Jin et al., 2008; Wee et al., 2011). Various beneficial pharmacological effects of ginseng or its components have been reported, including anti-cancer, anti-allergy, anti-inflammatory, anti-fatigue, anti-stress, and immunomodulatory activities (Kim et al., 2002; Park et al, 2004; Sumiyoshi et al., 2010; Lee and Cho, 2011). Red ginseng is produced by steaming fresh ginseng at 95°C to 100°C for a reasonable time (Kim et al., 2009). It is known to have greater pharmaceutical efficacy and functionality than white ginseng, due to its high saponin content (Kim and In, 2010). Also, red ginseng may be useful for the treatment of hypertension and pulmonary vascular obstruction (Han et al., 2005) and has immune stimulatory and antioxidant activity as well as vasorelaxing effects in several arterial vessels (Gillis, 1997; Shin et al., 2000). Previous studies with animals have demonstrated the potential effects of ginseng as an immunization agent against various pathogens (Hu et al., 2003; Rivera et al., 2003). For example, dietary supplementation of red ginseng extracts has been shown to improve egg production of laying hens (Jang et al., 2007; Kim et al., 2015). Many studies have indicated that saponin is a potentially bioactive ingredient of red ginseng (Ao et al., 2011). The extraction of red ginseng creates by-product has a potential bioactive

* Corresponding Author: C. H. Kim.
E-mail: kch8059@korea.kr

and immunostimulant additive to animal feed. To the best of our knowledge, limited research has been published on the effects of dietary red ginseng marc on immune system and the intestinal microflora of laying hens. Therefore, this study was performed to investigate the effects of dietary red ginseng by-product (RGB) supplementation on egg production, egg quality, blood biochemistry, serum immunoglobulins and microbial population in laying hens.

MATERIALS AND METHODS

The protocol for this experiment was reviewed and approved by the Institutional Animal Care and Welfare Committee of the National Institute of Animal Science, Rural Development Administration, Korea.

Preparation of red ginseng by-product

The RGB was obtained from Korean Ginseng Nonghyup (Geumsan, Korea). RGB was first dried and ground finely using a Wiley mill (J-NCM, Jisico Co. Ltd., Seoul, Korea). RGB saponins were extracted according to the method described in the previous research (Sung and Lee, 2008). The nutrient composition of the RGB was analyzed in duplicate for dry matter (AOAC, 1990; method 934.01), crude ash (AOAC, 1990; method 934.01), crude fat (AOAC, 1990; method 942.05), crude protein (AOAC, 1990; method 988.05), and crude fiber (AOAC, 1990; method; 978.01), and the results are presented in Table 1.

Birds and experimental design

A total of 120, 75-week-old Hy-Line Brown laying hens were randomly allotted to 1 of 3 dietary treatments. Each treatment had 4 replicates with 5 cages and 2 hens per cage (30×37×40 cm = width×length×height). A commercial type basal diet was formulated to meet or exceed nutrient recommendations of the National Research Council (NRC, 1994) for laying hens (Table 2). Two additional diets were prepared by supplementing 5.0 or 10.0 g/kg of RGB to the basal diet at the expense of corn. The experimental period was 4 weeks. During the experiment, hens were provided with feed and water *ad libitum* and were exposed to a 16-h:8-h light:dark lighting schedule. The temperature and humidity of the laying house was maintained at 18°C±3°C and 65% to 70%, respectively, during the experiment.

Table 1. Analyzed composition of red ginseng by-product

Items	Red ginseng by-product[1]
Dry matter (g/kg)	966.8
Crude fat (g/kg)	12.0
Crude fiber (g/kg)	153.6
Crude protein (g/kg)	142.5
Crude ash (g/kg)	29.5
Total saponin (mg/g)	0.62

[1] Nutrient composition was analyzed in duplicate for dry matter (AOAC, 1990; method 934.01), crude fat (AOAC, 1990; method 920.39), crude fiber (AOAC, 1990; method 978.10), crude protein (AOAC 1990; method 988.05).

Table 2. Composition and nutrient content of experimental diets

Items	RGB 0.0	RGB 5.0	RGB 10.0
Ingredients (g/kg)			
Corn	411.5	406.5	401.5
Wheat	150.0	150.0	150.0
Soybean meal	250.0	250.0	250.0
DDGS	50.0	50.0	50.0
Canola meal	20.0	20.0	20.0
Tallow	5.0	5.0	5.0
Molasses	5.0	5.0	5.0
Dicalcium phosphate	7.0	7.0	7.0
Limestone	97.0	97.0	97.0
Sodium chloride	2.0	2.0	2.0
Red ginseng by-product	-	5.0	10.0
Vitamin premix[1]	1.5	1.5	1.5
Mineral premix[2]	1.0	1.0	1.0
Total	1,000.0	1,000.0	1,000.0
Energy and nutrient content[3]			
ME_n (MJ/kg)	11.4	11.4	11.4
Crude protein (g/kg)	142.0	142.0	142.0
Calcium (g/kg)	45.0	45.0	45.0
Available P (g/kg)	3.3	3.3	3.3
Lysine (g/kg)	7.5	7.5	7.5
Methionine (g/kg)	3.6	3.6	3.6

RGB, red ginseng by-product; DDGS, distillers dried grains with soluble; ME_n, nitrogen-corrected metabolizable energy.

[1] Provided per kilogram of the complete diet: vitamin A (vitamin A acetate), 12,500 IU; vitamin D_3, 2,500 IU; vitamin E (DL-α-tocopheryl acetate), 20 IU; vitamin K_3, 2 mg; vitamin B_1, 2 mg; vitamin B_2, 5 mg; vitamin B_6, 3 mg; vitamin B_{12}, 18 µg; calcium pantothenate, 8 mg; folic acid, 1 mg; biotin, 50 µg; niacin, 24 mg.

[2] Provided per kilogram of complete diet: Fe ($FeSO_4 \cdot 7H_2O$), 40 mg; Cu ($CuSO_4 \cdot H_2O$), 8 mg; Zn ($ZnSO_4 \cdot H_2O$), 60 mg; Mn ($MnSO_4 \cdot H_2O$) 90 mg; Mg (MgO) as 1,500 mg.

[3] Nutrient contents in all diet were calculated.

Laying performance and egg quality

Hen-day egg production rate and egg weight were recorded daily, whereas feed intake and feed conversion ratio were recorded weekly. Ten eggs per replicate were randomly collected at the end of the each week to measure eggshell strength, eggshell thickness, eggshell color, and Haugh units (HU). Eggshell strength was measured by the Texture Systems compression test cell (model T2100C, Food Technology Co., Ltd., Rockville, MD, USA) and was expressed as units of compression force exposed to units of eggshell surface area (kg/cm^2). Eggshell thickness (without the inner and outer shell membrane) was determined at three different points (top, middle, and bottom) using a dial pipe gauge (model 7360, Mitutoyo Co. Ltd., Kawasaki, Japan) without cracking the eggshell. Egg yolk color was evaluated by the Roche color fan (Hoffman-La Roche Ltd., Basel, Switzerland; 15 = dark orange; 1 = light pale). The HU values were calculated using a micrometer (model S-8400, Ames, Waltham, MA, USA) from the following formula described by Eisen et al. (1965): HU = 100 log (H–$1.7W^{0.37}$+7.6), where W is egg weight and H is albumen height.

Sample collection

At the end of the 28-day feeding trial, 2 birds/replicate with a body weight close to the average (i.e., 8 birds per treatment) were selected to be euthanized by cervical dislocation. Immediately after death, a 5-mL blood sample was collected from the jugular vein of each bird using ethylenediaminetetraacetic acid vacuum tubes (Becton Dickinson, Franklin Lakes, NJ, USA), and stored on ice. Serum samples, obtained by centrifuging the samples for 20 min at 25,000×*g* and 4°C, were stored at –15°C. Aspartate aminotransferase (AST), alanine aminotransferase (ALT), triglyceride, and total cholesterol in the serum were quantified using an ADVIA 1650 chemistry system (Bayer Diagnostic, Putraux, France). The cecal contents were collected from the euthanized chickens and used for analyzing bacterial populations. The cecum was ligated at both sides and removed from the gastrointestinal tract, and the contents were aseptically collected into a 2-mL Eppendorf tube. The cecal contents were immediately frozen at –80°C before analysis.

Measurement of serum IgG and IgM concentration

The plasma samples were used to measure the concentrations of immunoglobulin (IgG) and IgM isotypes by using chicken IgG, IgA, and IgM enzyme-linked immunosorbent assay quantification kits, respectively (Bethyl Laboratories, Montgomery, TX, USA). Briefly, flat-bottomed microtiter plates were coated for 60 min with capture antibody (purified goat anti-chicken IgG or IgM) and coating buffer (0.05 M carbonate-bicarbonate, pH 9.6). They were washed 3 times with washing solution (50 mM Tris-buffered saline, 0.14 M NaCl, 0.05% Tween 20, pH 8.0); subsequently, blocking solution (50 mM Tris-buffered saline, 0.14 M NaCl, 1% bovine serum albumin (BSA), pH 8.0) was added to the wells for 30 min, and then the wells were rinsed 3 times with washing solution. The calibrator (chicken reference plasma) and sample-conjugate diluent (50 mM Tris-buffered saline, 0.14 mM NaCl, 1% BSA, 0.05% Tween 20, pH 8.0) were used to prepare standards, whereas plasma samples, which were thawed at 4°C overnight, were diluted at 1:1,000 in the sample-conjugate diluent. Subsequently, they were added to the wells for 60 min and washed 5 times with washing solution. The detection antibody horseradish peroxidase (goat anti-chicken IgG, or IgM) diluted in sample-conjugate diluent was added to the wells, incubated for 60 min, and rinsed 5 times with washing solution. Next, the enzyme substrate (3,3′,5,5′-tetramethyl benzidine peroxidase substrate and peroxidase solution B) was added, and the samples were incubated for 15 min (IgM) or 30 min (IgG). Finally, 2 M H_2SO_4 was used to stop the enzyme substrate reaction. A microtiter plate reader (Spectramax 190; Molecular Device, Salt Lake City, UT, USA) was used to measure the absorbance at 450 nm. The immunoglobulin (IgG or IgM) titers were determined by developing a four-parameter logistic curve fit by using the Hy-line brown laying hen reference plasma absorbance.

Measurement of microbial populations

The composite cecal material sample (1 g) of each hen was diluted with 9 mL of 0.9% saline solution and mixed using a vortex. Viable counts of bacteria in the cecal samples were then estimated by plating serial 10-fold dilutions (in 1% peptone solution) on lactobacilli de Man, Rogosa, and Sharpe (Lactobacilli MRS), MacConkey, and *Salmonella shigella* agar plates (Difco Laboratories, Becton Dickinson, Franklin Lakes, NJ, USA) in order to isolate *Lactobacillus*, *Escherichia coli*, and *Salmonella*, respectively. The Lactobacilli MRS plates were then incubated for 48 h at 37°C under anaerobic conditions, and the MacConkey and *Salmonella shigella* plates were incubated for 24 h at 37°C under aerobic conditions. *Lactobacillus*, *E. coli*, and *Salmonella* colonies were counted immediately after removal from the incubator.

Statistical analysis

All data were analyzed by one-way analysis of variance as a completely randomized design using the PROC MIXED procedure (SAS Institute Inc., Cary, NC, USA). Outlier data were identified by the UNIVARIATE procedure of SAS, but no outliers were found. Least squares means were calculated and the means among treatments were compared by the PDIFF option with the Tukey adjustment. Significance was set at $p<0.05$.

RESULTS AND DISCUSSION

There were no differences in feed intake, egg weight, and feed conversion ratio during the 4 weeks of the feeding trial among groups. However, hen-day egg production was greater ($p<0.05$) for RGB treatment groups than that for basal treatment group, and no significant differences were observed among the RGB treatment groups (Table 3). It was previously suggested that ginseng may improve physiological function and immunity, and exerts various pharmacological effects (Kiefer and Pantuso, 2003). Therefore, a beneficial influence of RGB on performance was expected. Jang et al. (2007) and Kim et al. (2015)

Table 3. Laying performance of layers fed the diet containing red ginseng by-product[1]

Items	Dietary treatment[2]			SEM	p-value
	RGB 0.0	RGB 5.0	RGB 10.0		
Hen-day egg production (%)	75.9[b]	82.1[a]	82.6[a]	0.30	0.04
Feed intake (g/d/hen)	150.3	151.8	156.6	1.69	0.82
Egg weight (g)	64.4	63.7	66.1	0.32	0.59
Feed conversion ratio (g/g)	3.07	3.03	3.01	0.04	0.49

RGB, red ginseng by-product; SEM, pooled standard error of means.
[1] Data are least squares means of 4 observations per treatments.
[2] Basal diet was supplemented at the level of red ginseng by-product 5.0 or 10.0 g/kg to the diets, respectively.
[a,b] Values with different superscripts in the same row are significantly different ($p<0.05$).

Table 5. Blood biochemical parameters of layers fed the diet containing red ginseng by-product (RGB)[1]

Items	Dietary treatment[2]			SEM	p-value
	RGB 0.0	RGB 5.0	RGB 10.0		
Total cholesterol (mg/dL)	251.3[a]	238.7[b]	235.1[b]	1.35	0.04
Triglyceride (mg/dL)	266.7	265.1	258.5	1.02	0.19
AST (U/L)	141.7	142.5	140.8	0.59	0.31
ALT (U/L)	14.7	14.7	14.1	0.39	0.28

SEM, pooled standard error of means.
[1] Data are least squares means of 4 observations per treatments.
[2] Basal diet was supplemented at the level of red ginseng by-product 5.0 or 10.0 g/kg to the diets, respectively.
[a,b] Values with different superscripts in the same row are significantly different ($p<0.05$).

reported that fermented wild ginseng culture by-product or red ginseng extracts could increase egg production, which may be attributed to the improvement in the health status of birds fed diets supplemented with ginseng. Jenkins and Atwal (1994) suggested that dietary saponins (main bioactive compounds in ginseng) had adverse effects on growth performance and feed intake of chicks due to their bitter taste. In contrast, no reduction of feed intake due to RGB supplementation was observed in the current experiment. Due to the limited experiments on ginseng in livestock, further studies are needed to be conducted.

In addition, the supplementation of RGB did not have an effect on eggshell strength, eggshell thickness, egg yolk color, and HU during the feeding trial (Table 4). Similar to the present study, Ao et al. (2011) reported that egg qualities were not affected by the supplementation level of red ginseng extracts. However, Jang et al. (2007) observed that supplemented red ginseng could improve egg quality compared to the basal treatment. These conflicting results may be due to different strains, ginseng sources, or methods of their preparation, as reported by Ao et al. (2011). There were no differences in triglyceride, AST, and ALT levels during the 4 weeks of the feeding trial (Table 5). However, the serum IgG and IgM concentrations of the chickens fed RGB supplemented diet were 10.5% and 29.14% higher than those in the basal treatment group respectively, whereas no significant differences were observed among the RGB treatment groups (Table 6). Hu et al. (2003) reported that RGB could have a beneficial effect on immune function. This study also showed that RGB may have positive effect on the immune system of laying hens.

In the current experiment, reduced serum cholesterol by RGB supplementation in the diets is likely caused by the inhibition of cholesterol or bile acid absorption as was observed with ginseng feeding (Oakenfull and Sidhu, 1990). Some saponins form insoluble complexes with cholesterol in the digesta and inhibit the intestinal absorption of endogenous and exogenous cholesterol (Rao and Gurfinkel, 2000). Previous experiments suggested that dietary ginseng impaired avian hepatic cholesterogensis and reduced serum total cholesterol (Qureshi et al., 1983; Muwalla and Abuirmelieh, 1990). Jang et al. (2007) reported that total cholesterol was decreased by supplementation of red ginseng extracts in the diet of laying hens. The concentration of *Lactobacillus* was greater ($p<0.05$) in the RGB treatments groups than that in the basal treatment

Table 4. Eggshell quality of layers fed the diet containing red ginseng by-product (RGB)[1]

Items	Dietary treatment[2]			SEM	p-value
	RGB 0.0	RGB 5.0	RGB 10.0		
Eggshell strength (kg/cm²)	3.40	3.30	3.40	0.11	0.25
Eggshell thickness (μm)	351.0	352.0	360.9	4.35	0.45
Egg yolk color	8.0	8.8	8.6	0.10	0.68
Haugh unit	83.5	83.8	82.9	0.84	0.50

SEM, pooled standard error of means.
[1] Data are least squares means of 4 observations per treatments.
[2] Basal diet was supplemented at the level of red ginseng by-product 5.0 or 10.0 g/kg to the diets, respectively.

Table 6. Serum immunoglobulin (Ig) of layers fed the diets containing red ginseng by-product (RGB)[1]

Items	Dietary treatment[2]			SEM	p-value
	RGB 0.0	RGB 5.0	RGB 10.0		
IgM (μg/mL)	173.75[b]	190.28[a]	192.13[a]	10.25	0.41
IgG (μg/mL)	78.34[b]	100.16[a]	101.17[a]	8.13	0.32

SEM, pooled standard error of means.
[1] Data are least squares means of 4 observations per treatments.
[2] Basal diet was supplemented at the level of red ginseng by-product 5.0 or 10.0 g/kg to the diets, respectively.
[a,b] Values with different superscripts in the same row are significantly different ($p<0.05$).

Table 7. Intestinal microflora of layers fed the diets containing red ginseng by-product (RGB)[1]

Items	Dietary treatment[2]			SEM	p-value
	RGB 0.0	RGB 5.0	RGB 10.0		
	------------ log 10 cfu/g ----------				
Lactobacillus	5.17[b]	5.49[a]	5.63[a]	0.13	0.04
Salmonella	4.66	4.55	4.56	0.45	0.56
Escherichia coli	4.23	4.25	4.30	0.35	0.19

SEM, pooled standard error of means.

[1] Data are least squares means of 4 observations per treatments.

[2] Basal diet was supplemented at the level of red ginseng by-product 5.0 or 10.0 g/kg to the diets, respectively.

[a,b] Values with different superscripts in the same row are significantly different ($p<0.05$).

group, whereas no significant differences were observed among the RGB treatment groups (Table 7). The concentrations of *Salmonella* and *E. coli* in the ileum were not affected by inclusion of RGB in the diets.

The results of this study indicated that dietary supplementation of RGB improves laying performance and decreases serum cholesterol levels. The results also showed that RGB may be utilized as an immunostimulant for laying hens. Therefore, RGB can be considered as a potential functional ingredient to improve egg production and immune response of laying hens.

ACKNOWLEDGMENTS

This work was carried out with the support of the "Cooperative Research Program for Agriculture Science & Technology Development (Project title: Development of technical skills to improve older laying hens meat quality for export invigoration, Project No. PJ008409)", Rural Development Administration, Republic of Korea.

REFERENCES

AOAC. 1990. Official Methods of Analysis. 15th edn. Association of Official Analytical Chemists, Aarlington, VA, USA.

Ao, X., T. X. Zhou, H. J. Kim, S. M. Hong, and I. H. Kim. 2011. Influence of fermented red ginseng extract on broilers and laying hens. Asian Australas. J. Anim. Sci. 24:993-1000.

Cheung, S. C. M., Y. T. Szeto, and I. F.F. Benzie. 2007. Antioxidant protection of edible oils. Plant Foods Hum. Nutr. 62:39-42.

Eisen, E. J., B. B. Bohren, and H. E. McKean. 1965. The haugh unit as a measure of egg albumen quality. Poult. Sci. 41:1461-1468.

Ernst, E. 2010. Panax ginseng: An overview of the clinical evidence. J. Ginseng Res. 34:259-263.

Gillis, C. N. 1997. Panax ginseng pharmacology: A nitric oxide link? Biochem. Pharmacol. 54:1-8.

Han, K., I. C. Shin, K. J. Choi, Y. P. Yun, J. T. Hong, and K. W. Oh. 2005. Korea red ginseng water extract increases nitric oxide concentrations in exhaled breath. Nitric Oxide 12:159-162.

Hu, Y. J., Y. C. Lin, G. L. Zhou, and D. Q. Yu. 2003. Effect of Chinese extracts on performance and T lymphocyte cell subset of yellow broilers. China Poult. 12:14-17.

Jang, H. D., H. J. Kim, J. H. Cho, Y. J. Chen, J. S. Yoo, B. J. Min, J. C. Park, and I. H. Kim. 2007. Effect of dietary supplementation of fermented wild-ginseng culture by-products on egg productivity, egg quality, blood characteristics and ginsenoside concentration of yolk in laying hen. Korean J. Poult. Sci. 34:271-278.

Jenkins, K. J. and A. S. Atwal. 1994. Effects of dietary saponins on fecal bile acids and neutral sterols, and availability of vitamins A and E in the chick. J. Nutr. Biochem. 5:134-137.

Jin, Y., V. S. Kotakadi, L. Ying, A. B. Hofseth, X. Cui, P. A. Wood, A. Windust, L. E. Matesic, E. A. Pena, C. Chiuzan, N. P. Singh, M. Nagarkatti, P. S. Naharkatti, M. J. Wargovich, and L. J. Hofseth. 2008. American ginseng suppresses inflammation and DNA damage associated with mouse colitis. Carcinogenesis 29:2351-2359.

Kang, K. S., T. Yokozawa, N. Yambe, H. Y. Kim, and J. H. Park. 2007. ESR study on the structure and hydroxyl radical-scavenging activity relationships of ginsenosides isolate from Panax Ginseng C. A. Meyer. Biol. Pharm. Bull. 30:917-921.

Kiefer, D. and T. Pantuso. 2003. Panax ginseng. Am. Fam. Physician 68:1539-1542.

Kim, C. S., J. B. Park, K. J. Kim,S. J. Chang, S. W. Ryoo, and B. H. Jeon. 2002. Effect of Korea red ginseng on cerebral blood flow and superoxide production. Acta Pharmacol. Sin. 23:1152-1156.

Kim, D. C. and M. J. In. 2010. Production of hydrolyzed red ginseng residue and its application to lactic acid bacteria cultivation. J. Ginseng Res. 34:321-326.

Kim, H., I, Oh, K. H. Park, N. M. Kim, J. H. Do, and Y. Cho. 2009. Stimulatory effect of dietary red ginseng on epidermal hydration and ceramide levels in ultraviolet-irradiated hairless mice. J. Med. Food 12:746-754.

Kim, Y. J., G. D. Lee, and I. H. Choi, 2015. Effects of dietary red ginseng marc on egg production, egg quality and blood characteristics of laying hens. J. Appl. Anim. Res. 43:242-246.

Lee, J. H. and S. H. Cho. 2011. Korean red ginseng extract ameliorates skin lesions in NC/N ga mice: an atopic dermatitis model. J. Ethnopharmacol. 133:810-817.

Muwalla, M. M. and N. M. Abuirmeileh. 1990. Suppression of avian hepatic cholesterogenesis by dietary ginseng. J. Nutr. Biochem. 1:518-521.

NRC. 1994. Nutrient Requirements of Poultry. 9th rev. edn. National Academy Press, Washington, DC, USA.

Oakenfull, D. G. and G. S. Sidhu. 1990. Could saponins be a useful treatment for hypercholesterolaemia? Eur. J. Clin. Nutr. 44:79-88.

Park, E. K., M. K. Choo, M. J. Han, and D. H. Kim. 2004. Ginsenoside Rh 1 possesses antiallergic and anti-inflammatory activities. Int. Arch. Allergy Immunol. 133:113-120.

Qureshi, A. A., Z. Z. Din, N. Abuirmeleh, W. C. Burger, Y. Ahmad, and C. E. Elson. 1983. Suppression of avian hepatic lipid metabolism by solvent extracts of garlic: Impact on serum lipids. J. Nutr. 113:1746-1755.

Rao, A. V. and D. M. Gurfinkel. 2000. The bioactivity of saponins: Triterpenoid and steroidal glycosides. Durg Metabol. Drug Interact. 17:211-235.

Rivera, E., S. Hu, and C. Concha. 2003. Ginseng and aluminum hydroxide act synergistically as vaccine adjuvants. Vaccine 21:1149-1157.

Shin, H. R., J. Y. Kim, T. K. Yun, G. Morgan, and H. Vainio. 2000. The cancer-preventive potential of Panax ginseng: A review of human and experimental evidence. Cancer Causes Control 11:565-576.

Sumiyoshi, M., M. Sakanaka, and Y. Kimura. 2010. Effects of red ginseng extract on allergic reactions to food in balb/c mice. J. Ethnopharmacol. 132:206-212.

Sung, W. S. and D. G. Lee. 2008. The combination effect of Korean Red Ginseng Saponins with Kanamycin and Cefotaxime against Methicillin-Resistant Staphylococcus aureus. Biol. Pharm. Bull. 31:1614-1617.

Szeto, Y. T., J. W. M. Wong, S. C. Y. Wong, S. C. Park, and I. F. F. Benzie. 2011. DNA protective effect of ginseng and the antagonistic effect of Chinese turnip: A preliminary study. Plant Foods Hum. Nutr. 66:97-100.

Thomas, D. V., A. L. Molan, and A. V. Ravindran. 2010. The ability of green tea to positively modify the gut microflora in broiler chickens. Aust. Poult. Sci. Symp. 21:203-206.

Wee, J. J., M. K. Park, and A. S. Chung. 2011. Biological activities of ginseng and its application to human health. In: Herbal Medicine: Biomolecular and Clinical Aspects (Eds. I. F. F. Benzie and S. Wachel-Galor). CRC Press, Boca Raton, FL, USA. pp. 157-174.

24

Effects of Dietary Calcium Levels on Productive Performance, Eggshell Quality and Overall Calcium Status in Aged Laying Hens

S. H. An[1], D. W. Kim[2], and B. K. An*

Department of Animal Science and Technology, Konkuk University, Seoul 05029, Korea

ABSTRACT: This study was conducted to investigate the effects of diets with varying levels of calcium on egg production, shell quality and overall calcium status in aged laying hens. A total of five hundred 70-wk-old Hy-Line Brown layers were divided five groups and fed one of the five experimental diets with 3.5%, 3.8%, 4.1%, 4.4%, or 4.7% Ca, for 10 weeks. There were no significant differences in feed intake, egg production and egg weight among groups. The cracked eggs were linearly reduced as dietary Ca levels increased to 4.7% ($p<0.01$). A significant linear improvement for eggshell strength and thickness were determined with increasing dietary Ca levels ($p<0.01$). The contents of serum Ca and phosphorus were not affected by dietary Ca levels. With increase in dietary Ca levels, the tibial breaking strength slightly increased. There were no significant differences in the tibial contents of ash, Ca and phosphorus among groups. In conclusion, eggshell quality, as measured by appearance, strength and thickness of eggshell, were influenced by dietary Ca content as expected ($p<0.05$). These results suggested that aged laying hens require relatively higher level of Ca than required levels from current Korean feeding standards for poultry. (**Key Words:** Dietary Calcium Levels, Cracked Eggs, Eggshell Strength, Tibial Breaking Strength, Aged Laying Hens)

INTRODUCTION

The eggshell quality continues to be a major concern of the egg industry. Eggs with inferior shell quality are a leading economical loss to poultry producers (Roberts, 2004). It has been reported that the average of eggs cracked and lost prior to point of consumption ranged from 13% to 20% (Roland, 1988).

The increased incidence of cracked eggs occurs mainly in late laying period. Decrease in eggshell quality of aged laying hens might be attributed to reduced intestinal Ca uptake and increased egg size (Al-Batshan et al., 1994). Egg size and weight increased with increasing hen age, but it is generally not accompanied by a proportional increase in shell weight, which leads to a decrease in the shell weight to egg weight ratio. Elaroussi et al. (1994) suggested that the increase in cracked eggs seen in aged layers could be a result of disturbances related with the Ca homeostasis.

Calcium is one of the key nutrients required for production and optimal eggshell quality of laying hens (Ahmed et al., 2013). Most research reported that a linear improvement in eggshell quality was evident with increasing dietary Ca levels. Roland (1987) also suggested that a linear increase in eggshell quality when feeding dietary Ca above 4.35 g/d. On the other hands, Leeson et al. (1993) did not find any effect of higher levels of dietary Ca on eggshell quality and concluded that 3.4 g of daily Ca intake was enough for brown egg layers. These discrepancies may be attributed to differences in strains, environmental factors and other nutrients such as phosphorus, which can affect Ca requirement (Garlich et al., 1984).

To our knowledge, a considerable amount of research has been conducted on the effect of feeding various Ca levels during early, mid or total laying stage, but only limited information is available on overall Ca requirement

* Corresponding Author: B. K. An.
E-mail: abk7227@hanmail.net
[1] Team of An Educational Program for Specialists in Global Animal Science, Brain Korea 21 Plus Project, Konkuk University, Seoul 05029, Korea.
[2] Animal Nutrition & Physiology Team, National Institute of Animal Science, RDA, Wanjugun 55365, Korea.

in aged laying hens. This experiment was conducted to investigate the effects of dietary Ca levels with equal in the contents of energy and other nutrients, including available phosphorus, on eggshell quality and overall Ca status in aged laying hens.

MATERIALS AND METHODS

Animals, diets and managements

Five hundred 70-wk-old Hy-Line Variety Brown hens were used and allotted in the experimental windowless house. The layers were divided into five dietary treatments with 10 replicates of 10 birds per each and two hens at a time were put into one wire cage (35×40 cm). The layers were fed one of the five experimental diets with 3.5%, 3.8%, 4.1%, 4.4%, or 4.7% Ca, respectively. All diets were formulated to meet and exceed the nutrients requirements of NRC (1994) and Korean Feeding Standard for poultry (2012), except for Ca, as shown in Table 1. The level of available phosphorus was equally set at 0.23% due to addition of conventional phytase. Proximal composition of formulated diets are shown in Table 2. The analyzed values of Ca were slightly lower than calculated composition. The experiment lasted 10 wk and during which diets and water were provided for *ad libitum* intake. A room temperature of 25°C±5°C and a photoperiod of 16/8 h light/dark cycle were maintained throughout the experimental period. The diets were freshly added everyday and feed intake of each replicate was recorded weekly. The protocol for the experiment was approved by the Institutional Animal Care and Use Committee at Konkuk University.

Table 1. Ingredient composition of experimental diets, as-fed basis

Items	Level of Ca (%)				
	3.5	3.8	4.1	4.4	4.7
Ingredients (%)					
Corn	59.55	60.02	60.5	60.98	61.47
Soybean meal	14.54	14.74	14.95	15.16	15.36
Wheat bran	9.49	7.78	6.06	4.34	2.63
Limestone coarse	8.66	9.44	10.23	11.02	11.80
Canola meal	5.00	5.00	5.00	5.00	5.00
Soybean oil	1.00	1.00	1.00	1.00	1.00
Corn gluten meal	0.37	0.59	0.80	1.01	1.22
Dicalcium phosphate	0.71	0.74	0.76	0.79	0.82
Salt	0.30	0.30	0.30	0.30	0.30
Mineral mixture[1]	0.12	0.12	0.12	0.12	0.12
Vitamin mixture[2]	0.10	0.10	0.10	0.10	0.10
DL-methionine, 98%	0.07	0.07	0.07	0.07	0.07
Phytase	0.05	0.05	0.05	0.05	0.05
$NaHCO_3$	0.03	0.03	0.04	0.04	0.04
Choline-Cl, 50%	0.01	0.02	0.02	0.02	0.02
Calculated nutrient content					
CP (%)	14.50	14.50	14.50	14.50	14.50
Ca (%)	3.50	3.80	4.10	4.40	4.70
Avail. P (%)	0.23	0.23	0.23	0.23	0.23
Total Lys (%)	0.65	0.65	0.65	0.65	0.65
Total Met (%)	0.33	0.33	0.33	0.33	0.33
Total TSAA (%)	0.58	0.58	0.58	0.58	0.58
TMEn (kcal/kg)	2,760	2,760	2,760	2,760	2,760

CP, crude protein; TSAA, total sulfur amino acid; TMEn, nitrogen-corrected true metabolizable energy.

[1] Mineral mixture provided following nutrients per kg of diet: Fe, 56 mg; Zn, 106 mg; Mn, 124 mg; Cu, 11.5 mg; I, 1.7 mg; Se, 0.54 mg; Cr, 0.24 mg.

[2] Vitamin mixture provided following nutrients per kg of diet: vitamin A, 8,666 IU; vitamin D_3, 2,666 IU; vitamin E, 20 IU; vitamin K_3, 2 mg; vitamin B_1, 2 mg; vitamin B_2, 4.6 mg; vitamin B_6, 3.3 mg; vitamin B_{12}, 0.013 mg.

Egg production and qualities

In this experiment, the egg production was recorded daily and mean egg weight was determined by daily average weight of egg, excluding abnormal eggs. The percentages of cracked eggs were calculated by replicate (number of soft-shell and broken eggs/number of eggs produced×100). At 6, 8, and 10 wks of experiment, five eggs from each replicate were collected, weighed individually and stored overnight at room temperature for subsequent measurements.

The breaking strength of uncracked eggs was measured with an eggshell strength tester (FHK, Fujihara Ltd., Tokyo, Japan). Eggshell thickness without shell membrane was tested by micrometer (Digimatic micrometer, Series 293-330, Mitutoyo, Japan). Eggshell color and albumin height were measured by using Egg multi tester made by TSS (Technical Services and Supplies Ltd., York, England). Haugh unit, along with albumen height and egg weight, was calculated as previously described (An et al., 2010). Egg yolk color was measured by comparing with Roche yolk color fan (Hoffman-La Roche Ltd., Basel, Switzerland).

Sampling and measurements

At the end of experiment, 10 birds were randomly selected from each treatment. Thereafter, the blood was

Table 2. Analyzed nutrient composition of formulated diet, as-fed basis[1]

Composition (%)	Level of Ca (%)				
	3.5	3.8	4.1	4.4	4.7
Dry matter	91.1	91.6	91.8	91.7	92.1
Crude protein	15.1	15.7	15.3	15.3	15.1
Ether extract	2.1	2.2	3.7	4.1	3.7
Crude fiber	3.1	3.0	2.8	2.7	2.6
Crude ash	7.3	8.4	8.6	8.7	9.4
Ca	3.1	3.3	3.8	4.0	4.3
Total P	0.45	0.47	0.47	0.46	0.45

[1] Data are the mean of duplicate analysis of each diet.

drawn from wing vein and analyzed for concentrations of Ca and phosphorus. The concentrations of serum Ca and phosphorus were measured according to the colorimetric method using biochemical analyzer (Hitachi modular system, Hitachi Ltd., Tokyo, Japan). At euthanasia, the right legs were immediately collected and stored in the refrigerator for the determination of mechanical property and chemical composition of tibias. Bone breaking strength was measured on fresh tibias using an Instron (Model 3342, Instron Universal Testing Machine, Instron Corp., Norwood, MA, USA) with 50-kg-load cell as 50-kg load range with a crosshead speed of 50 mm per min with tibia supported on a 3.35 cm span. The graphs showed the plateau curve of applied maximal force (KN) to measure the tibial strength as expressed as energy stored in the bone. The sheared tibia pieces were collected and deffatted, after which the tibia samples were oven-dried at 100°C for 24 h and then weighted to obtain the dry weight. The tibia samples were ashed in a muffle furnace (Isotemp muffle furnace, Fisher Scientific, Pittsburgh, PA, USA) at 600°C for 24 h in crucibles. The contents of Ca and phosphorus in tibia were determined using AOAC methods (AOAC, 1995).

Statistical analysis

Data were analyzed using the general linear model procedures of SAS 9.2 (SAS Inst. Inc., Cary, NC, USA). The cage was considered the experimental unit. Linear, quadratic, or both compared using the orthogonal contrast coefficients. The NLIN procedure of SAS according to Robbins et al. (2006) was used to find optimum breakpoint of Ca level whenever linear and or quadratic effects were significant.

However, all variables only showed the linear effect that is cannot account for optimum breakpoint of Ca level and therefore proc NLIN procedure was not included in the predictive model. Results were considered significant if their p-values were <0.05.

RESULTS AND DISCUSSION

Egg production

The feed intake and egg production in aged laying hens fed diets with varying levels of Ca are presented in Table 2. There were no significant linear and quadratic trends of dietary Ca levels affecting feed intake, egg production and egg weight. With increasing dietary Ca levels from 3.5% to 4.7%, cracked eggs linearly reduced ($p<0.01$) from 3.6% to 2.1%.

A number of studies with laying hens have reported that laying performance was not influenced by dietary Ca levels. Cufadar et al. (2011) did not find any significant differences in egg production and egg weight among the Ca levels of 3.0%, 3.6%, or 4.2% of diets in aged laying hens. Frost and Roland (1991) and Keshavarz and Nakajima (1993) also reported that different levels of dietary Ca had no significant effect on egg production, egg weight and egg mass. However, an excess of dietary Ca exerted a negative effect on egg production as a result of reduced feed intake (Ousterhout, 1980; Pelicia et al., 2009).

In present study, the average daily feed intake ranged from 117.1 g to 120.5 g. The dietary Ca did not affect the total feed intake in aged layers and had no negative effects on laying performance. There have been contradictory findings in the relationship to feed intake after feeding diets with varying levels of Ca. Olver and Malan (2000) observed that the dietary Ca levels did not influence total feed intake during 16 to 80 wks of age. Contrary to this, Narvaez-Solarte et al. (2006) reported that daily feed intake was decreased as dietary Ca levels increased. While Chandramoni et al. (1998) found that with increasing dietary Ca levels, the daily feed intake tended to be increase, but not significantly. This discrepancy may be attributed to differences in age of bird, dietary energy density and feeding levels of Ca.

A significant linear decrease in incidence of cracked egg was evident with increasing dietary Ca (Table 3). This decrease in incidence of cracked egg might be associated with improvement in eggshell strength and thickness seen according to increasing dietary Ca. The intake of insufficient amounts of Ca may cause poor shell quality, leading to higher incidence of cracked eggs (Jiang et al., 2013). In this study, the feed intake was not affected by dietary Ca levels during overall experimental period, whereas the tentative total Ca intake was increased as

Table 3. Effect of graded levels of dietary calcium on production performance in the aged laying hens[1,2]

Item	Level of Ca (%)					SEM	p-value	
	3.5	3.8	4.1	4.4	4.7		Linear	Quadratic
Feed intake (g/d/bird)	120.5	117.5	117.8	117.1	118.6	2.06	0.517	0.319
Egg production (%)	75.1	76.0	75.2	75.0	79.1	1.43	0.134	0.197
Egg weight (g/egg)	60.4	60.3	61.0	60.8	61.2	0.64	0.223	0.486
Cracked egg (%)	3.6	3.4	2.3	2.2	2.1	0.46	0.007	0.475

SEM, standard error of the means.

[1] Data are least square of mean of 10 replicate with 5 cages with 2 birds per cage.

[2] Mean values from the overall experimental period.

Table 4. Effect of graded levels of dietary calcium on egg and eggshell qualities in aged laying hens[1,2]

Item	Level of Ca (%)					SEM	p-value	
	3.5	3.8	4.1	4.4	4.7		Linear	Quadratic
Eggshell color[3]	38.7	38.4	38.5	38.3	37.6	0.71	0.312	0.704
Yolk color[4]	5.0	5.1	5.3	5.4	5.5	0.14	<0.001	0.429
Eggshell strength[5]	2.25	2.31	2.37	2.37	2.46	0.05	0.003	0.994
Eggshell thickness[6]	34.8	35.1	35.1	36.6	36.0	0.52	0.006	0.890
Haugh unit[7]	76.55	72.64	75.07	72.59	73.89	1.44	0.237	0.305

SEM, standard error of the means; TSS, Technical Services and Supplies Ltd., York, England.

[1] Data are least square of mean of 10 replicate with 5 cages with 2 birds per cage. [2] Mean values from the 76 or 78 to 80 weeks of age.

[3] Eggshell color is measured by Egg multi tester made by TSS. [4] Yolk color is measured by Roche yolk color fan.

[5] Eggshell strength measurement is expressed as kg/cm^2. [6] Eggshell thickness measurement is expressed as 0.01 mm.

[7] Haugh unit value is determined using the procedure described by Haugh (1937). HU = 100×log (H-1.7×w$^{0.37}$+7.6), where: H = albumen height, mm; w = egg weight, g.

dietary Ca increased. To minimize the incidence rates of cracked eggs in aged layers, the diet must supply enough Ca due to the effect being linear.

Eggshell qualities

The egg and eggshell qualities in aged laying hens fed diets with varying levels of Ca are presented in Table 4. There were no significant linear and quadratic trends of dietary Ca levels affecting eggshell color and Haugh unit. Yolk color score was linearly increased as dietary Ca level increased, although the reason for the difference was not explainable. The eggshell quality was influenced by dietary Ca, as expected. The strength and thickness of eggshell were significantly increased ($p<0.01$) by dietary Ca levels in a linear manner (Table 4).

The available results about effect of dietary Ca levels on eggshell qualities are somewhat inconsistent. Jiang et al. (2013) found that layers on a diet with 2.62% Ca had a weaker eggshell breaking strength than those on a diet with 3.7% or 4.4% Ca. Roland (1987) suggested that the eggshell quality was linearly increased when dietary Ca levels were above 4.35 g per day. On the other hands, Keshavarz and Nakajima (1993) reported that increasing levels of dietary calcium from 3.5% to 5.5% did not have any beneficial effects on eggshell qualities in a long-term experiment. Cufadar et al. (2011) also noted that the level of dietary Ca had no significant effect on eggshell breaking strength and eggshell thickness.

The adequacy of recommended amounts of dietary Ca for optimal eggshell qualities is still being studied. But, based on the results obtained from previous studies, a constant increase in the level of dietary Ca has been associated with improvement of laying performance. Castillo et al. (2004) reported that the biological optimum level for maximum eggshell quality (as specific gravity) was 4.62% Ca in diet. An increase in Ca intake from 4.08 to 4.64 g/d improved the eggshell weight and eggshell thickness in aged Brown layers (Safaa et al., 2008), which is consistent with results of this study. Also, the research results led to the definition of a linear effect on dietary Ca with the eggshell quality. Pelicia (2009) reported that using 90 and 108 weeks of age laying hens, there was no effects of dietary Ca on eggshell strength and thickness; but the eggshell percentage and eggshell weight per surface area (ESWSA) was increased by increasing Ca concentration in the diet. And they obtained linear regression equation $y = 0.119x+8.9985$; $R^2 = 0.899$ in eggshell percentage and $y = 1.5879x+78.556$; $R^2 = 0.886$ in ESWSA. Likewise, the present study showed linear effect in the eggshell strength and eggshell thickness. The determined linear regression equations of the effect of dietary Ca on eggshell strength $y = 0.16x+1.70$; $R^2 = 0.941$ and on eggshell thickness $y = 1.31x+30.14$; $R^2 = 0.656$ showed that both eggshell strength and thickness linearly increase as dietary Ca intake increased. Through these results, we consider that dietary Ca has a strong linear relationship to eggshell strength.

The NRC (1994) suggested the Ca requirement of Brown layers to be 3.4% of dietary Ca for 110 g/d feed intake regardless of age, which seems inadequate for optimal eggshell qualities. More recently, the Korean feeding standard for poultry (2012) proposed the Ca requirement for aged Brown layers up to 4.1% at a feed intake of 110 g/d. The maximum requirement for calcium based on eggshell qualities is uncertain due to the effect being linear in present study. Obviously, aged Brown layers require considerably higher level of Ca to optimize eggshell quality than suggested levels in previous studies.

The inclusion of conventional phytase in layer diets has been greatly increased, in response to reduce the feed and production costs and to minimize phosphorus excretion. There is evidence that phytase positively influences the digestion and absorption of Ca, although the available results about dietary phytase did not have any consistent effects on the eggshell qualities. Punna and Roland (1999) observed a beneficial effect on eggshell quality of phytase inclusion, but others did not find any effect (Parsons, 1999). The possibility of positive effect by dietary phytase should not be precluded and further study is needed to clarify

Table 5. Effect of graded levels of dietary calcium on overall calcium status in serum and tibia[1,2]

Item	Level of Ca (%)					SEM	p-value	
	3.5	3.8	4.1	4.4	4.7		Linear	Quadratic
Serum (mg/dL)								
Calcium	28.9	30.2	29.6	27.2	29.0	1.20	0.488	0.561
Phosphorus	6.55	6.53	6.37	5.99	6.11	0.41	0.274	0.815
Tibia								
Length (cm)	11.61	11.88	11.78	11.66	11.98	0.13	0.233	0.876
Strength[3]	16.15	17.50	17.47	17.58	18.43	0.99	0.148	0.818
Ash (%)	48.83	45.85	45.86	46.57	46.84	1.24	0.408	0.912
Calcium (%)	17.79	17.26	17.68	18.16	18.25	0.48	0.234	0.917
Phosphorus (%)	9.19	8.69	8.55	9.27	9.33	0.24	0.253	0.317

SEM, standard error of the means.
[1] Data are least square of mean of 10 replicate with 1 hen per each replicate. [2] Mean values at 80 weeks of age.
[3] Breaking strength measurements is expressed as kg/mm^2.

dietary Ca levels on eggshell quality, depending on whether or not conventional phytase.

Overall calcium status

There were no significant linear and quadratic trends of dietary Ca levels affecting concentration of serum Ca and phosphorus (Table 5). Contrary to this, Frost and Roland (1991) reported that the level of plasma ionized Ca was significantly increased in a linear manner by increasing dietary Ca levels from 2.75% to 4.25%, but not plasma total calcium.

With increase in dietary Ca levels, the tibial breaking strength tended to be increased, but not significantly. There were no significant linear or quadratic trends of dietary Ca affecting ash, Ca and phosphorus contents in tibia among groups (Table 5). This result is consistent with that of Jiang et al. (2013), who reported that the hens fed diet with 4.4% Ca had similar bone density and strength as compared with those of diet with 3.7% Ca. Contrary to these results, a study has shown that increasing dietary Ca level linearly increased bone strength (Roland et al., 1996). Koutoulis et al. (2009) also suggested that increasing dietary Ca levels from 3.5% to 4.0% significantly increased tibial breaking strength in Brown layers at 72 wks of age. The reason for this discrepancy among authors with respect to bone status is not apparent, but might be attributed to differences in age, strain, dietary Ca levels and nutrient specification of experimental diets.

On the basis of present results, the dietary Ca levels did not affect on the total feed intake and laying performance in aged laying hens. But, the eggshell quality can be improved by ingesting more Ca, up to 4.7%, during last third of total laying period. In summary, our results indicate that aged Brown layers require relatively higher level of Ca to reduce cracked eggs and to maximize eggshell qualities than required levels, 4.1% of diet, from current Korean feeding standards for poultry.

ACKNOWLEDGMENTS

This paper was supported by the KU Research Professor Program of Konkuk University. The results were obtained with financial support of National Institute of Animal Science, RDA.

REFERENCES

Ahmed, N. M., K. A. Abdel Atti, K. M. Elamin, K. Y. Dafalla, H. E. E. Malik, and B. M. Dousa. 2013. Effect of dietary calcium sources on laying hens performance and egg quality. J. Anim. Prod. Adv. 3:226-231.

Al-Batshan, H. A., S. E. Sceideler, B. L. Black, J. D. Garlich, and K. E. Anderson. 1994. Duodenal calcium uptake, femur ash and eggshell quality decline with age and increase following molt. Poult. Sci. 73:1590-1596.

An, B. K., H. S. Kwon, B. K. Lee, J. Y. Kim, S. J. You, J. M. Kim, and C. W. Kang. 2010. Effects of dietary skullcap (*Scutellaria baicalensis*) extract on laying performance and lipid oxidation of chicken eggs. Asian Australas. J. Anim. Sci. 23:772-776.

AOAC (Association of Official Analytical Chemists) International. 1995. Official Methods of Analysis of AOAC International, 16th edn. AOAC International, Gaithersburg, MD, USA.

Castillo, C., M. Cuca, A. Pro, M. González, and E. Morales. 2004. Biological and economic optimum level of calcium in White Leghorn laying hens. Poult. Sci. 83:868-872.

Chandramoni, S. B. Jadhao, and R. P. Sinha. 1998. Effect of dietary calcium and phosphorus concentrations on retention of these nutrients by caged layers. Br. Poult. Sci. 39:544-548.

Cufadar, Y., O. Olgun, and A. O. Yildiz. 2011. The effect of dietary calcium concentration and particle size on performance, eggshell quality, bone mechanical properties and tibia mineral contents in moulted laying hens. Br. Poult. Sci. 52:761-768.

Elaroussi, M. A., L. R. Forte, S. L. Eber, and H. V. Biellier. 1994. Calcium homeostasis in the laying hen. 1. Age and dietary calcium effects. Poult. Sci. 73:1581-1589.

Frost, T. J. and D. A. Roland, Sr. 1991. The influence of various calcium, and phosphorus levels on tibia strength, and eggshell quality of pullets during peak production. Poult. Sci. 70:963-969.

Garlich, J., J. Brake, C. R. Parkhurst, J. P. Thaxton, and G. W. Morgan. 1984. Physiological profile of caged layers during one production year, molt and postmolt: Egg production, eggshell quality, liver, femur, blood parameters. Poult. Sci. 63:339-343.

Haugh, R. R. 1937. The Haugh unit for measuring egg quality. US Egg Poult. Mag. 43:552-573.

Jiang, S., L. Cui, C. Shi, X. Ke, J. Luo, and J. Hou. 2013. Effects of dietary energy and calcium levels on performance, egg shell quality and bone metabolism in hens. Vet. J. 198:252-258.

Keshavarz, K. and S. Nakajima. 1993. Re-evaluation of calcium and phosphorus requirements of laying hens for optimum performance and eggshell quality. Poult. Sci. 72:144-153.

Korean Feeding Standard for Poultry. 2012. Nutrient Requirement of Poultry. National Institute of Animal Science, RDA, Suwon, Korea.

Koutoulis, K. C., I. Kyriazakis, G. C. Perry, and P. D. Lewis. 2009. Effect of different calcium sources and calcium intake on shell quality and bone characteristics of laying hens at sexual maturity and end of lay. Int. J. Poult. Sci. 8:342-348.

Leeson, S., J. D. Summers, and L. Caston. 1993. Response of brown-egg strain layers to dietary calcium or phosphorus. Poult. Sci. 72:1510-1514.

Narvaez-Solarte, W., H. S. Rostagno, P. R. Soares, L. F. Uribe-Velasquez, and M. A. Silva. 2006. Nutritional requirement of calcium in white laying hens from 46 to 62 wk of age. Int. J. Poult. Sci. 5:181-184.

NRC (National Research Council). 1994. Nutrient Requirements of Poultry. 9th edn. National Academy Press, Washington, DC, USA.

Olver, M. D. and D. D. Malan. 2000. The effect of choice-feeding from 7 weeks of age on the production characteristics of laying hens. S. Afr. J. Anim. Sci. 30:110-114.

Ousterhout, L. E. 1980. Effect of calcium and phosphorus levels on egg weight and eggshell quality in laying hens. Poult. Sci. 59:1480-1484.

Parsons, C. M. 1999. The Effect of Dietary Available Phosphorus and Phytase Level on Long-term Performance of Laying Hens. BASF Corp., Florham Park, NJ, USA. pp. 24-33.

Pelicia, K., E. Gracia, C. Mori, A. B. G. Faitarone, A. P. Silva, A. B. Molino, F. Vercese, and D. A. Berto. 2009. Calcium levels and limestone particle size in the diet of commercial layers at the end of the first production cycle. Rev. Bras. Cienc. Avic. 11:87-94.

Punna, S. and D. A. Roland, Sr. 1999. Influence of supplemental microbial phytase on first cycle laying hens fed phosphorus-deficient diets from day one of age. Poult. Sci. 78:1407-1411.

Robbins, K. R., A. M. Saxton, and L. L. Southern. 2006. Estimation of nutrient requirements using broken-line regression analysis. J. Anim. Sci. 84:E155-E165.

Roberts, J. R. 2004. Factors affecting egg internal quality and eggshell quality in laying hens. J. Poult. Sci. 41:161-177.

Roland, Sr., D. A. 1987. Calcium and other factors involved in maintaining eggshell quality in commercial Leghorns. In: Proceedings of the Arkansas Nutrition Conference, Little Rock, AR, USA. pp. 96-100.

Roland, Sr., D. A. 1988. Research note: Egg shell problems: Estimates of incidence and economic impact. Poult. Sci. 67:1801-1803.

Roland, Sr., D. A., M. M. Bryant, and H. W. Rabon. 1996. Influence of calcium and environmental temperature on performance of first cycle (Phase 1) commercial leghorns. Poult. Sci. 75:62-68.

Safaa, H. M., M. P. Serrano., D. G. Valencia, M. Frikha, E. Jimenez-Moreno, and G. G. Mateos. 2008. Productive performance and egg quality of brown egg-laying hens in the late phase of production as influenced by level and source of calcium in the diet. Poult. Sci. 87:2043-2051.

25

Influence of Kaolinite Clay Supplementation on Growth Performance and Digestive Function in Finishing Calf-fed Holstein Steers

José Ortiz[1], Martín Montaño[1], Alejandro Plascencia[1,*], Jaime Salinas[2], Noemí Torrentera[1], and Richard A. Zinn

Department of Animal Science, University of California, Davis, CA 95616, USA

ABSTRACT: Two experiments were conducted to examine the influence of kaolinite clay supplementation (0%, 1%, or 2% diet dry matter [DM] basis) on characteristics of digestion (Trial 1) and growth performance (Trial 2) in calf-fed Holstein steers fed a finishing diet. In Trial 1, 6 Holstein steers (539±15 kg) with ruminal and duodenal cannulas were used to evaluate treatment effects on characteristics of digestion. Kaolinite clay supplementation decreased total tract DM digestion (linear effect, $p<0.01$) without effects ($p\geq0.10$) on site and extent of digestion of organic matter, neutral detergent fiber, starch and N, or ruminal microbial efficiency. There were no treatment effects on ruminal pH, volatile fatty acids molar proportions or estimated methane production. In Trial 2, 108 Holstein steers (132.4±5.6 kg) were used in a 308-d study to evaluate growth performance and carcass characteristics. There were no treatment effects ($p>0.10$) on average daily gain (ADG) and gain efficiency (ADG/dry matter intake). Kaolinite supplementation tended (linear effect, $p = 0.08$) to increase dietary net energy (NE) during the initial 112-d period. However, the overall (308-d) effect of supplementation dietary NE was not appreciable ($p>0.20$). However, due to the inertness of kaolinite, itself, the ratio of observed-to-expected dietary NE increased with kaolinite supplementation. This effect was more pronounced (linear effect, $p\leq0.03$) during the initial 224 d of the study. Overall (308 d), kaolinite supplementation tended to increase (linear effect, $p = 0.07$) dietary NE by 3% over expected. Kaolinite supplementation did not affect carcass weight, yield grade, longissimus area, kidney, pelvic and heart fat, and quality grade, but decreased (linear effect, $p = 0.01$) dressing percentage. It is concluded that kaolinite supplementation up to 2% of diet DM may enhance energetic efficiency of calf-fed Holstein steers in a manner independent of changes in characteristics of ruminal and total tract digestion. (**Key Words:** Kaolinite, Feedlot, Supplementation, Cattle, Growth, Digestion)

INTRODUCTION

Clay minerals (bentonite, kaolinite, zeolite) are ubiquitous in nature. Unique structural properties lend to their usefulness as feed additives. Zeolites are a family of minerals of volcanic origin that are composited of crystalline aluminosilicates. Their dimensional structures are characterized by an ability to lose and gain water reversibly and to exchange cations without major change of their structure (Trckova et al., 2004). Bentonite and kaolinite clay (or kaolin) on the other hand, are members of phillosilicates. Bentonite is a rock constituted of highly colloidal plastic clays composed mainly by montmorillonite (Safaei et al., 2014). The particular characteristic of bentonite is its capability to form gel with water and its high cation exchange capacity (about half the CEC of zeolite clay). Kaolinite is formed by the weathering of aluminous minerals such a feldspar, a plastic clay mineral (Owen et al., 2012), kaolinite has a low shrink–swell capacity and a low cation-exchange capacity (1 to 15 meq/100 g). Recognized as safe for feeding to livestock (EFSA, 2016), kaolinite is widely used as a binder for pelleted feeds, for anti-caking, as an anti-diarrheal and for aflatoxin reduction (Spotti et al., 2005). Due to its chemical, and antimicrobials properties, kaolinite supplementation has been shown to also enhance average

* Corresponding Author: Alejandro Plascencia.
E-mail: aplas_99@yahoo.com

[1] Instituto de Investigaciones en Ciencias Veterinarias, UABC, Mexicali, Baja California 21380, México.

[2] Facultad de Medicina Veterinaria y Zootecnia UAT, Cd. Victoria, Tamaulipas 87000, México.

daily gain (ADG), feed efficiencies, and carcass yields in non-ruminants species (Trckova et al., 2004). However, very little work has been reported that evaluates the effects of kaolinite supplementation on growth-performance of feedlot cattle is limited. The objective of this experiment was to examine the influence of kaolinite clay supplementation (0%, 1%, or 2% diet dry matter [DM] basis) on characteristics of digestion, feedlot growth performance and carcass characteristics of calf-fed Holstein steers fed a steam-flaked corn-based diet.

MATERIALS AND METHODS

All procedures involving animal care and management were in accordance with and approved by the University of California, Davis, Animal Use and Care Committee.

Trial 1: Characteristics of digestion and ruminal fermentation

Animals, sampling and treatments: Six Holstein steers (539±15 kg) with cannulas in the rumen (3.8 cm internal diameter) and proximal duodenum were used in a replicated 3×3 Latin square experiment to study treatment effects on characteristics of digestion. Dietary treatments consisted of a steam-flaked corn-based growing-finishing diet supplemented (diet DM basis) with 0%, 1%, or 2% of kaolinite clay (Ione Minerals Inc., Ione, CA, USA). Composition of experimental diets is shown in Table 1. Chromic oxide (3.0 g/kg of diet air-dry basis) was used as an indigestible marker to estimate nutrient flow and digestibility. Steers were maintained in individual pens (4 m^2) with automatic waterers. Diets were fed at 0800 and 2000 h daily. In order to avoid feed refusals, dry matter intake (DMI) was restricted to 10.3 kg/d (equivalent to 1.9% of live weight [LW]). Experimental periods consisted of a 10-d diet adjustment period followed by a 4-d collection period. Between each experimental period, steers were allowed a 7-d recovery period during which all steers were fed the control diet. During collection, duodenal and fecal samples were taken twice daily as follows: d 1, 0750 and 1350 h; d 2, 0900 and 1500 h; d 3, 1050 and 1650 h, and d 4, 1200 and 1800 h. Individual samples consisted of approximately 700 mL of duodenal chime and 200 g (wet basis) of fecal material. Samples from each steer within each collection period were composited for analysis. During the final day of each collection period, ruminal samples were obtained from each steer via ruminal cannula 4 h after feeding. Ruminal fluid pH was determined on fresh samples. Samples were strained through 4 layers of cheesecloth. Two milliliters of freshly prepared 25% (wt/vol) meta-phosphoric acid was added to 8 mL of strained ruminal fluid. Samples were then centrifuged (17,000×*g* for 10 min), and supernatant fluid was stored at –20°C for volatile fatty acids (VFA) analysis by gas chromatograph. Upon completion of the experiment, ruminal fluid was obtained via the ruminal cannula from all steers and composited for isolation of ruminal bacteria by differential centrifugation (Bergen et al., 1968).

Table 1. Composition of experimental diets fed to steers (DM basis)

Item	Kaolinite level (%)		
	0	1	2
Ingredient composition (%)			
Steam-flaked corn	67.53	66.53	65.53
Distillers dried grains plus solubles	10.00	10.00	10.00
Sudan grass hay	12.00	12.00	12.00
Molasses cane	4.00	4.00	4.00
Yellow grease	3.00	3.00	3.00
Urea	1.20	1.20	1.20
Limestone	1.80	1.80	1.80
Magnesium oxide	0.15	0.15	0.15
Trace mineral salt[1]	0.30	0.30	0.30
Kaolinite clay	-	1.00	2.00
Monensin (g/T)	35.00	35.00	35.00
Nutrient composition (DM basis)[2]			
Net energy (Mcal/kg)			
Maintenance	2.21	2.19	2.17
Gain	1.54	1.53	1.51
Crude protein (%)	14.0	13.9	13.8
Calcium (%)	0.79	0.79	0.80
Phosphorus (%)	0.34	0.33	0.33
Potassium (%)	0.75	0.75	0.74
Magnesium (%)	0.29	0.29	0.29
Sulfur (%)	0.16	0.16	0.16

DM, dry matter.

[1] Trace mineral salt contained (%): $CoSO_4$, 0.068; $CuSO_4$, 1.04; $FeSO_4$, 3.57; ZnO, 1.24; $MnSO_4$, 1.07; KI, 0.052; NaCl, 92.96.

[2] Based on tabular values for individual feed ingredients (NRC,1996).

Sample analysis and calculations: Feed, duodenal and fecal samples were subject to the following analysis: DM (oven drying at 105°C until no further weight loss; method 930.15; AOAC, 2000); ash (method 942.05; AOAC, 2000), Kjeldahl N (method 984.13; AOAC, 2000); aNDFom [Van Soest et al., 1991, corrected for neutral detergent fiber (NDF)-ash, incorporating heat stable α-amylase (Ankom Technology, Macedon, NY, USA) at 1 mL per 100 mL of NDF solution (Midland Scientific, Omaha, NE, USA)]; chromic oxide (Hill and Anderson, 1958), and starch (Zinn, 1990). Microbial organic matter (OM) and nitrogen leaving the abomasum were calculated using purines as a microbial marker (Zinn and Owens, 1986). The OM fermented in the rumen was considered equal to OM intake minus the difference between the amount of total OM reaching the duodenum and microbial OM reaching the duodenum. Feed N escape to the small intestine was considered equal to total N leaving the abomasum minus ammonia-N, microbial N

(MN) and endogenous N, assuming endogenous N is equivalent to 0.195 $W^{0.75}$ (Orskov et al., 1986). Ruminal microbial efficiency was estimated as duodenal MN, g/kg OM fermented in the rumen and N efficiency represent the duodenal non-ammonia N, g/g N intake. Methane production (mol/mol glucose equivalent fermented) was estimated based on the theoretical fermentation balance for observed molar distribution of VFA (Wolin, 1960).

Statistical design and analysis: Treatment effects on characteristics of digestion in cattle were analyzed as a replicated 3×3 Latin square design using the MIXED procedure (SAS Inst. Inc., Cary, NC, USA). Treatments effects on digestion and fermentation variables were tested by means of polynomial contrasts (SAS Inst.; Version 9.3). Contrasts were considered significant when the p-value was ≤0.05, and tendencies were identified when the p-value was >0.05 and ≤0.10.

Trial 2: Growth performance and carcass characteristics

Animals and diets: One hundred eight Holstein steers (132.4±5.6 kg) were used in a 308-d experiment to evaluate the influence of kaolinite clay supplementation on growth performance, dietary net energy (NE), and carcass characteristics. Cattle originated from Tulare, California, and were received at the University of California Desert Research Center, El Centro, on April 29, 2014. Upon arrival, steers were treated for parasites (Dectomax Injectable, Zoetis, New York, NY, USA), and injected subcutaneously with tuluthramycin (Draxxin, Pfizer, New York, NY, USA), and 500,000 IU vitamin A (Vital E-A+D, Stuart Products, Bedford, TX, USA). Steers were balanced by weight and assigned within weight groupings to 18 pens (6 steers/pen). Pens were 78 m^2 with 33 m^2 of overhead shade, automatic waterers, and fence-line feed bunks. Dietary treatments were the same as in Trial 1. Composition of experimental diets is shown in Table 1. Diets were prepared at weekly intervals and stored in plywood boxes located in front of each pen. Steers were allowed *ad libitum* access to their experimental diets. Fresh feed was provided twice daily. On day 112 and 224, all steers were injected subcutaneously with 500,000 IU vitamin A (Vital E-A+D, Stuart Products, USA) and implanted with Revalor-S (Intervet, Millsboro, DE, USA).

Estimation of dietary net energy: For calculating steer performance measures of live weight were reduced 4% to account for digestive tract fill. Final shrunk LW was carcass-adjusted by dividing hot carcass weights (HCW) by the decimal fraction of the average dressing percentage (0.618). Daily energy gain (EG; Mcal/d) was calculated by the equation: $EG = ADG^{1.097}\ 0.0557W^{0.75}$, where W is the mean shrunk weight (kg; NRC, 1984). Maintenance energy (EM) was calculated by the equation: $EM = 0.084W^{0.75}$ (Garrett, 1971). Dietary net energy for gain (NE_g) was derived from net energy for maintenance (NE_m) by the equation: $NE_g = 0.877NE_m - 0.41$ (Zinn et al., 2008). DMI is related to energy requirements and dietary NE_m according to the equation: $DMI = EG/(0.877NE_m - 0.41)$, and can be resolved for estimation of dietary NE by means of the quadratic formula: $x = (-b - \sqrt{b^2 - 4ac})/2c$, where: $x = NE_m$, $a = -0.42EM$, $b = 0.887EM + 0.41DMI + EG$, and $c = -0.887DMI$ (Zinn et al., 2008).

Carcass data: The HCW were obtained at time of slaughter. After carcasses chilled for 48 h, the following measurements were obtained: Longissimus (LM) area (cm^2) by direct grid reading of the muscle at the 12th rib; subcutaneous fat (cm) over the LM at the 12th rib taken at a location 3/4 the lateral length from the chine bone end (adjusted by eye for unusual fat distribution); kidney, pelvic and heart fat (KPH) as a percentage of HCW; marbling score (USDA, 1997; using 3.0 as minimum slight, 4.0 as minimum small, 5.0 as minimum modest, 6.0 as minimum moderate, etc.), and estimated retail yield of boneless, closely trimmed retail cuts from the round, loin, rib and chuck (% of HCW; Murphey et al., 1960) = 52.56–1.95×subcutaneous fat–1.06×KPH+0.106×LM area–0.018× HCW.

Statistical design and analysis: Performance (gain, gain efficiency, and dietary energetics) and carcass data were analysed as a randomised complete block design. The experimental unit was the pen. The MIXED procedure of SAS (SAS Institute, 2004) was used to analyse the variables. The fixed effect consisted of treatment, and pen as the random component. Treatment effects were tested by means of orthogonal polynomials (SAS Inst.; Version 9.3). Contrasts were considered significant when the p-value was ≤0.05, and tendencies were identified when the p-value was >0.05 and ≤0.10.

RESULTS AND DISCUSSION

Trial 1: Characteristics of digestion and ruminal fermentation

Treatment effects on characteristics of ruminal and total tract digestion of experimental diets are shown in Table 2. There were no treatment effects (p≥0.10) on ruminal microbial efficiency (g MN/kg OM fermented) or ruminal and total tract digestion N. Very limited information has been reported regarding the effects of kaolinite supplementation on site and extent of digestions of steers fed high-grain finishing diets. Previous studies evaluating supplemental zeolite clay at levels comparable to that of the present study (McCollum and Galyean, 1983; Cole et al., 2007) likewise did not show an effect of supplementation on N digestion in finishing diets fed to cattle.

There were no treatment effects (p≥0.10) on ruminal and total tract digestion of OM, starch and NDF. Because their sorbent properties that increase the ruminal fluid viscosity (Spotti et al., 2005), it has been postulated that inclusion of clay could decrease fluid dilution rate, possibly enhancing

Table 2. Influence of supplementation level of kaolinite clay on characteristics of apparent ruminal and total tract digestion in Holstein steers

Item	Kaolinite level (% diet DM)			SEM	p-value	
	0	1	2		Linear	Quadratic
Steer replicates	6	6	6			
Intake (g/d)						
DM	10,278	10,278	10,278			
OM	9,606	9,606	9,605			
NDF	1,595	1,661	1,728			
N	213	212	211			
Starch	5,371	5,298	5,225			
Flow to duodenum, g/d						
OM	5,250	5,036	5,392	151	0.53	0.16
NDF	1,059	958	1,053	60	0.94	0.22
Starch	955	993	1123	88	0.21	0.69
Non-ammonia N	239	235	236	9.5	0.85	0.82
Microbial N	140	145	141	5.6	0.85	0.59
Feed N	77	68	72	8.0	0.72	0.52
Ruminal digestion (%)						
OM	59.91	62.64	58.57	1.50	0.54	0.10
NDF	33.26	41.99	30.31	3.48	0.25	0.22
Starch	82.22	81.27	78.55	1.65	0.15	0.67
Feed N	64.15	68.00	65.61	3.79	0.79	0.52
MN efficiency [1]	24.47	24.19	25.29	1.23	0.65	0.66
N efficiency [2]	1.12	1.11	1.12	0.05	0.96	0.82
Fecal excretion (g/d)						
DM	2,392	2,648	2,684	52	<0.01	0.12
OM	1,969	2,099	2,096	56	0.15	0.36
NDF	913	972	965	40	0.39	0.53
Starch	86.9	83.6	88.7	16.4	0.94	0.84
N	60.1	64.3	64.2	2.6	0.29	0.52
Total tract digestion (%)						
DM	76.7	74.3	73.9	0.5	<0.01	0.13
OM	79.5	78.2	78.2	0.6	0.15	0.37
NDF	42.7	41.0	44.2	2.4	0.68	0.44
Starch	98.4	98.4	98.3	0.3	0.88	0.84
N	71.8	69.7	69.5	1.2	0.21	0.53

DM, dry matter; SEM, standard error of the mean; OM, organic matter; NDF, neutral detergent fiber; MN, microbial N.

[1] Microbial nitrogen, g/kg organic matter fermented.

[2] Non-ammonia nitrogen flow to the small intestine as a fraction of nitrogen intake.

the extent of ruminal digestion. However, the effect of zeolites and bentonite on ruminal passage rate has been inconsistent (McCollum and Galyean, 1983). Decreased total tract DM digestion with no effect on OM digestion has been reported previously with the inclusion of bentonite in sorghum-based (Martin et al., 1696) and corn silage-based (Ivan et al., 1992) diets. Likewise, Dinius et al. (1970) observed that total tract DM digestion decreased linearly with increasing kaolinite supplementation. Decreased total tract DM digestion with clays supplementation is expected, due to indigestibility of clay, itself.

Kaolinite supplementation did not affect (p>0.10) ruminal pH (averaging 5.82), VFA molar proportions or estimated methane production. Based on diet formulation, expected ruminal pH was 5.73 (NRC, 1996), in reasonably good agreement with observed. Tate et al. (2015) likewise did not observe effects of kaolinite supplementation (7.5 and 15 g/kg DM) on ruminal pH (determined *in vitro*). The lack of effects of kaolinite supplementation on ruminal pH and VFA profiles is consistent with non-appreciable effects on measures of ruminal digestion of OM, starch, NDF, and N (Table 3).

Trial 2: Growth performance and carcass characteristics

The effects of kaolinite clay supplementation on 308-d feedlot growth-performance of calf-fed Holstein steers are shown in Table 4. There were no treatment effects (p>0.10)

Table 3. Influence of supplementation level of kaolinite clay on ruminal pH, volatile fatty acid profile and estimate methane production

Item[1]	Kaolinite level (% diet DM)			SEM	p-value	
	0	1	2		Linear	Quadratic
Ruminal pH	5.87	5.74	5.86	0.09	0.94	0.30
Ruminal VFA (mol/100 mol)						
Acetate	53.2	50.8	50.8	1.8	0.52	0.19
Propionate	36.6	36.6	35.8	2.8	0.83	0.91
Butyrate	10.2	12.6	9.4	1.8	0.75	0.24
Acetate/propionate	1.61	1.53	1.63	0.17	0.95	0.68
Estimated methane[2]	0.40	0.40	0.42	0.03	0.73	0.70

DM, dry matter; SEM, standard error of the mean; VFA, volatile fatty acids.

[1] Measured at 4-h posprandium (morning meal).

[2] Estimated methane based on VFA molar proportions as mol/mol glucose equivalent fermented (Wolin, 1960).

Table 4. Influence of supplementation level of kaolinite clay on growth performance of feedlot steers

Item	Kaolinite level (% diet DM)			SEM	p-value	
	0	1	2		Linear	Quadraticc
Days on test	308	308	308			
Pen replicates	6	6	6			
Live weight (kg)[1]						
Initial	132.4	132.3	132.5	0.86	0.97	0.92
Final[2]	576.8	584.1	580.09	4.83	0.56	0.39
ADG (kg)						
1 to 112 d	1.39	1.40	1.42	0.01	0.11	0.73
112 to224 d	1.53	1.53	1.53	0.02	0.99	1.0
224 to 308 d	1.40	1.47	1.40	0.04	0.93	0.13
1 to 308 d	1.44	1.47	1.46	0.02	0.58	0.40
DMI (kg/d)						
1 to 112 d	5.42	5.35	5.43	0.05	0.95	0.25
112 to 224 d	8.31	8.18	8.32	0.11	0.98	0.32
224 to 308 d	10.45	10.67	10.44	0.11	0.55	0.06
1 to 308 d	7.82	7.83	7.85	0.05	0.70	0.99
ADG/DMI (kg/kg)						
1 to 112 d	0.257	0.262	0.263	0.002	0.59	0.377
112 to 224 d	0.184	0.187	0.184	0.002	0.97	0.222
224 to 308 d	0.135	0.138	0.134	0.003	0.87	0.349
1 to 308 d	0.185	0.187	0.186	0.001	0.62	0.205
Dietary NE (Mcal/kg)						
Maintenance						
1 to 112 d	2.06	2.09	2.10	0.01	0.08	0.38
112 to224 d	2.18	2.21	2.19	0.02	0.62	0.23
224 to 308 d	2.12	2.13	2.11	0.03	0.97	0.64
1 to 308 d	2.15	2.18	2.16	0.02	0.58	0.21
Gain						
1 to 112 d	1.40	1.43	1.43	0.01	0.08	0.38
112 to224 d	1.50	1.53	1.51	0.02	0.62	0.23
224 to 308 d	1.45	1.46	1.44	0.03	0.97	0.64
1 to 308 d	1.47	1.50	1.48	0.01	0.58	0.21
Observed:expected dietary NE						
Maintenance						
1 to 112 d	0.93	0.96	0.97	0.006	<0.01	0.41
112 to224 d	0.99	1.01	1.01	0.008	0.03	0.24
224 to 308 d	0.96	0.98	0.98	0.014	0.32	0.66
1 to 308 d	0.97	1.00	1.00	0.007	0.07	0.22
Gain						
1 to 112 d	0.91	0.95	0.96	0.008	<0.01	0.41
112 to224 d	0.98	1.02	1.02	0.011	0.03	0.24
224 to 308 d	0.95	0.97	0.97	0.018	0.33	0.66
1 to 308 d	0.97	1.00	1.00	0.009	0.07	0.22

DM, dry matter; SEM, standard error of the mean; ADG, average daily gain; DMI, dry matter intake; NE, net energy.

[1] Initial weight is the shrunk off truck arrival weight. Interim and final weights were reduced 4% to account for digestive tract fill.

[2] Final shrunk weight was adjusted for carcass weight by dividing the carcass weight by the decimal fraction of the average dressing percentage (0.618).

Table 5. Influence of supplementation level of kaolinite clay on carcass characteristics of Holstein steers

Item	Kaolinite level (% diet DM)			SEM	Contrast p-value	
	0	1	2		Linear	Quadratic
Pen replicates	6	6	6			
Hot carcass weight (kg)	356.25	360.8	358.8	3.0	0.56	0.39
Dressing percentage	62.1	61.8	61.4	0.2	0.01	0.90
Longissimus area (cm^2)	77.6	80.4	79.9	2.8	0.56	0.64
Fat thickness (cm)	0.76	0.89	0.82	0.04	0.35	0.07
KPH (%)	2.43	2.39	2.45	0.05	0.77	0.45
Yield grade (%)[1]	51.8	52.1	52.0	0.3	0.64	0.72
Quality grade[2]	4.93	5.08	4.73	0.22	0.52	0.37

DM, dry matter; SEM, standard error of the mean; KPH, kidney, pelvic and heart fat; LM, lgissimus.

[1] Kidney, pelvic, and heart fat as a percentage of carcass weight.

[2] Assessment of external 12th-rib fat thickness, KPH, LM area, lean and skeletal maturity, lean color, and marbling were used to determine a quality and yield grade for each carcass (USDA, 1997).

on ADG, and gain efficiency (ADG/DMI). Effects of clay inclusion on DMI and ADG in growing-finishing ruminants has not been consistent. Colling et al. (1979) observed decreased ADG and DMI in finishing steers fed high-moisture and steam-flaked corn-based diets supplemented with 2.5% bentonite. In other studies (Cammack et al., 2010), supplementation with 2% to 5% clay (as bentonite or zeolite) did not affect ADG, or gain efficiency. Mendel (1971) did not detect differences in DMI between controls and supplemented steers fed 2% to 4% montmorillonite clay. However, ADG was 8.9% and 13.2% greater with 2% and 4% supplemental montmorillonite, respectively. Berthiaume et al. (2007) observed greater ADG in steers supplemented with 2% bentonite in a silage-based diet.

Kaolinite supplementation tended (linear effect, $p = 0.08$) to increase dietary NE during the initial 112-d period. However, the overall (308-d) effect of supplementation on dietary NE was not appreciable ($p>0.20$). Kaolinite replaced steam-flaked corn in the basal diet (Table 1). Considering the indigestibility of kaolinite (contains no OM), the ratio of observed-to-expected dietary NE increased with kaolinite supplementation. This effect was more pronounced (linear effect, $p\leq0.03$) during the initial 224 d of the study. Overall (308 d), kaolinite supplementation tended to increase (linear effect, $p = 0.07$) dietary NE by 3% over expected. Based on the apparent absence of effects of supplemental kaolinite on ruminal and total tract digestion, and ruminal fermentation parameters, the positive effect of clay on the dietary energetic may operate in a manner independent of changes in characteristics of ruminal and total tract digestion. However, increases on feed efficiency have been observed in lambs and steers fed diets supplemented with bentonite (Britton et al., 1978).

The effects of supplemental kaolinite on carcass characteristics are shown in Table 5. As expected, there were no treatment effects on HCW, yield grade, LM area, KPH, and quality grade. Kaolinite supplementation linearly decreased ($p<0.01$) dressing percentage. The negative effect of supplemental kaolinite on dressing percentage could be more apparent than real. It has been argued that clay essentially is not absorbed and is excreted with the feces. Because of its density, it is reasonable that the clay particles would accumulate along the digestive tract, particularly in the forestomach regions.

It is concluded that kaolinite supplementation up to 2% of diet DM may enhance energetic efficiency of calf-fed Holstein steers in a manner independent of changes in characteristics of ruminal and total tract digestion.

REFERENCES

AOAC (Association Official Analytical Chemists). 2000. Official methods of analysis. 17th edn. Association Official Analytical Chemists. Gaithersburg, MD, USA.

Bergen, W. G., D. B. Purser, and J. H. Cline. 1968. Effect of ration on the nutritive quality of rumen microbial protein. J. Anim. Sci. 27:1497-1501.

Berthiaume, R., M. Ivan, and C. Lafreniere. 2007. Effects of sodium bentonite supplements on growth performance of feedlot steers fed direct-cut or wilted grass silage based diets. Can. J. Anim. Sci. 87:631-638.

Britton, R. A., D. P. Cooling, and T. J. Klopfenstein. 1978. Effect of complexing sodium bentonite with soybean meal or urea *in vitro* ruminal ammonia release and nitrogen utilization in ruminants. J. Anim. Sci.46:1738-1747.

Cammack, K. M., C. L.Wright, K. J. Austin, P. S. Johnson, R. R. Cockrum, K. L. Kessler, and K. C. Olson. 2010. Effects of high-sulfur water and clinoptilolite on health and growth performance of steers fed forage-based diets. J. Anim. Sci. 88:1777-1785.

Cole, N. A., R. W. Todd, and D. B. Parker. 2007. Use of fat and zeolite to reduce ammonia emissions from beef cattle feedyards. In: Proceedings of the Air Quality Waste Management Agriculture. Broomfield, CO, USA; ASABE Publication Number 701P0907cd.

Colling, D. P., R. A. Britton, S. D. Farlin, and M. K. Nielsen. 1979.

Effects of adding sodium bentonite to high grain diets for ruminants. J. Anim. Sci. 48:641-648.

Dinius, D. A., A. D. Peterson, T. A. Long, and B. R. Baumgardt. 1970. Intake and digestibility by sheep or rations containing various roughage substitutes. J. Anim. Sci. 30:309-312.

EFSA (European Food Safety Authority). 2016. Safety and efficacy of a natural mixture of illite, montmorillonite and kaolinite (Argile Verte du Velay) as a feed additive for all animal species. EFSA J. 14:4342-4360.

Garrett, W. N. 1971. Energetic efficiency of beef and dairy steers. J. Anim. Sci. 32:451-456.

Hill, F. N. and D. L. Anderson. 1958. Comparison of metabolizable energy and productive energy determinations with growing chicks. J. Nutr. 64:587-603.

Ivan, M., M. S. Dayrell, and M. Hidiroglou. 1992. Effects of bentonite and monensin on selected elements in the stomach and liver of fauna-free and faunated sheep. J. Dairy Sci. 75:201-208.

Martin, L. C., A. J. Clifford, and A. D. Tillman. 1969. Studies on sodium bentonite in ruminant diets containing urea. J. Anim. Sci. 29:777-782.

McCollum, F. T. and M. L. Galyean. 1983. Effects of clinoptilolite on rumen fermentation, digestion and feedlot performance in beef steers fed high concentrate diets. J. Anim. Sci. 56:517-524.

Mendel, V. E. 1971. Montmorillonite clay in feed lot rations. J. Anim. Sci. 33:891-894.

Murphey, C. E., D. K. Hallett, W. E. Tyler, and J. C. Pierce Jr. 1960. Estimating yields of retail cuts from beef carcasses. In: the 62nd Meeting of the American Society of Animal Production, Chicago, IL, USA. pp. 1-12.

NRC (National Research Council). 1984. Nutrient Requirements of Beef Cattle. 6th edn. National Academy of Press, Washington, DC, USA.

NRC (National Research Council). 1996. Nutrient Requirements of Beef Cattle. 7th edn. National Academy of Press, Washington, DC, USA.

Orskov, E. R., N. A. MacLeod, and D. J. Kyle. 1986. Flow of nitrogen from the rumen and abomasum in cattle and sheep given protein-free nutrients by intragrastric infusion. Br. J. Nutr.56:241-248.

Owen, O. J., M. B. Nodu, U. A. Dike, and H. M. Ideozu. 2012. The effect of dietary kaolin (clay) as feed additive on the growth performance of broiler chickens. Greener J. Agric. Sci. 6:233-236.

Safaei, M., F. Boldaji, B. Dastar, S. Hassani, M. S. A. Mutalib, and R. Rezaei. 2014. Effects of inclusion kaolin, bentonite and zeolite in dietary on chemical composition of broiler chicken meat. Asian J. Anim. Vet. Adv. 9:56-63.

SAS (Statistical Analysis System) Institute Inc. 2004. User's Guide: Statistics, version 9. SAS Inst. Cary, NC, USA.

Spotti, M., M. L. Fracchiola, F. Arioli, F. Canoni, and G. Pompa. 2005. Aflatoxin B_1 binding to sorbents in bovine ruminal fluid. Vet. Res. Commun. 29: 507-515.

Tate, K., G. Youan, B. Theng, G. Churchman, J. Sing, and P. Berben. 2015. Can geophagy mitigate enteric methane emission from cattle? J. Prelim. Res. 2:1-8.

Trckova, M., L. Matlova, L. Dvorska, and I. Pavlik. 2004. Kaolin, bentonite and zeolites as feed supplements for animals: health advantages and risks. Vet. Med. Czech. 49:389-399.

USDA (United States Department of Agriculture). 1997. United States Standards for Grading of Carcass Beef. Agricultural Marketing Service, USDA Washington, DC, USA.

Van Soest, P. J., J. B. Robertson, and B. A. Lewis. 1991. Methods for dietary fiber, neutral detergent fiber, and nonstarch polysaccharides in relation to animal nutrition. J. Dairy Sci. 74:3583-3597.

Wolin, M. J. 1960. A theoretical rumen fermentation balance. J. Dairy Sci. 43:1452-1459.

Zinn, R. A. 1990. Influence of flake density on the comparative feeding value of steam-flaked corn for feedlot cattle. J. Anim. Sci. 68:767-775.

Zinn, R. A. and F. N. Owens. 1986. A rapid procedure for purine measurement and its use for estimating net ruminal protein synthesis. Can. J. Anim. Sci. 66:157-166.

Zinn, R. A., A. Barreras, F. N. Owens, and A. Plascencia. 2008. Performance by feedlot steers and heifers: daily gain, mature weight, dry matter intake and dietary energetics. J. Anim. Sci. 86:2680-2689.

26

Use of Lysozyme as a Feed Additive on *In vitro* Rumen Fermentation and Methane Emission

Ashraf A. Biswas[a], Sung Sill Lee[1,a], Lovelia L. Mamuad, Seon-Ho Kim, Yeon-Jae Choi, Gui-Seck Bae[2], Kichoon Lee[3], Ha-Guyn Sung[4], and Sang-Suk Lee*

Ruminant Nutrition and Anaerobe Laboratory, Department of Animal Science and Technology, Sunchon National University, Suncheon 540-742, Korea

ABSTRACT: This study was conducted to determine the effect of lysozyme addition on *in vitro* rumen fermentation and to identify the lysozyme inclusion rate for abating methane (CH_4) production. An *in vitro* ruminal fermentation technique was done using a commercial concentrate to rice straw ratio of 8:2 as substrate. The following treatments were applied wherein lysozyme was added into 1 mg dry matter substrate at different levels of inclusion: Without lysozyme, 2,000, 4,000, and 8,000 U lysozyme. Results revealed that, lysozyme addition had a significant effect on pH after 24 h of incubation, with the highest pH ($p<0.01$) observed in 8,000 U lysozyme, followed by the 4,000 U, 2,000 U, and without lysozyme. The highest amounts of acetic acid, propionic acid ($p<0.01$) and total volatile fatty acid (TVFA) ($p<0.05$) were found in 8,000 U after 24 h of incubation. The CH_4 concentration was the lowest in the 8,000 U and the highest in the without lysozyme addition after 24 h of incubation. There was no significant differences in general bacteria, methanogen, or protozoan DNA copy number. So far, addition of lysozyme increased the acetate, propionate, TVFA, and decreased CH_4 concentration. These results suggest that lysozyme supplementation may improve *in vitro* rumen fermentation and reduce CH_4 emission. (**Key Words:** *In vitro*, Lysozyme, Methane, Ruminant, Volatile Fatty Acid)

INTRODUCTION

Lysozyme is a universal enzyme found in living organisms having diverse role starting from digestion to immune response. Lysozymes (e.g. 1, 4-β-N-acetylmuramidase) have both enzymatic and bacteriolytic activity. They cleave the glycosidic linkage in the peptidoglycan component of gram-positive bacterial cell walls, which ultimately leads to cell death (Ellison and Giehl, 1991). Interestingly, lysozymes lyse the microbial cells from the outside, in addition to inducing microbial autolysis (Ibrahim et al., 2001). Gram-positive bacteria are susceptible to the action of lysozyme but most of the gram-negative bacteria are not for their thick outer membrane beyond the peptidoglycan layer. In the animal production sector, antimicrobials, such as antibiotics, are used at low doses as growth promoters for their bactericidal effect. But, the adverse effects of the indiscriminate use of antibiotics (such as antibiotic resistance) on the animal as well as environment make the animal industry in vulnerable condition. So, animal production people as well as animal scientists are inspired to identify the alternative of antibiotic for sustainable animal production system. For the enzymatic and bacteriolytic characteristics of lysozyme, scientists are encouraging the application of lysozymes for enhancing growth performance and as an antimicrobials (Nyachoti et al., 2012). Lysozymes have been shown to reduce diarrhoea incidence during the preweaning

* Corresponding Author: Sang-Suk Lee.
E-mail: rumen@sunchon.ac.kr
[1] Division of Applied Life Science (BK21+, IALS), Gyeongsang National University, Jinju 52828, Korea.
[2] Department of Animal Science and Technology, Chung-Ang University, Anseong 456-756, Korea.
[3] Department of Animal Sciences, The Ohio State University, Columbus, OH 43210, USA.
[4] Department of Animal Science & Technology, Sangji University, Wonju 26339, Korea.
[a] These authors contributed equally to the work.

period of calves and tend to increase average daily gain (ADG) (Goncu et al., 2012). In addition, they have been shown to have an ADG-increasing capability in pigs (May et al., 2012). However, there is no information available about the effects of lysozymes on *in vitro* rumen fermentation and methane (CH_4) emission in ruminants. Therefore, this study aimed to evaluate the effects of lysozymes on *in vitro* fermentation parameters and to identify the inclusion rate of lysozymes for abating *in vitro* CH_4 production.

MATERIALS AND METHODS

An *in vitro* ruminal fermentation technique was adopted using late fattening commercial pelleted concentrate (Nonghyup co., Haman, Korea) and rice straw (8:2) as substrate. In this experiment, we used liquid lysozyme collected from Celltech co., Ltd., Eumseong, Korea containing 150,000 units/mL. In addition to the control (no added lysozyme), the following treatments were applied: 2,000, 4,000, and 8,000 U lysozyme (hereafter referred to as T1, T2, and T3, respectively), which were added to 1 g dry matter (DM) substrate. Inclusion level of lysozyme as U/g DM substrate were based on the previous research work done by Nyachoti et al. (2012). On the basis of inclusion level (U) and lysozyme content (U/mL) of liquid lysozyme, we have used 13.33 μL, 26.66 μL, and 53.33 μL liquid lysozyme in 1 g DM substrate for T1, T2, and T3 treatments, respectively.

In vitro experiment

Ruminal contents were obtained from a 48-month old rumen-cannulated Holstein cow (650 kg) fed rice straw and commercial pelleted concentrate (8:2 ratio) twice a day. The supplied rice straw contained 4.45% crude protein (CP) and 38.29% total digestible nutrients (TDN), whereas the concentrate contained 12% CP and 72% TDN. The pooled ruminal contents were squeezed and the extracted fluids were collected in a glass bottle after straining through cheese cloth that had been folded four times. The bottle was subsequently capped and immediately transported to the laboratory while maintaining the temperature at 39°C. In the laboratory, the bottle was placed in a water bath at 39°C. The bottle was shaken vigorously by hand, prior to mixing with a buffer. The particle-free rumen fluid was transferred to a buffer medium (pH 6.9) at a 1:3 rumen fluid: buffer ratio. The buffer medium was prepared according to the method described by Asanuma et al. (1999). Under a constant flow of N_2 gas, 100 mL of buffered rumen fluid was anaerobically transferred to 160 mL 96 serum bottles containing 1 g substrate (1 mm particle size) together with the different concentrations of lysozyme. The serum bottles were subsequently sealed with rubber septum stoppers and aluminium caps followed by incubation at 39°C for 0, 6, 12, and 24 h in a 120 rpm shaking incubator as described by Hattori and Matsui (2008). For each incubation time, three replicates per experimental treatment were used.

pH, total gas (TG), CH_4, ammonia nitrogen (NH_3-N), acetic acid, propionic acid, butyric acid and total volatile fatty acid (VFA) were analysed at each incubation time. Fifteen ml samples were collected from each of the serum bottles into the falcon tube and kept at –80°C until further analysis of VFA, NH_3-N, and molecular study. The gas produced inside the bottle was estimated for TG and gas sample trapped inside vacuum tubes and stored in a refrigerator until CH_4 determination. *In vitro* DM and organic matter (OM) disappearance were estimated after 0, 6, 12, and 24 h of incubation.

Analyses of *in vitro* fermentation parameters

Fermentation parameters were checked at the end of each incubation time. pH was determined using a Pinnacle series M530p meter (Schott Instruments, Mainz, Germany). Before opening serum bottle, an EA-6 (Sun Bee instrument, Inc., Seoul, Korea) pressure sensor meter was used to measure TG production from each of the serum bottles. Duplicates of 1 mL fermentation samples from each of the serum bottles were immediately centrifuged at 13,000 rpm for 10 min at 4°C using a Micro 17TR centrifuge (Hanil Science Industrial Co. Ltd., Gimpo, Korea). The supernatant was transferred to two 1.5 mL Eppendorf tubes and deep-frozen at –80°C until NH_3-N and VFA analysis also remaining pellet were preserved in the same temperature for DNA copy number estimation. At the time of NH_3-N and VFA analysis, the samples contained in the Eppendorf tubes were thawed at room temperature and used for measuring NH_3-N, individual VFA i.e. acetic acid, propionic acid, butyric acid and total VFA. The measurements of NH_3-N were performed according to the methods developed by Chaney and Marbach (1962), at an optical density of 630 nm using a Libra S22 spectrophotometer (Biochrom Ltd., CB40FJ, Cambourne, England). The VFA measurements followed the methods described by Han et al. (2005) and Tabaru et al. (1988). We used high performance liquid chromatography (Agilent Technologies 1200 series, Waldbronn, Baden-Wurttemberg, Germany) using a UV detector set at 210 and 220 nm and a METACARB87H (Varian, Palo Alto, CA, USA) column with 0.0085 N H_2SO_4 solvent as buffer at a flow rate of 0.6 mL/min. The individual VFA concentration in mM was calculated as parts per million divided by the molecular weight. Total VFA were calculated with sum of individual VFA and acetate/propionate ratio (A/P) value was calculated as acetic acid divided by propionic acid.

Determination of *in vitro* dry matter and organic matter disappearance

At the start of the *in vitro* rumen fermentation, the

Table 1. Primer used in quantitative polymerase chain reaction

Target microbes	Primer sequences	Annealing temp. (°C)	Product size	Reference
General bacteria	1114F: 5′-CGG CAA CGA GCG CAA CCC-3′ 1275R: 3′-CCA TTG TAG CAC GTG TGT AGC C-5′	60	130	Denman and McSweeney, 2006
Methanogens	qmcraA-F: 5′-GGA TTA GAT ACC CSG GTA GT-3′ qmcrA-R: 3′-GTT GAR TCC AAT TAA ACC GCA-5′	60	140	Denman and McSweeney, 2006
Protozoa	F: 5′-GCT TTC GWT GGT AGT GTA TT-3′ R: 3′-CTT GCC CTC YAA TCG TWC T-5′	60	223	Denman and McSweeney, 2006

substrate DM and OM were determined by drying at 105°C for 16 h and ashing at 550°C for 12 h, respectively. The resulting percent DM and OM were used to compute the initial DM (DM_i) and OM (OM_i) content (g) of the substrate. After each of the specified incubation periods, every treatment bearing three replicates fermentation samples from each serum bottle were drained in dried, pre-weighed nylon bags and knotted using nylon thread. They were then rinsed in flowing water until the turbidity of the rinse water disappeared. The final DM (DM_f) and OM (OM_f) of the substrate were determined using the same conditions as applied when determining DM_i and OM_i. The DM and OM disappearance (%) were calculated as $([DM_i–DM_f]/DM_i)\times 100$ and $([OM_i–OM_f]/OM_i)\times 100$, respectively.

Methane estimation

The samples contained in the vacuum tube were analysed for CH_4, using gas chromatography (Agilent Technologies HP 5890, Germany) using a TCD detector with a Carboxen 1006PLOT capillary column 30 m×0.53 mm (Supelco, Bellefonte, PA, USA).

Quantitative real-time polymerase chain reaction

General bacteria, methanogens, and protozoan DNA copies were estimated using the protocol of Denman and McSweeney (2006). The DNA concentration from the extracted gDNA from each pelleted sample was determined using an Epoch Microplate Spectrophotometer (Biotek, Winooski, VT, USA). Target-specific forward and reverse polymerase chain reaction (PCR) primers were used (Table 1). Amplification was done using Eco Real-Time PCR (Illumina, Inc., San Diego, CA, USA) in a total reaction volume of 20 μL per reaction mixture containing 10 μL of 2X qPCRBIO SyGreen Mix, 0.8 μL each forward and reverse PCR primers, and 8.4 μL template DNA of 50 ng/μL in sterile distilled water.

Statistical analysis

Data were analysed by analysis of variance using a general linear model for a complete randomised design. All treatments were conducted in triplicate. Duncan's multiple range test and orthogonal polynomial contrast were used to identify differences between and among treatments and control. A $p\leq 0.05$ was considered to indicate statistical significance. All analyses were carried out using Statistical Analysis System (SAS) (SAS, 2004) (version 9.1; SAS Inst. Inc., Cary, NC, USA).

RESULTS

There was no significant difference in pH after 6 and 12 h of incubation. After 24 h of incubation, the pH in the control group was the lowest ($p<0.01$), subsequently increase in T1 then T2 and the highest pH was in the T3 treatment group (Table 2). In the case of TG, no significant differences were observed after 12 and 24 h. However, a tendency towards higher TG in treatment groups compared to the control was observed after 12 and 24 h (Table 2). We found no differences in NH_3-N after 12 or 24 h of incubation. Nevertheless, at 6 h of incubation, the NH_3-N differed significantly among the treatments ($p = 0.05$) with the highest value observed in the T3 treatment (Table 2).

There were significant differences in acetic acid at all incubation periods except after 6 h of incubation (Table 3). The highest ($p<0.01$) acetic acid was observed in T3 (41.61 mM) after 24 h of incubation followed by T2, T1, and the control. Likewise, we found a significant difference ($p<0.01$) in propionic acid after 12 and 24 h of incubation with 12.21 mM and 12.01 mM, respectively. The highest and lowest propionic acid concentrations were found in T3 and control, respectively, after 12 and 24 h of incubation. In the case of butyric acid, there was no significant difference after 24 h but there was a tendency towards T3 having the highest butyric acid. Significant differences in total VFA ($p<0.05$) were found in all treatments except at 6 h of incubation. The highest total VFA was observed in T3 after 12 and 24 h of incubation (Table 3). There were no significant differences in A/P value after 6, 12, or 24 h of incubation, although there was a tendency towards the highest A/P value in the control after 12, and 24 h of incubation (Table 3).

There were no significant differences in DM or OM disappearance, but there was a tendency towards higher DM disappearance in T3 compared to the control after 24 h of incubation (Table 4). Likewise, after 24 h of incubation the tended to highest OM disappearance was found in T3 (Table 4). In the case of CH_4 concentration, significant differences

Table 2. pH, total gas and ammonia nitrogen (NH_3-N) production from *in vitro* rumen fermentation by the addition of lysozyme

Parameters/time	Treatments[1]				SEM	p value[2]	
	Control	T1	T2	T3		All	C vs T
pH value							
0 h	6.16[d]	6.20[c]	6.23[b]	6.26[a]	0.008	<0.001	<0.001
6 h	5.56	5.53	5.61	5.59	0.019	0.097	0.354
12 h	5.34	5.37	5.37	5.37	0.008	0.152	0.032
24 h	5.20[c]	5.21[bc]	5.23[b]	5.25[a]	0.005	0.002	0.002
Total gas (mL)							
0 h	3.75[a]	3.17[b]	3.67[a]	3.75[a]	0.124	0.010	0.144
6 h	51.33[b]	54.20[ab]	53.20[b]	57.00[a]	1.097	0.020	0.019
12 h	87.25	88.75	87.75	88.50	1.475	0.911	0.578
24 h	96.17	99.00	98.83	98.83	1.307	0.431	0.106
NH_3-N (mM)							
0 h	31.28	32.09	34.28	34.72	1.405	0.697	0.440
6 h	34.65[b]	38.46[ab]	37.13[b]	44.57[a]	2.081	0.050	0.072
12 h	36.67	38.73	42.93	45.82	2.159	0.203	0.157
24 h	41.58	44.59	43.40	45.99	2.424	0.697	0.357

SEM, standard error of the mean.
[1] Control: without lysozyme, T1: 2,000 unit lysozyme, T2: 4,000 unit lysozyme, T3: 8,000 unit lysozyme.
[2] C vs T is the comparison between control and treatment.
[a,b,c] Means in the same row with different superscript are significantly different ($p<0.05$).

Table 3. Volatile fatty acid (VFA) production (mM) and A/P ratio from *in vitro* fermentation by the addition of lysozyme

Parameters/time	Treatments[1]				SEM	p value[2]	
	Control	T1	T2	T3		All	C vs T
Acetic acid							
0 h	26.30[b]	26.76[a]	26.72[a]	26.87[a]	0.068	0.002	0.001
6 h	37.20	37.04	36.98	38.18	0.274	0.109	0.622
12 h	37.77[d]	38.63[c]	39.35[b]	40.40[a]	0.168	<0.001	<0.001
24 h	39.17[c]	39.92[bc]	40.48[b]	41.61[a]	0.237	0.003	0.004
Propionic acid							
0 h	7.82	7.78	7.77	7.76	0.023	0.460	0.145
6 h	10.61	10.62	10.21	11.36	0.181	0.078	0.711
12 h	11.04[b]	11.39[b]	11.98[a]	12.21[a]	0.134	0.002	0.001
24 h	10.67[b]	11.57[a]	11.50[a]	12.01[a]	0.175	0.008	0.002
Butyric acid							
0 h	4.65	4.60	4.58	4.47	0.027	0.012	0.015
6 h	9.06	9.46	8.76	9.28	0.207	0.191	0.676
12 h	15.42[a]	15.23[a]	13.71[b]	13.52[b]	0.188	<0.001	0.001
24 h	16.14	15.91	15.99	16.19	0.306	0.934	0.798
Total VFA							
0 h	38.77[b]	39.13[a]	39.06[a]	39.09[a]	0.071	0.046	0.008
6 h	56.87	57.11	56.94	58.82	0.701	0.185	0.340
12 h	64.22[c]	65.25[ab]	65.04[bc]	66.13[a]	0.273	0.010	0.005
24 h	66.29[b]	67.40[b]	67.98[ab]	69.81[a]	0.512	0.039	0.040
A/P ratio							
0 h	3.36[b]	3.44[a]	3.44[a]	3.48[a]	0.012	0.003	0.001
6 h	3.51	3.49	3.63	3.36	0.037	0.073	0.864
12 h	3.42	3.39	3.29	3.31	0.042	0.239	0.143
24 h	3.57	3.45	3.52	3.47	0.041	0.329	0.149

A/P, acetate/propionate; SEM, standard error of the mean.
[1] Control: without lysozyme, T1: 2,000 unit lysozyme, T2: 4,000 unit lysozyme, T3: 8,000 unit lysozyme.
[2] C vs T is the comparison between control and treatment.
[a,b,c] Means in the same row with different superscript are significantly different ($p<0.05$).

Table 4. Dry matter and organic matter disappearance (%) from as influenced by the addition of lysozyme

Parameters /time	Treatments[1]				SEM	p value[2]	
	Control	T1	T2	T3		All	C vs T
DM disappearance							
0 h	29.18	30.95	31.85	32.52	1.631	0.682	0.301
6 h	47.10	47.68	47.08	48.37	1.084	0.854	0.706
12 h	57.25	57.66	55.80	56.33	0.713	0.341	0.438
24 h	57.10	60.12	59.13	59.40	1.738	0.722	0.323
OM disappearance							
0 h	25.58	26.95	27.69	28.38	1.351	0.696	0.321
6 h	42.39	42.39	42.68	43.86	1.011	0.708	0.673
12 h	51.79	52.23	50.65	50.92	0.624	0.318	0.472
24 h	52.61	53.61	54.26	54.37	1.602	0.843	0.456

SEM, standard error of the mean; DM, dry matter; OM, organic matter.
[1] Control: without lysozyme, T1: 2,000 unit lysozyme, T2: 4,000 unit lysozyme, T3: 8,000 unit lysozyme.
[2] C vs T is the comparison between control and treatment.
Means in the same row with different superscript are significantly different ($p<0.05$).

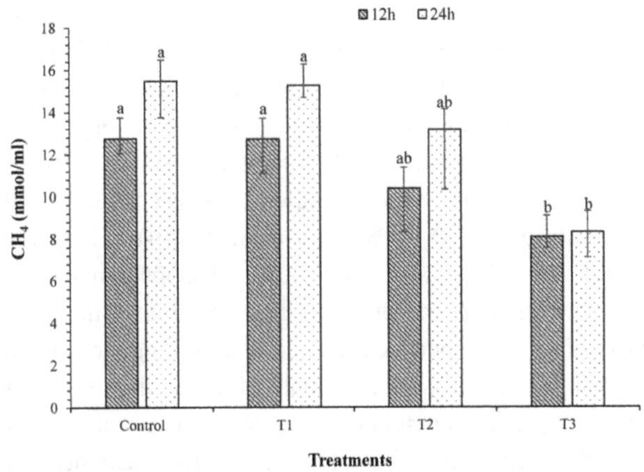

Figure 1. Methane concentration (mmol/mL) from as influenced by the addition of lysozyme. Control: without lysozyme, T1: 2,000 unit lysozyme, T2: 4,000 unit lysozyme, T3: 8,000 unit lysozyme. [a,b] Treatments with different letters are different at $p<0.05$.

were observed at all incubation times, except after 6 h of incubation. After 24 h, the lowest CH_4 concentration was observed in the T3 treatment, followed by T2 and T1 and highest CH_4 concentration was found in the control (Figure 1). Although non-significant, the general bacterial DNA copy number tended to be lower in the treatment groups (T1 to T3), compared to the control. Likewise, there were no significant differences in the case of methanogen DNA copy number. The protozoan DNA copy numbers were not significantly different either, but the lowest DNA copy numbers were found in the T3 treatment (Table 5).

DISCUSSION

Lysozyme is an antimicrobial enzyme that can lyses bacterial cell walls by hydrolysing the polysaccharide component (Salton, 1957). It cleaves the gram-positive bacterial glycosidic linkage in the peptidoglycan component (Ellision and Giehl, 1991) but most of gram negative bacteria are not susceptible to lysozyme for their protective outer membrane. Also, lysozyme has enzymatic activity on digestion process. According to Sahoo et al. (2012), lysozyme has been recruited as a digestive enzyme for some of animals. They have also mentioned that lysozyme has diverse role including digestion. In the current study, we found that the pH after 24 h of incubation was the lowest in the control group with sequentially increasing from low to high inclusion level of lysozyme, which is suggested that the high pH found in the treatment groups containing higher amounts of lysozyme, was due to the high pH (8.99) of the lysozyme used in this experiment. This high pH can reduce the ruminal acidosis (Owens et al., 1998) which will be helpful for ruminant production especially high yielding animal.

Significantly higher contents of acetic acid, propionic acid, and total VFA of this study after 24 h of incubation supported by Giraldo et al. (2007) who observed increased total VFA using fibrolytic enzyme (cellulase and xylanase) addition in *in vitro* trials. This increased VFA could be the effect of enzymatic degradation of feed material through hydrolytic action. Current experimental increased total VFA, which ultimately reflects better performance, also agrees with the findings of May et al. (2012) and Goncu et al. (2012), who reported tended to increase in daily gain of calves fed 10 mg lysozyme/L of milk. In addition, Nyachoti et al. (2012) also reported a tendency to an ADG increase in piglets, when 0.1% and 0.2% of lysozyme were added to the basal diet.

Digestibility of different feed ingredients is an important factor for the CH_4 production (Kim et al., 2013). We found no effect of lysozyme on DM or OM disappearance,

Table 5. Microbial DNA copies from as influenced by the addition of lysozyme after 24 h of incubation

Target microbes	Treatment[1]				SEM	p value[2]	
	Control	T1	T2	T3		All	C vs T
General bacteria	7.05E+08	5.40E+08	4.55E+08	6.07E+08	3.23E+08	0.965	0.704
Methanogens	4.23E+05	6.70E+05	2.84E+05	3.75E+05	9.76E+04	0.265	0.886
Protozoa	5.53E+04	3.58E+05	3.59E+05	2.59E+05	4.66E+04	0.181	0.055

SEM, standard error of the mean.
[1] Control: without lysozyme, T1: 2,000 unit lysozyme, T2: 4,000 unit lysozyme, T3: 8,000 unit lysozyme.
[2] C vs T is the comparison between control and treatment.

irrespective of incubation period is consistent with the findings of Quinn et al. (2009) and Zinn et al. (1997) where they observed no difference of DM and OM disappearance with the ionophores, antibiotics and sulphur supplementation in their *in vitro* digestion trial. This non-significant different DM and OM disappearance results are also consistent with the results observed by Soriano et al. (2014), where they elicited *in vitro* DM and OM disappearance with *Lactobacillus mucosae* on dried brewers grain. The results observed in our experiment tended towards higher DM and OM disappearance, whose ultimate effect on higher *in vitro* total VFA production supported by Van Dung et al. (2014), where they reported higher DM and OM disappearance with concentrate:roughage (8:2) substrate along with higher *in vitro* total VFA production. We found a significant decrease in CH_4 concentration after 12 and 24 h of incubation which is in concordance with the result of Mamuad et al. (2014), where significantly lower *in vitro* CH_4 concentration was revealed. Current experimental decreasing CH_4 concentration was found with increasing lysozyme inclusion, suggesting a negative correlation between lysozyme amount and CH_4 concentration. Present experimental lower CH_4 concentration is also negatively correlated with the higher total VFA. Nyachoti et al. (2012) reported a tendency to increase propionate in piglets, when 0.1% and 0.2% of lysozyme were added to the basal diet. Current experimental lysozyme with high concentrate based substrate can produce higher propionate by the utilization of hydrogen (H_2) through the fumarate to succinate production pathway. On the other hand, methanogen generate CH_4 from CO_2 and H_2 (Miller et al., 1986). So, lysozyme which facilitate H_2 utilization can be the methanogen competitors. Ultimately, lysozyme inhibit CH_4 production, presumably which reflected as lower CH_4 concentration in this experiment. However, further *in vivo* study should be conducted to draw rigid conclusion.

We observed no differences in the general bacterial DNA copy number. The same non-significant results were found by Zhou et al. (2011) in an *in vitro* experiment using 2-nitroethanol, sodium nitrate, and ethyl 2 butynote with the result of lower CH_4 concentration. Thus, the lack of difference in general bacterial DNA copy number observed in the present study may be linked to the non-significant difference in DM and OM disappearance and TG production after 12 and 24 h of incubation period. However, we did see a tendency towards a lower general bacterial DNA copy number in the treatment groups compared to the control group. This result is supported by Salton (1957), in which the lysozyme digested bacterial cell wall as well as inducing microbial autolysis (Ibrahim et al., 2001) resulting in DNA autolysis, ultimately decreasing the DNA copy number. We found no differences in methanogen DNA copy number. Likewise, Zhou et al. (2011) found no difference in methanogen DNA copies log number among control and treatment groups, although compared to the control groups, a lower CH_4 concentration was found in groups treated with a methanogen inhibitor (propynoic acid). In this connection, Ungerfeld et al. (2004) explained that different types of methanogens differ in their CH_4 production potential and display different sensitivities to methanogen inhibitors. The amount of methanogen DNA copies log number found in our experiment is supported by the findings of Stiverson et al. (2011) who used samples from the adhering portion of sheep rumen and observed similar methanogen DNA log copy number as current experiment. We did not find any significant differences in protozoan DNA copy number with lower CH_4 concentration in treatments group is in concordance with the result of Karnati et al. (2009), where they revealed a tendency to be lower CH_4 concentration with non-significant different of protozoal N flow, using a continuous *in vitro* culture treatment with monensin. As seen in Table 5, however, protozoan DNA copy number is nearly 1 log difference but result is statistically non-significant, this type of result also elicited by Zhou et al. (2011) where 1 log difference DNA copy showed non-significant difference. The protozoan DNA copies log number observed in the current study is supported by those found by Anantasook et al. (2013) who used urea treated rice straw and concentrate mixture with and without rain tree pod meal in an *in vivo* experiment using dairy steers.

CONCLUSION

Addition of 8,000 U lysozyme per g of DM feed substrate increased total VFA and decreased CH_4 concentration which indicates that it is capable to improving *in vitro* rumen fermentation and reducing CH_4 emission. Further *in vitro* studies should be conducted using higher doses of lysozyme, and *in vivo* trials are needed to draw rigid conclusions.

ACKNOWLEDGMENTS

This research was supported by the Korea Institute of Planning and Evaluation for Technology in Food, Agriculture, Forestry and Fisheries (Project number 314075-3). Authors also thankful to Dr. Byung-Hern Moon. Managing Director, CellTech for their technical support to this experiment.

REFERENCES

Anantasook, N., M. Wanapat, A. Cherdthong, and P. Gunun. 2013. Changes of microbial population in the rumen of dairy steers as influenced by plant containing tannins and saponins and roughage to concentrate ratio. Asian Australas. J. Anim. Sci. 26:1583-1591.

Asanuma, N., M. Iwamoto, and T. Hino. 1999. Effect of the addition of fumarate on methane production by ruminal microorganisms *in vitro*. J. Dairy Sci. 82:780-787.

Chaney, A. L. and E. P. Marbach. 1962. Modified reagents for determination of urea and ammonia. Clin. Chem. 8:130-132.

Denman, S. E. and C. S. McSweeney. 2006. Development of a real-time PCR assay for monitoring anaerobic fungal and cellulolytic bacterial populations within the rumen. FEMS Microbiol. Ecol. 58:572-582.

Ellison, R. T. 3rd and T. J. Giehl. 1991. Killing of gram-negative bacteria by lactoferrin and lysozyme. J. Clin. Invest. 88:1080-1091.

Giraldo, L. A., M. D. Carro, M. J. Ranilla, and M. L. Tejido. 2007. Influence of fibrolytic enzymes on *in vitro* methane production and rumen fermentation of a substrate containing 60% of grass hay. Livest. Res. Rural Dev. 19:257-261.

Goncu, S., M. Gorgulu, and G. Gokce. 2012. The effect of additional lyzozyme to milk on growth performances of Holstein calves. J. Anim. Vet. Adv. 11:3724-3727.

Han, S., S. Kim, and H. Shin. 2005. UASB treatment of wastewater with VFA and alcohol generated during hydrogen fermentation of food waste. Process Biochem. 40:2897-2905.

Hattori, K. and H. Matsui. 2008. Diversity of fumarate reducing bacteria in the bovine rumen revealed by culture dependent and independent approaches. Anaerobe 14:87-93.

Ibrahim, H. R., T. Matsuzaki, and T. Aoki. 2001. Genetic evidence that antibacterial activity of lysozyme is independent of its catalytic function. FEBS Lett. 506:27-32.

Karnati, S., J. Sylvester, C. Ribeiro, L. Gilligan, and J. Firkins. 2009. Investigating unsaturated fat, monensin, or bromoethanesulfonate in continuous cultures retaining ruminal protozoa. I. Fermentation, biohydrogenation, and microbial protein synthesis. J. Dairy Sci. 92:3849-3860.

Kim, S., L. L. Mamuad, C. Jeong, Y. Choi, S. S. Lee, J. Ko, and S. Lee. 2013. *In vitro* evaluation of different feeds for their potential to generate methane and change methanogen diversity. Asian Australas. J. Anim. Sci. 26:1698-1707.

Mamuad, L., S. H. Kim, C. D. Jeong, Y. J. Choi, C. O. Jeon, and S. Lee. 2014. Effect of fumarate reducing bacteria on *in vitro* rumen fermentation, methane mitigation and microbial diversity. J. Microbiol. 52:120-128.

May, K. D., J. E. Wells, C. V. Maxwell, and W. T. Oliver. 2012. Granulated lysozyme as an alternative to antibiotics improves growth performance and small intestinal morphology of 10-day-old pigs. J. Anim. Sci. 90:1118-1125.

Miller, T. L., M. J. Wolin, H. X. Zhao, and M. P. Bryant. 1986. Characteristics of methanogens isolated from bovine rumen. Appl. Environ. Microbiol. 51:201-202.

Nyachoti, C. M., E. Kiarie, S. K. Bhandari, G. Zhang, and D. O. Krause. 2012. Weaned pig responses to *Escherichia coli* K88 oral challenge when receiving a lysozyme supplement. J. Anim. Sci. 90:252-260.

Ownes, F. N., D. S. Secrist, W. J. Hill, and D. R. Gill. 1998. Acidosis in cattle: A review. J. Anim. Sci. 76:275-286.

Quinn, M. J., M. L. May, K. E. Hales, N. DiLorenzo, J. Leibovich, D. R. Smith, and M. L. Galyean. 2009. Effects of ionophores and antibiotics on *in vitro* hydrogen sulfide production, dry matter disappearance, and total gas production in cultures with a steam-flaked corn-based substrate with or without added sulfur. J. Anim. Sci. 87:1705-1713.

Sahoo, N. R., P. Kumar, B. Bhusan, T. K. Bhattacharya, S. Dayal, and M. Sahoo. 2012. Lysozyme in livestock: A guide to selection for disease resistance: A review. J. Anim. Sci. Adv. 2:347-360.

Salton, M. R. J. 1957. The properties of lysozyme and its action on microorganisms. Bacteriol. Rev. 21:82-100.

SAS (Statistical Analysis System) Institute Inc. 2004. SAS/STAT User's Guide. version 9.1. SAS Institute Inc. Cary, NC, USA.

Soriano, A. P., L. L. Mamuad, S. H. Kim, Y. J. Choi, C. D. Jeong, G. S. Bae, M. B. Chang, and S. S. Lee. 2014. Effect of *Lactobacillus mucosae* on *in vitro* rumen fermentation characteristics of dried brewers grain, methane production and bacterial diversity. Asian Australas. J. Anim. Sci. 27:1562-1570.

Stiverson, J., M. Morrison, and Z. Yu. 2011. Populations of select cultured and uncultured bacteria in the rumen of sheep and the effect of diets and ruminal fractions. Int. J. Microbiol. 2011:750613.

Tabaru, H., E. Kadota, H. Yamada, N. Sasaki, and A. Takeuchi. 1988. Determination of volatile fatty acids and lactic acid in bovine plasma and ruminal fluid by high performance liquid chromatography. Jpn. J. Vet. Sci. 50:1124-1126.

Ungerfeld, E. M., S. R. Rust, D. R. Boone, and Y. Liu. 2004. Effects of several inhibitors on pure cultures of ruminal methanogens. J. Appl. Microbiol. 97:520-526.

Van Dung, D., W. Shang, and W. Yao. 2014. Effect of crude protein levels in concentrate and concentrate levels in diet on *in vitro* fermentation. Asian Australas. J. Anim. Sci. 27:797-805.

Zhou, Z., Q. Meng, and Z. Yu. 2011. Effects of methanogenic inhibitors on methane production and abundances of methanogens and cellulolytic bacteria in *in vitro* ruminal cultures. Appl. Environ. Microbiol. 77:2634-2639.

Zinn, R. A., E. Alvarez, M. Mendez, M. Montano, E. Ramirez, and Y. Shen. 1997. Influence of dietary sulfur level on growth performance and digestive function in feedlot cattle. J. Anim. Sci. 75:1723-1728.

27

Effects of Combining Feed Grade Urea and a Slow-release Urea Product on Performance, Dietary Energetics and Carcass Characteristics of Feedlot Lambs Fed Finishing Diets with Different Starch to Acid Detergent Fiber Ratios

A. Estrada-Angulo, M. A. López-Soto[1], C. R. Rivera-Méndez[1], B. I. Castro, F. G. Ríos, H. Dávila-Ramos, A. Barreras[1], J. D. Urías-Estrada[1], R. A. Zinn[2], and A. Plascencia[1,*]

Veterinary and Animal Science School, University Autonomous of Sinaloa, Culiacán 1084, Sinaloa, México

ABSTRACT: Recent findings have shown that microbial nitrogen flow and digestible energy of diets are increased when urea is combined with a slow-release urea (SRU) in diets with a starch to acid detergent fibre ratio (S:F) 4:1. This affect is attributable to enhanced synchrony between ruminal N availability for microbial growth and carbohydrate degradation. To verify the magnitude of this effects on lamb performance, an experiment was conducted to evaluate the effects of combining urea and a SRU in diets containing S:F ratios of 3:1, 4:1, or 5:1 on performance, dietary energetics and carcass characteristics of finishing lambs. For that, 40 Pelibuey×Katahdin lambs (36.65±3 kg) were assigned to one of five weight groupings in 20 pens (5 repetition/treatments). The S:F ratio in the diet was manipulated by partially replacing the corn grain and dried distiller's grain with solubles by forage (wheat straw) and soybean meal to reach S:F ratios of 3:1, 4:1 or 5:1. An additional treatment of 4:1 S:F ratio with 0.8% urea as the sole source of non-protein nitrogen was used as a reference for comparing the effect of urea combination *vs.* conventional urea at the same S:F ratio. There were no treatment effects on dry matter intake (DMI). Compared the urea combination vs urea at the same S:F ratio, urea combination increased ($p<0.01$) average daily gain (ADG, 18.3%), gain for feed (G:F, 9.5%), and apparent energy retention per unit DMI (8.2%). Irrespective of the S:F ratio, the urea combination improved the observed-to-expected dietary ratio and apparent retention per unit DMI was maximal (quadratic effect, $p\leq0.03$) at an S:F ratio of 4:1, while the conventional urea treatment did not modify the observed-to-expected net energy ratio nor the apparent retention per unit DMI at 4:1 S:F ratio. Urea combination group tended (3.8%, $p = 0.08$) to have heavier carcasses with no effects on the rest of carcass characteristics. As S:F ratio increased, ADG, G:F, dietary net energy, carcass weight, dressing percentage and *longissimus thoracis* (LM) area increased linearly ($p\leq0.02$). Combining urea and a slow-release urea product results in positive effects on growth performance and dietary energetics, but the best responses are apparently observed when there is a certain proportion (S:F ratio = 4:1) of starch to acid detergent fibre in the diet. (**Key Words:** Slow-release Urea, Finishing Lambs, Growth Performance, Dietary Energetics, Carcass)

INTRODUCTION

A topic of interest in recent years in feedlots has been the search for strategies that optimise nutrient synchrony between N and carbohydrate compounds in the rumen in order to promote better nutrient utilisation and energy efficiency, and as a strategy for reducing the risk of environmental pollution (Hristov et al., 2011). The N retention in the rumen is mainly mediated by the rate of degradation of N compounds and carbohydrates, and by the energy available for the process of protein synthesis (Tedeschi et al., 2002). It has been observed that in high-grain diets (ratio of starch vs acid detergent fibre (ADF) greater than 5 to 1), urea can be supplemented at 50% higher than

* Corresponding Author: A. Plascencia.
E-mail: aplas_99@yahoo.com
[1] Research Institute of Veterinary Sciences, University Autonomous of Baja California, Tijuana 21100, México.
[2] Department of Animal Science, University of California, Davis, CA 95616, USA.

that recommended with positive effects on growth performance or in dietary energy utilisation (Milton et al., 1997; Zinn et al., 2003). The latter can be partially explained by the possible synchrony of ruminal degradation rates between feed-grade urea and starch. On the other hand, in cattle that were fed a high-forage diet (>10% ADF, i.e. rations for dairy and growing cattle), the use of slow-release urea products improved nutrient synchrony (Inostroza et al., 2010; Alvarez-Almora et al., 2011). Currently, as a result of the cost of corn grain, the replacement of corn grain by dried distillers grain with solubles (DDGS) in feedlot diets is a common practice (Klopfenstein et al., 2008). Although the energy value of DDGS is similar to corn grain (NRC, 2007; Estrada-Angulo et al., 2013), DDGS are lower in starch content (<6%) and higher in their content (>30%) of digestible fibre (Rosentrater, 2011; Carrasco et al., 2013). Therefore, depending on the replacement level, the starch:fibre ratio in finishing diets can be decreased (i.e. from 5.0 to 3.0). In growing–finishing diets, the few studies conducted in this field have been focused on evaluating the effect of SRU in direct substitution of high-protein ingredients (Pinos-Rodríguez et al., 2010; Bourg et al., 2012; Lascano et al., 2012) rather than as a strategy to promote synchrony, and no research has examined the role of the starch:fibre ratio of the finishing diets on the effects of the combination of both sources of urea on lambs growth performance and dietary energetics. Recent findings in a digestion trial showed that when conventional urea was combined with a SRU in diets with a ratio of starch-to-acid detergent fibre (S:F) of 4:1, the digestible energy (DE) was improved by 2% over the expected (p = 0.04) level; while, according to the expected DE values, the predicted DE was 1.00 time to the expected values with urea plus SRU in diets with lower (3:1) or greater (6:1) S:F ratios, and for those that were fed with only urea in diets with a similar ratio of 4:1 (López-Soto et al., 2014). Similarly, increases of 6% on net energy (NE) of diet was observed in feedlot cattle when were fed with a diet with a S:F ratio of 4.5 supplemented with a urea combination, while diets with conventional urea did not modify the observed-to-expected NE ratio when was included in diets with an identical S:F ratio (López-Soto et al., 2015). This is surprising, if is considers that different responses of animals can be caused not only by different sources of urea supplementation but also by dietary variations (i.e. rumen undegradable intake protein level); however, the differences on observed-to-expected DE and dietary NE obtained by Lopez-Soto et al. (2014; 2015) between SRU and for those that were fed with only urea in diets with a similar ratio of S:F justifies the need to confirm these results in a lamb performance trial. To test the findings of Lopez-Soto et al. (2014; 2015) on the impact on the dietary energetics, the objectives of this experiment was to examine, in feedlot lambs, the magnitude of the responses on dietary energetics with combining urea and a SRU in diets containing different (3:1, 4:1, or 5:1) S:F ratios.

MATERIALS AND METHODS

This experiment was conducted at the Universidad Autónoma de Sinaloa Feedlot Lamb Research Unit, located in Culiacán, Mexico (24° 46' 13"N and 107° 21' 14"W). Culiacán is about 55 m above sea level, and has a tropical climate. All animal management procedures were conducted within the guidelines of locally approved techniques for animal use and care (NOM-051-ZOO-1995: humanitarian care of animals during mobilisation of animals; NOM-062-ZOO-1995: technical specifications for the care and use of laboratory animals. Livestock farms, farms, centres of production, reproduction and breeding, zoos and exhibition halls, must meet the basic principles of animal welfare; NOM-024-ZOO-1995: animal health stipulations and characteristics during transportation of animals; and NOM-033-ZOO-1995: humanitarian care and animal protection during slaughter process).

Animals, diet, and experimental design

Fifty Pelibuey×Katahdin lambs were received at the research facility before initiation of the experiment. Upon arrival, the lambs were treated for parasites (Tasasel 5%, Fort Dodge, Animal Health, Mexico) and injected with 1×10^6 IU vitamin A (Synt-ADE, Fort Dodge Animal Health). For 2 weeks before the initiation of the experiment, lambs were fed the reference diet (without slow-release urea). Following a 2-week evaluation period, lambs were weighed individually before the morning meal (electronic scale; TORREY TIL/S: 107 2691, TORREY electronics Inc., Houston TX, USA) and 40 lambs (36.65±3 kg) were selected from the original group of 50 lambs for use in the study, based on the uniformity of weight and general condition and were assigned to one of five weight groupings in 20 pens, with two lambs per pen. Pens were 6 m^2 with overhead shade, automatic waterers and 1-m fence-line feed bunks. Dietary treatments were randomly assigned to pens within blocks. Four treatments were tested using urea and urea and slow-release urea combination in diets with different S:F ratios. The S:F ratio in the diet was manipulated by partially replacing the corn grain and DDGS by forage (wheat straw) and soybean meal to reach S:F ratios of 3:1, 4:1, or 5:1. The slow-release urea product used was a polymer-coated urea that contains 41% N (SRU, Optigen II; Alltech Mexico, Guadalajara, Jalisco, Mexico). Based on the hypothesis that a combination of feed-grade urea with slow-release urea in finishing diets promotes synchrony between ruminal N availability and carbohydrate digestion, the combination of urea and SRU (as a percentage of dry matter [DM] in the diet) was performed based on S:F ratios as follows: i) 0.80 U and 1.00% SRU for 3:1 S:F ratio

(U+SRU3); ii) 0.80 U and 0.80% SRU for 4:1 S:F ratio (U+SRU4); and iii) 1.00 U and 0.80% SRU for 5:1 S:F ratio (U+SRU5). An additional treatment of 4:1 S:F ratio with 0.8% urea (U4) as the sole source of non-protein nitrogen (NPN) was used as a reference for comparing urea combination vs conventional urea at the same S:F ratio.

The relative differences in protein concentration between the U4 diet and the U+SRU4 diet was 0.90% (14.01% vs 15.40% crude protein). Although, it is well recognized that when the diet contains more than 1.95 Mcal of net energy of maintenance (NE_m)/kg, increasing protein level above of 14% has no additional beneficial effects on the productive performance of finishing lambs (Ríos et al., 2014), it is important to consider that different responses of animals can be caused not only by different sources of urea supplementation but also by dietary variations (i.e. UIP level). Ingredients and chemical composition of dietary treatments are shown in Table 1. The experiment lasted 56 days and lambs were weighed at the beginning of the trial, at day 28 and in the end of the experiment. The initial body weight (BW) was converted to shrunk body weight (SBW) by reduction of 4% of BW to adjust for the gastrointestinal fill (Cannas et al., 2004), and all lambs were fasted (food but not drinking water was withdrawing) for 18 h before recording the final BW. Lambs were allowed *ad libitum* access to dietary treatments. Daily feed allotments to each pen were adjusted to allow minimal (<5%) feed refusals in the feed bunk. The amount of feed offered and of feed refused was weighed daily. Lambs were provided fresh feed twice daily at 0800 and 1400 hours. Feed bunks were visually assessed between 0740 and 0750 hours each morning, refusals were collected and weighed, and feed intake was determined. Adjustments to either increase or decrease daily feed delivery was provided at the afternoon feeding. Feed and refusal samples were collected daily for DM analysis, which involved oven drying the samples at 105°C until no further weight loss occurred (method 930.15; AOAC, 2000).

Feed analyses

Feed was subjected to the following analysis: DM (oven drying at 105°C until no further weight loss; method 930.15; AOAC, 2000); ash (method 942.05; AOAC, 2000), Kjeldahl N (method 984.13; AOAC, 2000); ADF (Van Soest et al., 1991); starch (Zinn, 1990); calcium (method 927.02; AOAC,

Table 1. Ingredients and composition of experimental diets

Item	Treatments[1]			
	U4	U+SRU3	U+SRU4	U+SRU5
Ingredient composition (% DMB)				
Steam flaked corn	60.00	55.00	60.00	65.00
DDGS	8.00	6.00	8.00	13.00
Soybean meal	5.00	5.00	4.00	0.00
Wheat straw	12.00	18.00	12.00	6.00
Urea	0.80	0.80	0.80	1.00
Optigen 1200[2]	-	1.00	0.80	0.80
Cane molasses	9.70	9.50	9.60	9.40
Yellow grease	2.20	2.50	2.50	2.50
Trace mineral salt[3]	0.50	0.50	0.50	0.50
Limestone	1.80	1.70	1.80	1.80
NE concentration[4] (Mcal/kg of DM basis)				
Maintenance	2.00	1.89	1.99	2.10
Gain	1.34	1.26	1.34	1.43
Nutrient composition (% of DM)[5]				
Crude protein	14.01	15.70	15.40	15.84
Starch	42.62	38.77	42.10	45.12
ADF	10.71	13.07	10.52	8.53
Calcium	0.78	0.76	0.80	0.79
Phosphorus	0.35	0.32	0.35	0.41

U, urea; SRU, slow-release urea; DMB, dry matter basis; DDGS, dried distillers grain with solubles; NE, net energy; DM, dry matter; ADF, acid detergent fibre; NE_m, net energy of maintenance; NE_g, net energy of gain.

[1] Please describe the treatments.

[2] Optigen-II. Alltech de México, Guadalajara Jalisco, Mexico.

[3] Trace mineral salt contained: $CoSO_4$, 0.068%; $CuSO_4$, 1.04%; $FeSO_4$, 3.57%; ZnO, 1.24%; $MnSO_4$, 1.07%; KI, 0.052%; NaCl, 92.96%.

[4] Based on tabular NE values for individual feed ingredients (NRC, 2007) with the exception of supplemental fat, which was assigned NE_m and NE_g values of 6.03 and 4.79, respectively (Zinn, 1988).

[5] Dietary composition was determined by analyzing subsamples collected and composited throughout the experiment. Accuracy was ensured by adequate replication with acceptance of mean values that were within 5% of each other.

2000) and phosphorus (method 964.06; AOAC, 2000).

Calculations

The estimations of dietary energetic and expected dry matter intake (DMI) were performed based on the average obtained of estimated initial SBW and observed final SBW. Average daily gains (ADG) were computed by subtracting the initial BW from the final BW and dividing the result by the number of days on feed. The efficiency of BW gain was computed by dividing ADG by the daily DMI. The estimation of expected DMI was performed based on observed ADG and SBW according to the following equation: Expected DMI, kg/d = $(EM/NE_m)+(EG/EN_g)$, where EM (energy required for maintenance, Mcal/d) = $0.056\times SBW^{0.75}$ (NRC, 1985), EG (energy gain, Mcal/d) = $0.276\times ADG\times SBW^{0.75}$ (NRC, 1985), NE_m and net energy of gain (NE_g) are energy concentrations of experimental diets (derived from tabular values based on the ingredient composition of the experimental diet; NRC, 1985). The apparent retention per unit of DM was estimated by dividing the observed DM intake over expected DMI. The coefficient (0.276) was estimated assuming a mature weight of 113 kg for Pelibuey×Katahdin male lambs (Canton and Quintal, 2007). From the derived estimates of energy required for maintenance and gain, the NE_m and NE_g values of the diet were obtained using the quadratic formula: $x = (-b-\sqrt{b^2-4ac}\ /2c$, where $a = -0.41EM$, $b = 0.877EM+0.41DMI+EG$, and $c = -0.877DMI$, and $NE_g = 0.877\ NE_m-0.41$ (Zinn et al., 2008).

Carcass data

The hot carcass weights (HCW) were obtained from all lambs at time of harvest. After carcasses (with kidneys and internal fat included) were chilled in a cooler at –2°C to 1°C for 48 h, the following measurements were obtained: i) body wall thickness (distance between the 12th and 13th ribs beyond the ribeye, five inches from the midline of the carcass); ii) fat thickness perpendicular to the *m. longissimus thoracis* (LM), measured over the centre of the ribeye between the 12th and 13th ribs; iii) LM surface area, measured using a grid reading of the cross-sectional area of the ribeye between the 12th and 13th ribs; and iv) kidney, pelvic and heart fat (KPH). The KPH was removed manually from the carcass, and then weighed and is reported as a percentage of the cold carcass weight (USDA, 1982).

Statistical analyses

Performance (gain, gain efficiency, and dietary energetics) and carcass data were analysed as a randomised complete block design. The experimental unit was the pen. The MIXED procedure of SAS (SAS Institute, 2004) was used to analyse the variables. The fixed effect consisted of treatment, and pen as the random component. Three contrasts were defined to answer: i) the effect of urea combination vs reference diet (urea at same S:F ratio, 4:1), ii) linear response of the S:F ratio in urea combination treatments, iii) quadratic response of the S:F ratio in urea combination treatments. F-test (numerator = 1 df, denominator = error df) was utilized to test contrasts. The analysis was carried out using SAS (SAS Inst., Inc., Cary, NC, USA; Version 9.1). Contrasts were considered significant when the p-value was ≤ 0.05, and tendencies were identified when the p-value was >0.05 and ≤ 0.10.

RESULTS

According to the determinations of starch and ADF obtained in the laboratory, the starch:ADF ratio reached 100%, 99%, 100% and 106% of that planned for each treatment (Table 1). Treatment effects on growth performance of feedlot lambs are shown in Table 2.

Across the entire 56-day period, the average observed-to-expected DMI of lambs fed the reference diet was 102% of the expected value, based on tabular (NRC, 2007) estimates of diet energy density and observed SBW and ADG values (Table 2), which supports the suitability of the prediction equations proposed by the NRC (1985) for the estimation of DMI in relation to SBW and ADG in feedlot lambs. We expect that dietary NE ratio (observed-to-expected) would be to 1 this mean that animals were performed as expected. Or stated differently, animals performance is consistent with DMI and dietary energy density (NRC). If ratio is greater than 1, the observed dietary NE represent a greater value (concentration) than expected according to NRC, therefore the energy was better utilized by the animal, thus, the efficiency was improved. In contrast, if ratio is less than 1, energetic efficiency was less than expected (contrary to the observed:expected DMI in which values greater than 1 represent lower efficiencies.

There were no effects of the urea combination or SF ratio on DM intake. Even when the diets that contain the same proportion of S:F ratio contained the same amount of available energy (Table 1), with urea combination the ADG, gain for feed, and apparent energy retention per unit DMI were increased ($p<0.01$) by 18.3%, 9.5%, and 8.2%, respectively.

Irrespective of the S:F ratio, the urea combination improved the observed-to-expected dietary ratio and apparent retention per unit DMI was maximal (quadratic effect, $p\leq 0.03$) at an S:F ratio of 4:1, while the urea treatment did not modify the observed-to-expected NE ratio nor the apparent retention per unit DMI at 4:1 S:F ratio. In contrast with lambs fed the reference diet (urea at 4 S:F ratio), lambs fed with dietary treatments containing combination of urea with SRU at the same S:F ratio (4 S:F), tended (3.8%, p = 0.08) to have heavier carcasses with no effects on carcass

Table 2. Influence of treatments on growth performance and dietary energy of lambs

Item	Treatments[1]				SEM	S:F ratio[2]		
	U4	U+SRU3	U+SRU4	U+SRU5		U4 vs U+USR4	Linear	Quadratic
Pen replicates	5	5	5	5				
Days on feed	56	56	56	56				
Weight (kg) [3]								
Initial	36.61	36.49	36.75	36.73	0.21	0.66	0.42	0.60
Final	49.89	49.30	52.34	52.42	0.64	0.02	<0.01	0.09
Average daily gain (kg)	0.235	0.229	0.278	0.280	0.013	0.04	0.02	0.15
Dry matter intake (kg)	1.237	1.257	1.335	1.295	0.046	0.16	0.57	0.31
Gain for feed (kg/kg)	0.190	0.180	0.208	0.216	0.006	<0.01	<0.01	0.03
Dietary net energy (Mcal/kg)[4]								
Maintenance	2.03	1.98	2.15	2.21	0.02	0.01	<0.01	0.03
Gain	1.37	1.33	1.48	1.53	0.02	0.01	<0.01	0.03
Observed to expected dietary ratio[5]								
Maintenance	1.02	1.04	1.08	1.05	0.01	<0.01	0.42	0.03
Gain	1.02	1.05	1.10	1.06	0.01	<0.01	0.60	0.02
Apparent energy retention per unit DMI[6]	0.98	0.94	0.90	0.94	0.01	<0.01	0.57	0.01

U, urea; SRU, slow-release urea; SEM, standard error of the mean; DMI, dry matter intake; BW, body weight; NE, net energy; ADG, average daily gain; DMI, dry matter intake; NE_m, net energy of maintenance; NE_g, net energy of gain.

[1] U4 = 0.80% U for 4 S:F ratio; U+SRU3 = 0.80 U and 1.00% SRU for 3 S:F ratio; U+SRU4 = 0.80 U: 0.80% SRU for 4 S:F ratio; U+SRU5 = 1.00 U and 0.80% SRU for 5 S:F ratio.

[2] Proportion of starch to fibre acid detergent in diet.

[3] The initial BW was reduced by 4% to adjust for the gastrointestinal fill, and all lambs were fasted (food but not drinking water was withdrawing) for 18 h before recording the final BW.

[4] The estimation of dietary NE was performed based on observed ADG, DMI and average shrunk weight (SBW) and was estimated by means of the quadratic formula: $x = (-b\pm\sqrt{b^2-4ac})/2c$, where $x = NE_m$, $a = -0.41EM$, $b = 0.877$ EM+0.41 DMI+EG, and $c = -0.877$ DMI, where EM = maintenance coefficient of 0.056 Mcal/$BW^{0.75}$ (NRC, 1985), EG is the daily energy deposited (Mcal/d) estimated by equation: EG = ([0.276×ADG]×SBW $^{0.75}$; NRC, 1985), and DMI is the average daily dry matter intake (Zinn et al., 2008).

[5] Observed to expected dietary NE ratio was computed by dividing NE observed between expected diet NE, which was estimated based on tabular values for individual dietary ingredients (NRC, 2007).

[6] Expected DMI was performed based on observed ADG, average shrunk weight (SBW) and the calculated NE diet and was computed as follows: DMI, kg/d = (EM/NE_m)+(EG/EN_g), where EM = maintenance coefficient of 0.056 Mcal/$BW^{0.75}$ (NRC, 1985) and EG is the daily energy deposited (Mcal/d) estimated by equation: EG = ([0.276×ADG]×SBW $^{0.75}$, NRC, 1985). The divisors NE_m and NE_g are the NE of diet (Table 1, calculated from tables of composition of feed [NRC, 2007]).

characteristics.

As energy concentration (S:F ratio) increased in diet, ADG, G:F, dietary NE, carcass weight, dressing percentage and LM area increased linearly ($p \leq 0.02$).

DISCUSSION

Urea combination effects on growth performance and dietary energy of diet

The absence of the effects on feed intake as a consequence of the supplementation of combination of urea plus SRU have been observed previously in finishing lambs when lambs were fed with a 50:50 forage:concentrate diet (Moura et al., 2014) and in steers when they were fed a finishing diet (>70% concentrate; Tedeschi et al., 2002; Pinos-Rodríguez et al., 2010; Castañeda-Serrano et al., 2013). However, a tendency for a reduction in the DMI has been observed in some studies when feedlot cattle were supplemented with protected urea or with combinations of SRU plus urea (Huff et al., 2000; Taylor-Edwards et al., 2009). The basis for the inconsistencies in the DMI responses to SRU supplementation is not clear, but may be related to the taste of SRU products and/or diet composition (i.e. inclusion of high levels of corn silage).

There is limited information concerning the effects of SRU on growth performance and dietary energetics in lambs; however, improvements in feed efficiency in finishing steers supplemented with SRU have been previously reported (Huff et al., 2000). Similarly, combining conventional urea with slow-release urea has been reported to promote milk production (Akay et al., 2004). Changes in productivity and/or energy efficiency can be partially explained by improvements in nutrient synchrony between N and carbohydrate compounds in the rumen and greater N retention (decreases in ruminal ammonia concentration and increases in the flow of microbial N to the duodenum). However this is not to be confused with the popular notion that rate of soluble feed N release to the rumen be in synchrony with carbohydrate fermentation. Numerous studies have proved the concept indefensible. Providing adequate ruminal available N, irregardless of the rate at which it is degraded or solubilized within the rumen, is the relevant factor affecting microbial protein synthesis. Irregardless of source (NPN or intact protein), microbial protein synthesis is maximal when degradable intake protein is roughly 10% of digestible organic matter intake (Zinn and Shen, 1998). This effect is due to N recycling to the rumen (Calsamiglia et al., 2010). Conversely, López-Soto et al. (2014) showed that steers fed a combination of urea and slow-release urea (using the same source of SRU) with an S:F ratio of 4:1 had higher ($p = 0.04$) flows of microbial N and DE of the diet than those fed urea at the same S:F ratio, or those fed urea plus SRU in diets with 3:1 and 6:1 S:F ratios. In studies conducted with steers (Tedeschi et al., 2002; Pinos-Rodríguez et al., 2010), the urea combination did not affect growth performance or digestibility of the diet. Based on the experimental diets of the study of Tedeschi et al. (2002), the estimated S:F ratio of their experimental diets was 14:1, while in the study conducted by Pinos-Rodríguez et al. (2010), the estimated S:F ratio of the diets was 5.4:1. Thus, the high S:F ratios of the diets used in the studies conducted by Tedeschi et al. (2002) and by Pinos-Rodriguez et al. (2010) could be a factor, as in the present experiment, in the lack or small effects on performance and feed efficiency of steers fed a urea combination.

Irrespective of the S:F ratio, the urea combination improved the observed-to-expected dietary ratio and apparent retention per unit DMI. According to the expected NE values (NRC, 2007), the observed dietary NE was 1.02 for lambs fed the reference diet, and 1.04, 1.08, and 1.05 times the expected values for the urea combinations at 3 S:F, 4 S:F, and 5 S:F ratios, respectively. At a 4 S:F ratio, the observed NE value in the urea combination treatment was improved on average by 4% compared to the rest of the urea combination treatments. It is important to consider that different responses of animals can be caused not only by different sources of urea supplementation but also by dietary variations (UIP level among others). However, in a growth-performance study conducted with feedlot steers, Lopez-Soto et al. (2015) with a similar urea combination as in the present experiment in a diet with a S:F ratio of 4.5 observed a 6% of increases on NE of diet and decreases of 6% on the apparent retention per unit DM, while diets with conventional urea did not modify neither the observed-to-expected NE ratio nor the apparent retention per unit DMI, when was included in diets with a S:F ratios of 3, 4.5, and 5.5. It has been observed that in high-grain diets (ratio of starch vs ADF greater than 5 to 1) urea can be supplemented at 50% higher than that recommended with positive effects on growth performance and in dietary energy utilisation (Milton et al., 1997; Zinn et al., 2003). Those researchers argued that those results can be partially explained by the possible synchrony of ruminal degradation rates between urea and starch. At lower S:F ratios it is expected a lower positive effects, therefore, absence of improves of observed NE ratio over expected in urea treatment at 4:1 S:F ratio is not aberrant (observed-to-expected DMI = 0.98), as mentioned above, absence of improvements of observed NE ratio over expected (averaging 0.98) with conventional urea supplementation in diets with S:F ratios of 3, 4.5, and 5.5 was previously reported (López-Soto et al., 2015). The observed-to-expected dietary energy and intake are an important and practical application of current standards for energetics in nutrition research (Zinn et al., 2008). Based on diet composition and growth performance, there is an expected energy intake and hence an expected of DMI (NRC, 1985). The estimation of dietary energy and the ratio of observed-to-expected DMI (apparent energy retention per unit DMI) revealed differences on the efficiency of energy utilisation of the diet itself. In the present experiment, the greatest improvement in the observed-to-expected DMI and dietary NE of the combination urea treatments was at 4:1 S:F ratio. Starch and fibre at these proportions provide an energetic advantage when they were supplemented with the urea combination. For example, if considering the same diet composition between the reference diet (U4) and U+SRU4 treatments (Table 1), then——compared with the reference diet—the energy improvement in the U+SRU4 treatment represents an equivalent increase of 5.3% ([2.15-2.03]/2.24) corn grain in the diet. This could support the theory that the S:F ratio is the most important factor that impacts on the synchrony when urea and SRU are combined, rather than the energy level *per se*.

S:F ratio effects on growth performance and dietary energy of diet

The energy level (S:F ratio) did not affect the DMI. In high-energy diets, ME intake, rather than physical fill, appeared to be the dominant factor influencing the DMI. Lu and Potchoiba (1990) observed a curvilinear response in goats when comparing three levels of energy (1.66, 1.86, and 2.06 Mcal NE_m/kg DM) in diets. However, consistent with our results, other studies (Mahgob et al., 2000; Sheridan et al., 2000; Loe et al., 2004) did not find any effect on DMI in finishing lambs when comparing diets from 1.90 up to 2.16 Mcal NE_m/kg, which is similar to the range of energy density for the three S:F ratio treatments used in the present study (Table 1).

Increases in feed efficiency have been a common response when comparing high-energy and low-energy diets (NRC, 2007; Kioumarzi et al., 2008; Adbel-Basset, 2009). However, the effects of increased dietary energy levels on the ADG have been less consistent. In some instances (Lu and Potchoiba, 1990; García et al., 2003), increasing the energy level had no effect on the ADG, whereas in others

Table 3. Treatment effects on carcass characteristics

Item	Treatments[1]				SEM[2]	S:F ratio[2]		
	U4	U+SRU3	U+SRU4	U+SRU5		U4 vs U+SRU4	Linear	Quadratic
Hot carcass weight (kg)	29.79	28.45	30.97	31.16	0.43	0.08	<0.01	0.05
Cold carcass weight (kg)	29.44	28.13	30.68	30.83	0.42	0.06	<0.01	0.04
Drip loss (%)	1.18	1.03	0.95	1.09	0.16	0.34	0.80	0.59
Dressing percent	59.66	57.66	59.17	59.44	0.45	0.46	0.02	0.28
Longissimus muscle area (cm^2)	15.21	14.63	14.88	16.82	0.51	0.65	0.02	0.20
Backfat thickness (mm)	2.52	2.53	2.66	2.61	0.25	0.69	0.82	0.78
Kidney-pelvic fat (%)	2.84	2.56	2.90	3.08	0.21	0.85	0.09	0.73
Body wall thickness (mm)	13.81	13.42	13.43	13.81	0.50	0.61	0.59	0.77

U, urea; SRU, slow_release urea; SEM, standard error of the mean.

[1] U4 = 0.80% U for 4 S:F ratio; U+SRU3 = 0.80 U and 1.00% SRU for 3 S:F ratio; U+SRU4 = 0.80 U: 0.80% SRU for 4 S:F ratio, and U+SRU5 = 1.00 U and 0.80% SRU for 5 S:F ratio.

[2] Proportion of starch to fibre acid detergent in diet.

(Kioumarzi et al., 2008; Adbel-Basset, 2009), an increase in energy level markedly increased the ADG. The latter could be explained by the strong relationship between DMI and the dietary energy density (Cannas et al., 2004).

Treatments effects on carcass characteristics

The treatments effects on the carcass characteristics are shown in Table 3. There is limited information concerning the effects of SRU on carcass characteristics in lambs, but, consistent with previous findings with steers (Duff et al., 2000; Pinos-Rodríguez et al., 2010), urea combinations that replace soybean meal did not affect carcass characteristics. The linear increases in HCW and dressing percentage as a result of an increased S:F ratio is likely to be due to a concomitant linear increase in the ADG (Block et al., 2001). In the same manner, increased LM area has been a consistent response to increased rate of ADG in steers (Zinn et al., 2007).

Combining urea and a slow-release urea product results in positive effects on growth performance and dietary energetics, but the best responses are apparently observed when there is a certain proportion (S:F ratio = 4:1) of starch to ADF in the diet. When the S:F ratio increases or decreases, the level of response decreases. Further studies are needed to determine the conditions of the finishing diet so that it is possible to get the best response from the use of slow-release urea.

ACKNOWLEDGMENTS

This experiment was supported by PROFAPI-UAS of México (project code: PROFAPI/050).

REFERENCES

Adbel-Baset, N. S. 2009. Effect of different dietary energy levels on the performance and nutrient digestibility of lambs. Vet. World 2:418-420.

Akay, V., J. Tikofsky, C. Holtz, and K. A. Dawson. 2004. Optigen 1200: Controlled release of non-protein nitrogen in the rumen. In: Proceedings of the Alltech's 20th Annual Simposium of Nutritional Biotechnology in the Feed and Food Industries. Nottingham University Press. Nottingham, UK. pp. 179-186.

Alvarez-Almora, E. G., G. B. Huntington, and J. C. Burns. 2011. Effects of supplemental urea sources and feeding frequency on ruminal fermentation, fiber digestion, and nitrogen balance in beef steers. Anim. Feed Sci. Technol. 171:136-145.

AOAC (Association of Official Analytical Chemists) International. 2000. Official Methods of Analysis. 17th edn. Association Official Analytical Chemists. Gaithersburg, MD, USA.

Block, H. C., J. J. McKinnon, A. F. Mustafa, and D. A. Christensen. 2001. Manipulation of cattle growth to target carcass quality. J. Anim. Sci. 79:133-140.

Bourg, B. M., L. O. Tedeschi, T. A. Wickersham, and J. M. Tricarico. 2012. Effects of a slow-release urea product on performance, carcass characteristics, and nitrogen balance of steers fed steam-flaked corn. J. Anim. Sci. 90:3914-3923.

Calsamiglia, S., A. Ferret, C. K. Reynolds, N. B. Christensen, and A. M. van Vuuren. 2010. Strategies for optimizing nitrogen use by ruminant. Anim. 7:1184-1196.

Cannas, A., L. O. Tedeschi, D. G. Fox, A. N. Pell, and P. J. Van Soest. 2004. A mechanistic model for predicting the nutrient requirements and feed biological values for sheep. J. Anim. Sci. 82:149-169.

Canton, J. G. and J. A Quintal. 2007. Evaluation of growth and carcass characteristics of pure Pelibuey sheep and their cross with Dorper and Katahdin breeds. J. Anim. Sci. 85 (Suppl. 1):581 (Abstr.).

Carrasco, R., A. A. Arrizon, A. Plascencia, N. G. Torrentera, and R. A. Zinn. 2013. Comparative feeding value of distillers dried grains plus solubles as a partial replacement for steam-flaked corn in diets for calf-fed Holstein steers: characteristics of digestion, growth-performance, and dietary energetic. J. Anim. Sci. 91:1801-1810.

Castañeda-Serrano, R. D., A. Ferriani-Branco, S. Teixeira, T. García-Díaz, and A. Diego-Sofiati. 2013. Slow release urea in beef cattle diets: digestibility, microbial synthesis and rumen kinetic. Agrociencia 47:13-24.

Huff, G. C., D. A. Walker, K. J. Malcom-Callis, M. W. Wiseman, J. D. Rivera, M. L. Galyean, and T. H. Montgomery. 2000. Effects of a slow-release urea product on feedlot performance and carcass characteristics of beef steers. In: Proceedings of the American Society of Animal Science Western Section. Baltimore, MD, USA. pp. 506-509.

Estrada-Angulo, A., Y. S. Valdés, O. Carrillo-Muro, B. I. Castro-Pérez, A. Barreras, M. A. López-Soto, A. Plascencia, H. Dávila-Ramos, F. G. Ríos, and R. A. Zinn. 2013. Effects of feeding different levels of chromium-enriched live yeast in hairy lambs fed a corn-based diet: effects on growth performance, dietary energetics, carcass traits and visceral organ mass. Anim. Prod. Sci. 53:308-315.

Garcia, C. A., C. Costa, A. L. Gomes, M. A. Neres, and G. J. M. Rosa. 2003. Energy levels on performance and carcass characteristics in lambs fed creep feeding. R. Bras. Zootec. 32:1371-1379.

Hristov, A. N., M. Hanigan, A. Cole, R. Todd, T. A. McAllister, P. M. Ndegwa, and A. Rotz. 2011. Review: ammonia emissions from dairy farms and beef feedlots. Can. J. Anim. Sci. 91:1-35.

Inostroza, J. F., R. D. Shaver, V. E. Cabrera, and J. M. Tricárico. 2010. Effect of diets containing a controlled-release urea product on milk yield, milk composition, and milk component yields in commercial Wisconsin dairy herds and economic implications. Prof. Anim. Sci. 26:175-180.

Kioumarsi, H., K. J. Khorshidi, M. Zahedifar, A. R. Zeidavi, S. Z. Mirhosseini, and M. R. Taherzadeh. 2008. The effect of dietary energy and protein level on performance, efficiency and carcass characteristics of Taleshi lambs. Asian J. Anim. Vet. Adv. 3:307-313.

Klopfenstein, T. J., G. E. Erickson, and V. R. Bremer. 2008. BOARD-INVITED REVIEW: Use of distillers by-products in the beef cattle feeding industry. J. Anim. Sci. 86:1223-1231.

Lascano, G. J., M. Vélez, J. M. Tricarico, and A. J. Heinrichs. 2012. Short communication: Nutrient utilization of fresh sugarcane-based diets with slow-release nonprotein nitrogen addition for control-fed dairy heifers. J. Dairy Sci. 95:370-376.

Loe, E. R., M. L. Bauer, G. P. Lardy, J. S. Caton, and P. T. Berg. 2004. Field pea (*Pisum sativum*) inclusion in corn-based lamb finishing diets. Small Rum. Res. 53:39-45.

López-Soto, M. A., C. R. Rivera-Méndez, J. A. Aguilar-Hernández, A. Barreras, J. F. Calderón-Cortés, A. Plascencia, H. Dávila-Ramos, A. Estrada-Angulo, and Y. S. Valdés-García. 2014. Effects of combining conventional urea and a slow-release urea product on characteristics of digestion, microbial protein synthesis and digestible energy in steers fed diets with different starch:ADF ratios. Asian Australas. J. Anim. Sci. 27:187-193.

López-Soto, M. A., J. A. Aguilar-Hernández, H. Dávila-Ramos, A. Estrada-Angulo, F. G. Ríos, J. D. Urías-Estrada, A. Barreras, J. F. Calderón, and A. Plascencia. 2015. Effects of a combining feed grade urea and a slow-release product on performance, dietary energetics and carcass characteristics of steers fed finishing diets. J. Appl. Anim. Res. 43:303-308.

Lu, C. D. and M. J. Potchoiba. 1990. Feed intake and weight gain of growing goats fed diets of various energy and protein levels. J. Anim. Sci. 68:1751-1759.

Mahgoub, O., C. D. Lu, and R. J. Early. 2000. Effects of dietary energy density on feed intake, body weight gain and carcass chemical composition of Omani growing lambs. Small Rum. Res. 37:35-42.

Milton, C. T., R. T. Brandt Jr., and E. C. Titgemeyer. 1997. Urea in dry-rolled corn diets: finishing steers performance, nutrient digestion and microbial protein production. J. Anim. Sci. 75:1415-1424.

Moura, E., D. Rodriguez, M. Amaral, E. J. Dos-Santos, M. L. Alburquerque, and M. Dos-Santos. 2014. Nitrogen metabolism and microbial synthesis in sheep fed diets containing slow release urea to replace the conventional urea. Acta Sci. 36:55-62.

NRC (National Research Council). 1985. Nutrient Requirement of Sheep. 6th Rev. edn. National Academy Press. Washington, DC, USA.

NRC (National Research Council). 2007. Nutrient Requirement of Small Ruminant. Sheep, Goats, Cervids, and New World Camelids. National Academy Press. Washington, DC, USA.

Pinos-Rodríguez, J. M., L. Y. Peña, S. S. González-Muñoz, R. Bárcena, and A. Salem. 2010. Effects of a slow-release coated urea product on growth performance and ruminal fermentation in beef steers. Ital. J. Anim. Sci. 9:16-19.

Rosentrater, K. A. 2011. Feeding DDGS in other animals. In: Distiller Grain, Production Properties and Utilization (Eds. K. Lui and K. A. Rosentrater). CRC Press, Boca Raton, FL, USA. pp. 391-398.

SAS (Statistical Analysis System) Institute Inc. 2004. User's Guide: Statistics, version 9. SAS Inst. Cary, NC, USA.

Sheridan, R., A. V. Ferreira, L. C. Hoffman, and S. J. Schoeman. 2000. Effect of dietary energy level on efficiency of SA Mutton Merino lambs and Boer goat kids under feedlot conditions. S. Afr. J. Anim. Sci. 30:122-123.

Ríos-Rincón, F. G., A. Estrada-Angulo, A. Plascencia, M. A. López-Soto, B. I. Castro-Pérez, J. J. Portillo-Loera, J. C. Robles-Estrada, J. F. Calderón-Cortes, and H. Dávila-Ramos. 2014. Influence of protein and energy level in finishing diets for feedlot hair lambs: Growth performance, dietary energetics and carcass characteristics. Asian Australas. J. Anim. Sci. 27:55-61.

Taylor-Edwards, C. C., G. Hibbard, S. E. Kitts, K. R. McLeod, D. E. Axe, E. S. Vanzant, N. B. Kristensen, and D. L. Harmon. 2009. Effects of slow-release urea on ruminal digesta characteristics and growth performance in beef steers. J. Anim. Sci. 87:200-208.

Tedeschi, L. O., M. J. Baker, D. J. Ketchen, and D. G. Fox. 2002. Performance of growing and finishing cattle supplemented with a slow-release urea product and urea. Can. J. Anim. Sci. 82:567-573.

USDA (United States Department of Agriculture). 1982. Official United States Standards for Grades of Lambs, Yearling Mutton, and Mutton Carcasses. Agriculture Marketing Service, USA.

Van Soest, P. J., J. B. Robertson, and B. A. Lewis. 1991. Methods for dietary fiber, neutral detergent fiber, and nonstarch polysaccharides in relation to animal nutrition. J. Dairy Sci. 74:3583-3597.

Zinn, R. A. 1990. Influence of steaming time on site of digestion of flaked corn in steers. J. Anim. Sci. 68:776-781.

Zinn, R. A. and Y. Shen. 1998. An evaluation of ruminally

degradable intake protein and metabolizable amino acid requirements of feedlot calves. J. Anim. Sci.76:1280-128.

Zinn, R. A., A. Barreras, F. N. Owens, and A. Plascencia. 2008. Performance by feedlot steers and heifers: Daily gain, mature body weight, day matter intake, and dietary energetics. J. Anim. Sci. 86:2680-2689.

Zinn, R. A., R. Barrajas, M. Montaño, and R. A. Ware. 2003. Influence of dietary urea level on digestive function and growth performance of cattle fed steam-flaked barley- based finishing diets. J. Anim. Sci. 81:2383-2389.

Zinn, R. A., J. F. Calderon, L. Corona, A. Plascencia, M. F. Montaño, and N. Torrentera. 2007. Phase feeding strategies to meet metabolizable amino acid requeriments of calf-fed Holstein steers. Prof. Anim. Sci. 23:333-339.

28

Effects of Dietary Methionine Levels on Choline Requirements of Starter White Pekin Ducks

Z. G. Wen[1], J. Tang, M. Xie, P. L. Yang[1], and S. S. Hou*
Institute of Animal Sciences, Chinese Academy of Agricultural Sciences, Beijing 100193, China

ABSTRACT: A 2×5 factorial experiment, using 2 dietary methionine levels (0.28% and 0.48%) and 5 dietary choline levels (0, 394, 823, 1,239, and 1,743 mg/kg), was conducted to study the effects of dietary methionine status on choline requirements of starter white Pekin ducks from 7 to 28 days of age. Four hundred eighty 7-d-old male White Pekin ducks were randomly allotted to ten dietary treatments, each containing 6 replicate pens with 8 birds per pen. At 28 d of age, weight gain, feed intake, and feed/gain were measured and the legs of all ducks from each pen were examined for incidence of perosis. Perosis and growth depression were observed in choline-deficient ducks and supplementation of choline reduced perosis and significantly increased weight gain and feed intake regardless of dietary methionine levels ($p<0.05$). In addition, significant positive effects of dietary methionine supplementation on weight gain, feed intake, and feed/gain were observed at any choline level ($p<0.05$). Supplementation of 1,743 mg/kg choline in diets alleviated the depression of weight gain and feed intake caused by methionine deficiency at 0.28% methionine level. The interaction between choline and methionine influenced weight gain and feed intake of ducks ($p<0.05$). At 0.28% methionine level, 1,743 mg/kg choline group caused 4.92% and 3.23% amount of improvement in weight gain and feed intake compared with 1,239 mg/kg choline group, respectively. According to the broken-line regression, the choline requirements of starter Pekin ducks for weight gain and feed intake were 1,472 and 1,424 mg/kg at 0.28% methionine level and 946 and 907 mg/kg at 0.48% methionine level, respectively. It suggested the choline recommendations of starter Pekin ducks on a semi-purified diet were 1448 mg/kg at 0.28% methionine level and 927 mg/kg at 0.48% methionine level, respectively. Compared with the adequate methionine level, menthionine deficiency markedly increased the choline requirements of ducks. (**Key Words:** Duck, Choline, Methionine, Growth Performance)

INTRODUCTION

Choline, a water-soluble vitamin, plays important roles in the synthesis of the membrane phospholipids (Hollenbeck, 2010), neurotransmitter synthesis (Blusztajn and Wurtman, 1983; Wessler et al., 2001) and methyl-group metabolism (Zeisel and Blusztajn, 1994). Poor growth, hepatic fatty infiltration, and perosis were all observed in choline-deficient Pekin ducks (Bernard and Demers, 1949; Wen et al., 2014). NRC (1994) provided choline recommendations for chickens, turkeys, geese, and quails, but the recommendation of this vitamin was missing for ducks. At present, although the effects of different choline levels on growth performance of Tsaiya ducks was reported by Lien and Jan (1999), the data about the requirement of this vitamin was not given in their study and new literature on choline requirements of ducks was lacking until now.

On the other hand, choline is the methyl donor that provides the one-carbon unit in the conversion of homocysteine to methionine (Zeisel and Blusztajn, 1994). The interrelationship between choline and methionine remains a subject of ongoing debate. Ketola and Nesheim (1974) suggested that the choline requirement of chicks is elevated by high dietary methionine levels. However, some researchers reported more dietary choline was required when methionine is deficient in poultry (Miles et al., 1983; Harms and Miles, 1984). At present, studies on the interrelationship between methionine and choline in Pekin ducks are scarce and the effect of dietary methionine levels

* Corresponding Author: S. S. Hou.
E-mail: houss@263.net
[1] Feed research institute, Chinese Academy of Agricultural Sciences, Beijing 100081, China.

on choline requirements of ducks is not clear. Therefore, the objective was to estimate the choline requirements of white Pekin ducks with deficient and adequate methionine levels and to evaluate the effects of methionine status on the requirements of choline for ducks.

MATERIALS AND METHODS

This study was approved by the animal care and use committee of the Institute of Animal Sciences of Chinese Academy of Agricultural Sciences, and all efforts were made to minimize the suffering of animals.

Birds and housing

A 2×5 factorial experiment, using 2 dietary methionine levels (0.28% and 0.48%) and 5 dietary choline levels (0, 394, 823, 1,239, and 1,743 mg/kg), was conducted to study the effects of methionine status on choline requirements in white Pekin ducks from 7 to 28 d of age. A total of 480 seven-d-old male White Pekin ducklings with average body weight of 132±3 g were randomly divided into ten dietary treatments, each containing 6 replicate pens with 8 birds per pen. These birds were kept in plastic-wire-floor pens (200 by 100 by 40 cm) in an environmentally controlled experimental building. The temperature was kept at 28°C (at floor level) from 7 to 10 d of age, and then it was reduced gradually to room temperature until 28 d of age. All birds had free access to water and feed at any time and lighting was continuous. Water was provided by drip-nipple water supply lines and feed was fed in pellet form.

Table 1. Ingredient composition and nutrient content of the basal diet (% as-fed basis)

Item	Content
Ingredient (%)	
Corn starch	63.0
Isolated soybean protein	25.7
Rice hull powder	5.0
Soybean oil	2.0
Dicalcium phosphate	2.2
Limestone meal	0.8
Premix[1]	1.0
Sodium chloride	0.3
Calculated values	
Metabolizable energy[2] (kcal/kg)	3,480
Ca (%)	0.90
Nonphytate P (%)	0.42
Analyzed values[3]	
Crude protein (%)	22.12
Total lysine (%)	1.29
Total threonine (%)	0.78
Total tryptophan (%)	0.20
Total methionine (%)	0.28
Choline[4] (mg/kg)	0

[1] Premix provided the following per kg of diets: Cu, 10 mg; Fe, 60 mg; Zn, 60 mg; Mn, 80 mg; Mg, 500 mg; K, 2,000 mg; Se, 0.3 mg; I, 0.2 mg; retinyl acetate, 10,000 IU; cholecalciferol, 3,000 IU; DL-α-tocopheryl acetate, 20 IU; menadione sodium bisulphate, 2 mg; thiamin, 2 mg; pyridoxine hydrochloride, 4 mg; cobalamin, 0.02 mg; calcium-D-pantothenate, 20 mg; nicotinic acid, 50 mg; folic acid, 1 mg; riboflavin 8 mg; biotin, 0.2 mg.

[2] Values were calculated according to the MEn of feedstuffs for poultry provided by NRC (1994).

[3] Analyzed in duplicate.

[4] Below the level of detection.

Experimental diets

The basal diet was formulated using corn starch and isolated soybean protein that were low in methionine and choline content (Table 1). To produce experimental diets, the basal diets were supplemented with different levels of *DL*-methionine (0 and 0.2%) and choline (0, 400, 800, 1,200 and 1,600 mg/kg). Choline was added in the form of choline chloride (silica carrier, purity≥50%, Hebei Be-Long Feed Additive Co., Ltd, Heibe, China). The choline concentration was analyzed by ion exchange chromatography according to Zhai et al. (2013). In brief, powdered feed samples (2 g) were extracted with 25 mL of ultrapure water in a 50-mL polypropylene centrifuge tube and homogenized for 1 min. Then feed samples were extracted by using ultrasound bath for 30 min at 70°C. The supernatant was transferred into a 50-mL volumetric flask after extracts were centrifuged. After repeating the procedure, the volume of the supernatant was increased to 50 mL with ultrapure water. The supernatant was loaded into a plastic 2.5-mL syringe and filtered by a 0.45-μm filter membrane. This solution was then used for choline quantification by an ion chromatograph. Choline standards were choline chloride (purity ≥98%, Sinopharm Chemical Reagent Co., Ltd., Shanghai, China). The analyzed levels of dietary choline were 0, 394, 823, 1,239, and 1,743 mg/kg diet, respectively. Contents of methionine in basal diet were analyzed by ion-exchange chromatography with an amino acid analyzer (L-8900, Hitachi, Tokyo, Japan) after acid hydrolysis according to the method recommended by the Standardization Administration of China (2000). The analyzed methionine concentration in the basal diet was 0.28%. Crude protein in the basal diet was determined by the Kjeldahl method (method 954.01; AOAC, 2000). All diets were firstly prepared in mash form and then all pelleted (2.5 mm in diameter).

Measurements

At 28 d of age, weight gain, feed intake and feed/gain of ducks from each pen were measured after 12 h of fasting. Feed intake and feed/gain were all corrected for mortality. In addition, the legs of all ducks from each pen were examined for incidence of perosis according to symptom descriptions of Evans et al. (1942). The symptoms of perosis were divided into four types: Enlarged hock joint; twisting, turning, or bowing leg; slipped tendon; crippled. When single or more symptoms were observed in one or both legs of ducks, the bird was considered to have perosis. The perosis of all ducks were examined by the same person every time.

Statistical analysis

Data were analyzed by two-way analysis of variance procedure of SAS (SAS Institute, 2003) with the pen being used as the experimental unit. Linear and quadratic polynomial contrasts were performed to determine the effects of dietary choline. The variability in the data was expressed as the standard error of the means and a probability level of $p<0.05$ was considered to be statistically significant. In study, broken-line regression model analysis (Robbins et al., 2006) was used to estimate dietary choline requirement at each supplemental methionine level using the NLIN procedure of SAS. The model was as follows: $y = l+u\ (r-x)$, where y = weight gain, x = dietary choline level, r = choline requirement, l = the response at x = r and u = the slope of model. In this model, y = l when x>r. In addition, according to the statistical method suggested by Sterling et al. (2003), a t-test was used to determine if a difference existed for the choline requirements established at the two methionine levels.

RESULTS AND DISCUSSION

In previous studies, a diet containing a large percentage of corn starch and isolated soybean protein had been successfully used to determine the choline requirement and the availability of choline in feedstuffs for broilers (Molitoris and Baker, 1976; Pesti et al., 1981; Emmert and Baker, 1997). Therefore, in order to reduce choline content in the basal diet, purified diets containing corn starch and isolated soybean protein were used in this study as they contained little choline. The choline concentration in basal diet was analyzed by ion exchange chromatography, but it was not detected. The reason may be associated with the detection limit, precision, and recovery. So the choline concentration in basal diet was given as zero. Previous studies have used measurements of perosis and growth depression as the characteristic symptom of choline-deficiency in ducks (Wen et al., 2014). In the present study, a lack of supplementation of choline caused perosis and depressed weight gain and feed intake, which indicated that the basal diet was choline deficient and was hence appropriate for estimation of choline requirements of Pekin ducks. Perosis caused by choline deficiency was reduced by increasing dietary choline level. The reason for perosis caused by choline deficiency may be that the content of phosphatidylcholine in the bone cell was decreased, because choline is a component of the membrane phospholipids (Hollenbeck, 2010). Additionally, choline deficiency could cause a decrease in choline kinase activity (Li et al., 2005); previous study has shown that choline kinase was required for normal endochondral bone formation (Li et al., 2014).

As shown in Table 2, supplementation of methionine significantly improved weight gain and feed intake and decreased feed/gain ($p<0.05$). In poultry, this positive effect of methionine on growth performance has been shown by many researchers (Bornstein and Lipstein, 1964; Blair et al., 1986; Ohta and Ishibashi, 1995). In addition, supplementation of choline caused significant improvement in weight gain and feed intake (Table 2). This result is in accordance with a previous study in chickens by Ketola and Nesheim (1974), who reported that the addition of choline could improve the body weight gain of chicks either in the presence or absence of methionine supplementation. In this study, the data also indicated that methionine had a greater growth-promoting effect than choline because increasing dietary methionine significantly decreased feed/gain ($p<0.05$), but it was not true for increasing choline. In chicks, excess choline was found to efficiently compensate for dietary methionine deficiency (Quillin et al., 1961; Pesti et al., 1980; Baker et al., 1983). This study found similar results. As shown in Table 2, at low methionine level, 1,743 mg/kg choline could improve weight gain and feed intake further. However, at normal methionine level, weight gain and feed intake showed no increase at the 1,743 mg/kg choline level. It indicated that supplementation of choline in diets could alleviate the depression of weight gain and feed intake caused by methionine deficiency. Methionine plays an important role in poultry growth, and it is the first-limiting amino acid in common poultry diets. The early growth of young birds is mainly due to the deposition of body protein, and feed intake is an important factor affecting body protein synthesis (Kita et al., 1996). Dietary methionine deficiency caused an imbalance in the diet's amino acid profile, which altered feed intake. This pattern in ducks has also been observed by Chen et al. (1991). This interaction between choline and methionine was also found in fish studies. In the absence of methionine, choline can provide a portion of the methionine requirement (Kasper et al., 2000; Wu and Davis, 2005). Because choline is a methyl donor and can donate a methyl to homocysteine to form methionine (Zeisel and Blusztajn, 1994), which was clearly shown by the interaction between choline and methionine in this study ($p<0.05$).

A linear increase in weight gain and feed intake was obtained as dietary choline was increased from 0 to 1,734 mg/kg at 0.28% methionine level (Table 2). However, at 0.48% methionine level, weight gain and feed intake increased significantly as dietary choline increased from 0

Table 2. Effect of dietary choline on growth performance and perosis of Pekin ducks from 7 to 28 d of age at two dietary methionine level[1]

Dietary treatment			Weight gain	Feed intake	Feed/gain	Incidence of
Dietary Met[2]	Analyzed choline[2]	Calculated choline	(g/d/bird)	(g/d/bird)	(g:g)	perosis[3] (%)
0.28	0	2	12.13[e]	29.49[e]	2.48	100
	394	402	13.30[e]	32.53[de]	2.46	45
	823	802	16.80[cd]	39.89[bc]	2.38	22
	1,239	1,202	21.13[ab]	49.23[a]	2.35	5
	1,743	1,602	22.17[a]	50.82[a]	2.32	0
0.48	0	2	15.94[d]	35.96[cd]	2.26	100
	394	402	19.15[bc]	42.01[b]	2.20	15
	823	802	22.05[a]	50.58[a]	2.31	4
	1,239	1,202	23.11[a]	52.17[a]	2.30	0
	1,743	1,602	22.99[a]	51.38[a]	2.26	0
Pooled SEM			0.57	1.19	0.04	
Main effect						
Dietary Met level	0.28		17.10[b]	40.39[b]	2.40[b]	
	0.48		20.65[a]	46.42[a]	2.26[a]	
Pooled SEM			1.29	2.71	0.03	
Dietary choline level	0		14.04[d]	32.73[d]	2.38	
	394		16.22[c]	37.27[c]	2.33	
	823		19.43[b]	45.24[b]	2.35	
	1,239		22.12[a]	50.70[a]	2.32	
	1,743		22.58[a]	51.09[a]	2.29	
Pooled SEM			1.65	3.66	0.01	
Probability						
Met			<0.001	<0.001	0.048	
Choline			<0.001	<0.001	0.896	
Met×choline			0.046	0.045	0.908	
Choline linear			<0.001	<0.001	0.584	
Choline quadratic			0.028	0.104	0.974	

SEM, standard error of the means.
[1] Values are the means of 6 replicates of 8 ducks. [2] Analyzed value.
[3] Incidence of perosis = the number of ducks with perosis per treatment/the total ducks number per treatment.
[a-c] Mean values in the same column with no common superscript differ significantly ($p<0.05$).

to 1,239 mg/kg and then reached a plateau at the ranges of dietary choline from 1,239 to 1,743 mg/kg. Pekin ducks showed different growth response with increasing choline level at two methionine levels. This data indicates that more choline was needed by ducks at the low methionine level. Broken-line regression analysis has been used to estimate the choline requirements in broilers (Pesti et al., 1980), ducks (Wen et al., 2014), and fish (Mai et al., 2009; Jiang et al., 2012). Therefore, the choline requirements of ducks at two methionine levels were estimated by broken-line regression (Table 3). According to this regression, the choline requirements of White Pekin ducks for weight gain and feed intake were 1,472 and 1,424 mg/kg at 0.28% methionine level and 946 and 907 mg/kg at 0.48% methionine level, respectively (Table 3). The t-test showed that the choline requirement at 0.28% methionine level was

Table 3. Choline requirements for weight gain and feed intake at different supplemental methionine levels according to broken-line regression

Methionine level (%)	Response criterion	Broken-line model	p-value	R^2	Choline requirement (mg/kg)[1]
0.28	Weight gain	Y = 22.17–0.00738×(1,472–x)	0.029	0.947	1,472±212[a]
	Feed intake	Y = 50.82–0.0161×(1,424–x)	0.024	0.959	1,424±187[a]
0.48	Weight gain	Y = 23.05–0.00741×(946–x)	0.002	0.982	946±31[b]
	Feed intake	Y = 51.775–0.0178×(907–x)	0.005	0.981	907±48[b]

[1] It was expressed as requirement±standard error.
[a,b] Values in the same column with no common superscript differ significantly ($p<0.05$).

much greater than the requirements of choline at 0.48% methionine level ($p<0.05$). Therefore, it was concluded that methionine deficiency markedly increased the choline requirements for ducks, and this reason for this may be that in the presence of low methionine there is greater reliance on any surplus choline above the requirement of the bird, as this can be oxidized to betaine which can provide methyl for conversion of homocysteine to methionine (Setoue et al., 2008).

CONCLUSION

In conclusion, choline deficiency could cause growth depression and perosis of starter Pekin ducks. In this study it was found that the choline requirements of male Pekin ducks from 7 to 28 d of age for weight gain and feed intake were 1,472 and 1,424 mg/kg at 0.28% methionine level and 946 and 907 mg/kg at 0.48% methionine level, respectively. It suggested the choline recommendations of starter Pekin ducks in semi-purified diet were 1,448 mg/kg at 0.28% methionine level and 927 mg/kg at 0.48% methionine level, respectively. Compared with the adequate methionine level, menthionine deficiency markedly increased the choline requirements of ducks.

ACKNOWLEDGMENTS

This research was supported by the Earmarked fund for the Modern Waterfowl Industry Technology System of China (CARS-43) and the Agricultural Science and Technology Innovation Program (ASTIP-ISS09).

REFERENCES

AOAC (Association of Official Analytical Chemists). 2000. Official Methods of Analysis. 17th edn. Association of Official Analytical Chemists, Gaithersburg, MD, USA.

Baker, D., K. M. Halpin, G. Czarnecki, and C. Parsons. 1983. The choline-methionine interrelationship for growth of the chick. Poult. Sci. 62:133-137.

Bernard, R. and J. M. Demers. 1949. Lipotropic activity of choline, betaine, and methionine in ducklings. Can. J. Res. 27:281-289.

Blair, M. E., L. M. Potter, B. A. Bliss, and J. R. Shelton. 1986. Methionine, choline, and sulfate supplementation of practical-type diets for young turkeys. Poult. Sci. 65:130-137.

Blusztajn, J. K. and R. J. Wurtman. 1983. Choline and cholinergic neurons. Science 221:614-620.

Bornstein, S. and B. Lipstein. 1964. Methionine supplementation of practical broiler rations: II. The value of added methionine in chick starter rations. Br. Poult. Sci. 5:175-186.

Chen, L. X., S. R. Zhang, and G. W. Yue. 1991. A study on the methionine requirement of Jianchang duck and Tianfu×Jianchang crossbred duck. J. Sichuan Agric. Univ. 4:630-642.

Emmert, J. L. and D. H. Baker. 1997. A chick bioassay approach for determining the bioavailable choline concentration in normal and overheated soybean meal, canola meal and peanut meal. J. Nutr. 127:745-752.

Evans, R. J., E. I. Robertson, M. Rhian, and L. A. Wilhelm. 1942. The development of perosis in turkey poults and its prevention. Poult. Sci. 21:422-429.

Harms, R. H. and R. D. Miles. 1984. Effect of supplemental methionine and potassium sulfate on the choline requirement of the turkey poult. Poult. Sci. 63:1464-1466.

Hollenbeck, C. B. 2010. The importance of being choline. J. Am. Diet. Assoc. 110:1162-1165.

Jiang, G. Z., M. Wang, W. B. Liu, G. F. Li, and Y. Qian. 2013. Dietary choline requirement for juvenile blunt snout bream, *Megalobrama amblycephala*. Aquac. Nutr. 19:499-505.

Kasper, C. S., M. R. White, and P. B. Brown. 2000. Choline is required by tilapia when methionine is not in excess. J. Nutr. 130:238-242.

Ketola, H. G. and M. C. Nesheim. 1974. Influence of dietary protein and methionine levels on the requirement for choline by chickens. J. Nutr. 104:1484-1489.

Kita, K., S. Matsunami, and J. Okumura. 1996. Relationship of protein synthesis to mRNA levels in the muscle of chicks under various nutritional conditions. J. Nutr. 126:1827-1832.

Li, Z., L. B. Agellon, and D. E. Vance. 2005. Phosphatidylcholine homeostasis and liver failure. J. Biol. Chem. 280:37798-37802.

Li, Z., G. Wu, R. B. Sher, Z. Khavandgar, M. Hermansson, G. A. Cox, M. R. Doschak, M. Murshed, F. Beier, and D. E. Vance. 2014. Choline kinase beta is required for normal endochondral bone formation. Biochim. Biophys. Acta 1840:2112-2122.

Lien, T. F. and D. F. Jan. 1999. The effect on the lipid metabolism of Tsaiya ducks when high levels of choline or methionine are added to the ducks' diet. Asian Australas. J. Anim. Sci. 12:1090-1095.

Mai, K., L. Xiao, Q. Ai, X. Wang, W. Xu, W. Zhang, Z. Liufu, and M. Ren. 2009. Dietary choline requirement for juvenile cobia, *Rachycentron canadum*. Aquaculture 289:124-128.

Miles, R. D., N. Ruiz, and R. H. Harms. 1983. The interrelationship between methionine, choline, and sulfate in broiler diets. Poult. Sci. 62:495-498.

Molitoris, B. A. and D. H. Baker. 1976. The choline requirement of broiler chicks during the seventh week of life. Poult. Sci. 55:220-224.

NRC (National Research Council). 1994. Nutrient Requirements of Poultry. 9th rev. edn. National Academy Press, Washington, DC, USA.

Ohta, Y. and T. Ishibashi. 1995. Effect of dietary glycine on reduced performance by deficient and excessive methionine in broilers. Nippon Kakin Gakkaishi 32:81-89.

Pesti, G. M., N. J. Benevenga, A. E. Harper, and M. L. Sunde. 1981. Factors influencing the assessment of the availability of choline in feedstuffs. Poult. Sci. 60:188-196.

Pesti, G. M., A. E. Harper, and M. L. Sunde. 1980. Choline/methionine nutrition of starting broiler chicks. Three models for estimating the choline requirement with economic considerations. Poult. Sci. 59:1073-1081.

Quillin, E. C., G. F. Combs, R. D. Creek, and G. L. Romoser. 1961. Effect of choline on the methionine requirements of broiler chickens. Poult. Sci. 40:639-645.

Robbins, K. R., A. M. Saxton, and L. L. Southern. 2006. Estimation of nutrient requirements using broken-line regression analysis. J. Anim. Sci. 84:E155-E165.

SAS Institute. 2003. SAS User's Guide: Statistics Version 9.0. SAS Institute Inc., Cary, NC, USA.

Setoue, M., S. Ohuchi, T. Morita, and K. Sugiyama. 2008. Choline deprivation induces hyperhomocysteinemia in rats fed low methionine diets. J. Nutr. Sci. Vitaminol. 54:483-490.

Sterling, K. G., G. M. Pesti, and R. I. Bakalli. 2003. Performance of broiler chicks fed various levels of dietary lysine and crude protein. Poult. Sci. 82:1939-1947.

Standardization Administration of China. 2000. Determination of Amino Acids in Feeds. Standards Press of China, Beijing, China.

Wen, Z. G., J. Tang, S. S. Hou, Y. M. Guo, W. Huang, and M. Xie. 2014. Choline requirements of white Pekin ducks from hatch to 21 days of age. Poult. Sci. 93:3091-3096.

Wessler, I., H. Kilbinger, F. Bittinger, and C. J. Kirkpatrick. 2001. The biological role of non neuronal acetylcholine in plants and humans. Jpn. J. Pharmacol. 85:2-10.

Wu, G. and D. A. Davis. 2005. Interrelationship among methionine, choline, and betaine in channel catfish *ictalurus punctutus*. J. World Aquac. Soc. 36:337-345.

Zeisel, S. H. and J. K. Blusztajn. 1994. Choline and human nutrition. Annu. Rev. Nutr. 14:269-296.

Zhai, Q. H., X. F. Dong. J. M. Tong, Y. M. Guo, and Y. E. Bao. 2013. Long-term effects of choline on productive performance and egg quality of brown-egg laying hens. Poult. Sci. 92:1824-1829.

29

Different Coefficients and Exponents for Metabolic Body Weight in a Model to Estimate Individual Feed Intake for Growing-finishing Pigs

S. A. Lee, C. Kong, O. Adeola[1], and B. G. Kim*

Department of Animal Science and Technology, Konkuk University, Seoul 05029, Korea

ABSTRACT: Estimation of feed intake (FI) for individual animals within a pen is needed in situations where more than one animal share a feeder during feeding trials. A partitioning method (PM) was previously published as a model to estimate the individual FI (IFI). Briefly, the IFI of a pig within the pen was calculated by partitioning IFI into IFI for maintenance (IFI_m) and IFI for growth. In the PM, IFI_m is determined based on the metabolic body weight (BW), which is calculated using the coefficient of 106 and exponent of 0.75. Two simulation studies were conducted to test the hypothesis that the use of different coefficients and exponents for metabolic BW to calculate IFI_m improves the accuracy of the estimates of IFI for pigs, and that PM is applied to pigs fed in group-housing systems. The accuracy of prediction represented by difference between actual and estimated IFI was compared using PM, ratio (RM), or averaging method (AM). In simulation studies 1 and 2, the PM estimated IFI better than the AM and RM during most of the periods ($p<0.05$). The use of 0.60 as the exponent and the coefficient of 197 to calculate metabolic BW did not improve the accuracy of the IFI estimates in both simulation studies 1 and 2. The results imply that the use of 197 kcal×kg $BW^{0.60}$ as metabolizable energy for maintenance in PM does not improve the accuracy of IFI estimations compared with the use of 106 kcal×kg $BW^{0.75}$ and that the PM estimates the IFI of pigs with greater accuracy compared with the averaging or ratio methods in group-housing systems. (**Key Words:** Feed Consumption, Group-housing, Growth Performance, Maintenance Energy, Swine)

INTRODUCTION

Estimation of feed intake (FI) for individual animals in a group-housed pen is often required due to mortality or abnormal weight gain of animals (Lindemann and Kim, 2007). Moreover, when a response of animals is related to total intake of a specific nutrient or energy rather than the concentration of the nutrient or energy (Baker, 1986), it is crucial to knowledge the FI of individual animals in a group-housed pen. A partitioning method (PM) was previously proposed as a procedure to estimate individual FI (IFI) for pigs group-housed in a pen (Lindemann and Kim, 2007). This method partitions the IFI of animals within the pen into IFI for maintenance (IFI_m) and IFI for growth (IFI_g). The IFI_m is calculated based on the metabolizable energy (ME) of a diet and the mean body weight (BW) of an individual animal during the feeding period using an equation suggested by NRC (1998).

The ME for maintenance (ME_m) has been estimated from various coefficients and exponents for metabolic BW (Kil et al., 2013). In NRC (1998), the use of ME_m values calculated based on the metabolic BW using an exponent of 0.75 and a mean estimate of 106 as the coefficient for growing-finishing pigs: ME_m, kcal/d = 106 kcal×kg $BW^{0.75}$. However, Noblet et al. (1999) suggested that the exponent of metabolic BW be expressed as 0.60 rather than 0.75 for growing-finishing pigs. In NRC (2012), the ME_m of growing-finishing pigs is described using the exponent of 0.60 and the mean coefficient of 197: ME_m, kcal/d = 197 kcal×kg $BW^{0.60}$. To our knowledge, the use of the coefficient and metabolic BW exponent for calculating ME_m suggested in NRC (2012) has not been tested for the estimation of IFI.

Lindemann and Kim (2007) proposed a model for IFI estimation and compared the model with 2 other methods

* Corresponding Author: B. G. Kim.
E-mail: bgkim@konkuk.ac.kr
[1] Department of Animal Sciences, Purdue University, West Lafayette, IN 47907, USA.

using simulation studies. These simulation studies used data from pigs fed individually. The application of PM to group-housed pigs has not been documented. Therefore, the objective was to test the hypotheses that the use of the coefficient and metabolic BW exponent suggested by NRC (2012) for ME_m to calculate IFI_m improves the accuracy of the estimates of IFI for pigs and that the PM estimates IFI of pigs fed in group-housing systems with greater accuracy compared with other methods.

MATERIALS AND METHODS

Simulation study 1

In simulation study 1, twenty-four barrows and 24 gilts with initial BW of 33.9 kg (standard deviation, SD = 2.1) were used and the animals were fed individually. Pigs were allowed *ad libitum* access to the diet (2,945 kcal of ME/kg) and water. Body weight of individual pigs and IFI were recorded weekly. Growth data on a bi-weekly basis were used to compare the actual and estimated IFI. The final BW of the pigs was 57.7 kg (SD = 3.7).

Within pens, the growth data of each pig were pooled, calculated, and compared according to the procedures described by Lindemann and Kim (2007). Two hundred and forty-four artificial pens were created by grouping 4 or 6 pigs in a pen. Complete randomization, randomization within blocks by BW without regard to sex, randomization within blocks by BW with balanced sex, and randomization within blocks by BW and sex were used to regroup the pigs. The experimental animal allotment program of Kim and Lindemann (2007) was used for the simulation. Lastly, the IFI estimated based on the new pens were compared with the actual IFI.

Simulation study 2

Twenty gilts with initial BW of 67.1 kg (SD = 6.1) were used in simulation study 2. Pigs had *ad libitum* access to the diet of 3,312 kcal of ME/kg and to water. Pigs were fed in group pens of practical farms, and individual BW of pigs and IFI records were kept daily using FIRE System (Osborne Industries Inc., Osborne, KS, USA) and Acema-128 (Acemo, Pontivy, France). The growth data on a bi-weekly basis were used to compare the actual and estimated IFI. The final BW of the pigs was 101.8 kg (SD = 7.4).

The growth data were pooled, calculated, and analyzed as described in simulation study 1. Thirty-two artificial pens were created by grouping 4, 5, or 6 pigs in a pen. Complete randomization was used for 5, 4, or 3 pens, which yielded the aforementioned grouping, respectively. The pigs were also blocked by BW and allotted to each pen to make 2, 3, 4, or 5 replications using the animal allotment program (Kim and Lindemann, 2007). Lastly, the IFI estimated based on the new pens were compared with the actual IFI.

Partitioning method to estimate individual feed intake

The model proposed by Lindemann and Kim (2007) partitions IFI into IFI_m and IFI_g. Based on the ME in feed (kcal/kg), IFI_m can be calculated as:

$$IFI_m\text{, kg} = (106\ \text{kcal} \times \text{kg BW}^{0.75} \times \text{d})/ME_f$$

where BW is the mean BW for the period of interest; d is the number of days in the period of interest; and ME_f is the ME concentration in the feed (kcal/kg). The coefficient of 197 and exponent of 0.60 for metabolic BW is also used according to NRC (2012):

$$IFI_m\text{, kg} = (197\ \text{kcal} \times \text{kg BW}^{0.60} \times \text{d})/ME_f$$

Sum of IFI_m for all pigs within the pen represented as the pen FI (PFI) for maintenance (PFI_m) is then subtracted from total PFI (kg), and the remainder is PFI for growth (PFI_g) as:

$$PFI_g\text{, kg} = \text{total PFI} - PFI_m$$

where PFI_m is the sum of IFI_m for all pigs in the pen, in kilograms.

Next step is to calculate the IFI_g by apportioning PFI_g equally to each kg of BW gain during the period of interest within the pen:

$$IFI_g\text{, kg} = PFI_g \times (\text{IBWG}/\text{PBWG})$$

where IBWG is the individual BW gain (kg); and PBWG is the pen BW gain whose value is the sum of IBWG for all pigs in the pen (kg).

Lastly, the sum of IFI_m and IFI_g is IFI for the pig in the pen as:

$$IFI\text{, kg} = IFI_m + IFI_g$$

The PM calculated based on the metabolic BW using the coefficient of 197 and exponent of 0.60 (NRC, 2012) was referred to as PM 1, and the PM calculated based on the coefficient of 106 and exponent of 0.75 (NRC, 1998) was referred to as PM 2.

Ratio method to estimate individual feed intake

The ratio method (RM) does not consider the IFI_m of pigs, but focuses of IBWG. This method only apportions the PFI equally to each kg of IBWG:

$$IFI\text{, kg} = \text{total PFI} \times (\text{IBWG}/\text{PBWG})$$

Averaging method to estimate individual feed intake

The averaging method (AM) does not reflect the individual's biological aspects to the FI of pigs. All pigs in a pen are believed to have the same FI during the feeding trials as:

IFI, kg = total PFI/the number of pigs

Comparison among methods to estimate individual feed intake

To describe the accuracy of estimation, we measured the difference using percentages. The difference between the actual and estimated IFI were calculated as:

$$\text{Difference, \%} = 100\times|\text{actual IFI} - \text{estimated IFI}| / \text{actual IFI}$$

In this study, a smaller value of the difference (%) was considered to be more accurate when there was a significant difference.

Statistical analysis

Analysis of variance and mean separation tests were performed using PROC general linear model of SAS (SAS Inst. Inc., Cary, NC, USA) with PDIFF option and Tukey's adjustment. The model included method as a fixed variable. The experimental unit was each difference between the actual and estimated IFI from a pig, and the significance level was set at p-value less than 0.05.

RESULTS

Metabolizable energy for maintenance

The absolute values of difference between the estimated ME_m (kcal/d) based on the different coefficients and exponents for metabolic BW ranged from 0 to 459.9 kcal/d for growing-finishing pigs (Figure 1). Point of intersection where the difference in values of ME_m equaled 0 was approximately 62.3 kg of BW.

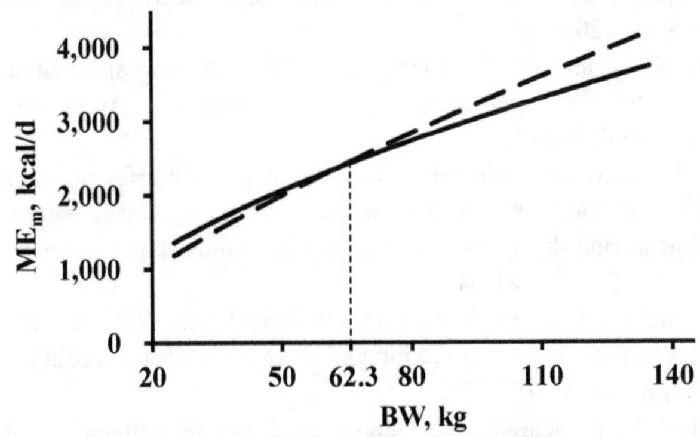

Figure 1. Metabolizable energy for maintenance (ME_m, kcal/d) calculated using different coefficients and exponents for metabolic body weight (BW). Metabolizable energy for maintenance was estimated based on the equations: 106 kcal×kg $BW^{0.75}$ (NRC, 1998) represented by dashed line, and 197 kcal×kg $BW^{0.60}$ (NRC, 2012) represented by solid line.

Table 1. Difference (%) between actual and estimated individual feed intake (IFI) using different methods for IFI estimation in simulation study 1 (n = 1,104)

Item	Method				Statistical parameter	
	PM 1	PM 2	AM	RM	SEM	p-value
d 0 to 14	9.26^b	9.33^b	10.7^a	11.5^a	0.27	<0.001
d 14 to 28	8.45^b	8.43^b	9.74^a	9.79^a	0.25	<0.001
d 0 to 28	8.23^b	8.25^b	9.74^a	8.90^{a,b}	0.24	<0.001

PM 1, partitioning method which estimates IFI based on the model proposed by Lindemann and Kim (2007), but calculated IFI for maintenance using the metabolic BW with a coefficient of 197 and an exponent of 0.60 (NRC, 2012); PM 2, partitioning method which estimates IFI based on the model proposed by Lindemann and Kim (2007); AM, averaging method; RM, ratio method; SEM, standard error of the mean.

[a,b] Within a row, means without a common superscript differ (p<0.05).

Simulation study 1

In simulation study 1, the PM 1 and 2 were able to estimate IFI better than AM during any period of interest (p<0.05; Table 1). The greater accuracy of PM 1 and 2 for IFI estimation was observed than RM during d 0 to 14 and d 14 and 28, but the accuracy did not differ between PM and RM during d 0 to 28. The accuracy did not differ between AM and RM during any period of interest. Lastly, the criterion also did not differ between PM 1 and 2 during all periods.

Simulation study 2

In simulation study 2, the PM 1 and 2 were able to estimate IFI better than AM during d 0 to 14, d 14 to 28, and d 0 to 42 (p<0.05; Table 2), but the criterion did not differ between PM 1 or 2 and AM during d 28 to 42. The

Table 2. Difference (%) between actual and estimated individual feed intake (IFI) using different methods for IFI estimation in group-housing systems (simulation study 2; n = 152)

Item	Method				Statistical parameter	
	PM 1	PM 2	AM	RM	SEM	p-value
d 0 to 14	4.89^b	4.86^b	6.90^a	7.61^a	0.41	<0.001
d 14 to 28	6.00^b	6.07^b	8.23^a	8.33^a	0.45	<0.001
d 28 to 42	8.39^b	8.14^b	8.10^b	11.3^a	0.58	<0.001
d 0 to 42	4.42^b	4.39^b	5.96^a	6.05^a	0.32	<0.001

PM 1, partitioning method which estimates IFI based on the model proposed by Lindemann and Kim (2007), but calculated IFI for maintenance using the metabolic BW with a coefficient of 197 and an exponent of 0.60 (NRC, 2012); PM 2, partitioning method which estimates IFI based on the model proposed by Lindemann and Kim (2007); AM, averaging method; RM, ratio method; SEM, standard error of the mean.

[a,b] Within a row, means without a common superscript differ (p<0.05).

PM 1 and 2 estimated IFI better than the RM during all periods ($p<0.05$). The RM showed less accuracy ($p<0.05$) in estimation of IFI than AM during d 28 to 42, but the accuracy did not differ between AM and RM during d 0 to 14, d 14 to 28, and d 0 to 42. Again, the accuracy of PM 1 and 2 for IFI estimation did not differ during any period.

DISCUSSION

The PM 1 and 2 consistently showed greater accuracy in estimation of IFI than AM and RM in both simulation studies 1 and 2. Lindemann and Kim (2007) validated the PM as a model to estimate IFI that is represented by PM 2 in the present study and found greater accuracy of PM than AM and RM in individual-housing system. This result was consistent with the present study. Feed intake for maintenance is an important factor to estimate IFI because PM 1 and 2 showed greater accuracy in most of the periods in simulation studies 1 and 2 than RM which calculates IFI based on only IBWG of pigs. Additionally, the results from the simulation study 2 showed that the PM may be applied to estimate IFI with fairly good accuracy in the group-housing systems. Turner et al. (2003) reviewed the previous studies of different group size (3<pigs/pen<100) and found no correlation between the average daily FI of growing-finishing pigs and the group size.

The IFI estimated based on AM may be attained with reliable values when the actual IFI are homogenous among the pigs within a pen. In the simulation study 2, however, the variation in FI among pigs during d 28 to 42 was similar to that of other periods (coefficient of variation = 10.1%, 11.1%, and 11.4%, respectively for d 0 to 14, d 14 to 28, and d 28 to 42). The different responses among the periods may be partly explained by the different physiological states such as a digestive physiology of pigs. Digestibility of the nutrients is one of the major factors affecting the accretion of proteins and lipids (Harris et al., 2012), but Kim et al. (2007) failed to find the effects of BW on the dry matter digestibility within the weanling, growing, and finishing stages. In this study, the growing and finishing pigs were used and whether the digestibility was affected by the different stage was not clear.

On the other hand, pigs showed greater variation in gain to feed ratio during d 28 to 42 than other periods in simulation study 2 (coefficient of variation = 10.7%, 11.7%, and 17.8%, respectively for d 0 to 14, d 14 to 28, and d 28 to 42). When the gain to feed ratio showed large variation among the pigs, accurate IFI may not be attained by PM because gain to feed ratio represents performance traits of animals such as BW gain and FI and because PM calculates the IFI based on the IBWG and FI of pigs. Indeed, greater difference between estimated and actual FI was observed during d 28 to 42 than other periods (Table 2).

The accuracy of PM 1 and 2 for IFI estimation did not differ for IFI of pigs during any period tested in the simulation studies 1 and 2. The ME_m estimated using the equation suggested by NRC (1998) becomes greater than ME_m estimated using the equation suggested by NRC (2012) as the BW of pigs exceeds 62.3 kg (Figure 1). The mean BW of pigs for the total period of simulation studies 1 and 2 were 48.6 and 84.4 kg, respectively. The absolute difference between the respective ME_m values of pigs with BW of 48.6 and 84.4 kg calculated based on the equations suggested by NRC (1998) and NRC (2012) were 74.0 and 132 kcal/d, which accounts for small portion of the daily FI when the ME_m was divided by ME in the diets.

CONCLUSION

The results from this study demonstrate that the use of 197 kcal×kg $BW^{0.60}$ as ME_m in PM does not improve the accuracy of IFI estimates compared with the use of 106 kcal×kg $BW^{0.75}$, and that partitioning the IFI into IFI_m and IFI_g provided the IFI of the pigs with greater accuracy compared with an AM or RM in group-housing systems.

ACKNOWLEDGMENTS

This work was supported in part by the Rural Development Administration (Republic of Korea; PJ009340).

REFERENCES

Baker, D. H. 1986. Problems and pitfalls in animal experiments designed to establish dietary requirements for essential nutrients. J. Nutr. 116:2339-2349.

Harris, A. J., J. F. Patience, S. M. Lonergan, C. J. M. Dekkers, and N. K. Gabler. 2012. Improved nutrient digestibility and retention partially explains feed efficiency gains in pigs selected for low residual feed intake. J. Anim. Sci. 90:164-166.

Kil, D. Y., B. G. Kim, and H. H. Stein. 2013. Invited review: Feed energy evaluation for growing pigs. Asian Australas. J. Anim. Sci. 26:1205-1217.

Kim, B. G. and M. D. Lindemann. 2007. A new spreadsheet method for experimental animal allotment. J. Anim. Sci. 85(Suppl. 2):112.

Kim, B. G., M. D. Lindemann, G. L. Cromwell, A. Balfagon, and J. H. Agudelo. 2007. The correlation between passage rate of digesta and dry matter digestibility in various stages of swine. Livest. Sci. 109:81-84.

Lindemann, M. D. and B. G. Kim. 2007. Technical note: A model to estimate individual feed intake of swine in group feeding. J. Anim. Sci. 85:972-975.

Noblet, J., C. Karege, S. Dubois, and J. van Milgen. 1999. Metabolic utilization of energy and maintenance requirements in growing pigs: effects of sex and genotype. J. Anim. Sci. 77:1208-1216.

NRC (National Research Council). 1998. Nutrient Requirements of Swine, 10th rev. edn. Natl. Acad. Press. Washington, DC, USA.

PERMISSIONS

All chapters in this book were first published in AJAS, by Asian-Australasian Association of Animal Production Societies; hereby published with permission under the Creative Commons Attribution License or equivalent. Every chapter published in this book has been scrutinized by our experts. Their significance has been extensively debated. The topics covered herein carry significant findings which will fuel the growth of the discipline. They may even be implemented as practical applications or may be referred to as a beginning point for another development.

The contributors of this book come from diverse backgrounds, making this book a truly international effort. This book will bring forth new frontiers with its revolutionizing research information and detailed analysis of the nascent developments around the world.

We would like to thank all the contributing authors for lending their expertise to make the book truly unique. They have played a crucial role in the development of this book. Without their invaluable contributions this book wouldn't have been possible. They have made vital efforts to compile up to date information on the varied aspects of this subject to make this book a valuable addition to the collection of many professionals and students.

This book was conceptualized with the vision of imparting up-to-date information and advanced data in this field. To ensure the same, a matchless editorial board was set up. Every individual on the board went through rigorous rounds of assessment to prove their worth. After which they invested a large part of their time researching and compiling the most relevant data for our readers.

The editorial board has been involved in producing this book since its inception. They have spent rigorous hours researching and exploring the diverse topics which have resulted in the successful publishing of this book. They have passed on their knowledge of decades through this book. To expedite this challenging task, the publisher supported the team at every step. A small team of assistant editors was also appointed to further simplify the editing procedure and attain best results for the readers.

Apart from the editorial board, the designing team has also invested a significant amount of their time in understanding the subject and creating the most relevant covers. They scrutinized every image to scout for the most suitable representation of the subject and create an appropriate cover for the book.

The publishing team has been an ardent support to the editorial, designing and production team. Their endless efforts to recruit the best for this project, has resulted in the accomplishment of this book. They are a veteran in the field of academics and their pool of knowledge is as vast as their experience in printing. Their expertise and guidance has proved useful at every step. Their uncompromising quality standards have made this book an exceptional effort. Their encouragement from time to time has been an inspiration for everyone.

The publisher and the editorial board hope that this book will prove to be a valuable piece of knowledge for researchers, students, practitioners and scholars across the globe.

LIST OF CONTRIBUTORS

P. Chanjula
Department of Animal Science, Faculty of Natural Resources, Prince of Songkla University, Songkhla 90112, Thailand

T. Raungprim, S. Poonko, S. Majarune and W. Maitreejet
Buffalo and Beef Production Research Center, Suwanvajokkasikit Animal Production Research and Development Institute, Kasetsart University, Kamphaeng Saen Campus 73140, Thailand

S. Yimmongkol
Department of Animal Science, Faculty of Agriculture, Kasetsart University, Kamphaeng Saen Campus 73140, Thailand

Chang Bon Choi
School of Biotechnology, Yeungnam University, Gyeongsan 38541, Korea

Hana Kwon
Department of Food and Nutrition, Yeungnam University, Gyeongsan 38541, Korea

Sung Il Kim
Department of Animal Science, Gyeongbuk Provincial College, Yecheon 36830, Korea

Un Mok Yang, Ju Hwan Lee and Eun Kyu Park
Woosung Feed Co., Ltd., Daejeon 34379, Korea

Chuang Xua, Cheng Xia, Hong-you Zhang and Jia-san Zheng
Department of Clinical Veterinary Medicine, Animal Science and Technology College, Heilongjiang August First Land Reclamation University, Daqing 163319, China

Ling-wei Sun
Jiangsu Engineering Technology Research Center of Meat Sheep & Goat Industry, Nanjing Agricultural University, Nanjing, 210095, China

Jun-song Wang
Center for Molecular Metabolism, School of Environmental and Biological Engineering, Nanjing University of Science Technology, Nanjing, 210095, China

F. Shahzad, M. N. Tahir and Z. Kamran
University College of Veterinary and Animal Sciences, The Islamia University of Bahawalpur, Bahawalpur 63100, Pakistan

M. Abdullah and J. A. Bhatti
Department of Livestock Production, University of Veterinary and Animal Sciences, Lahore 42000, Pakistan

A. S. Chaudhry
School of Agriculture, Food & Rural Development, Newcastle University, Newcastle, NE1 7RU, UK

M. A. Jabbar
Pakistan Agriculture Research Board, Lahore 42000, Pakistan

F. Ahmed T. Mehmood and M. Asim
Buffalo Research Institute, Pattoki 55300, Pakistan

S. Ahmed
Department of Animal Nutrition, University of Veterinary and Animal Sciences, Lahore 42000, Pakistan

I. Irshad
Department of Pathology, University of Veterinary and Animal Sciences, Lahore 42000, Pakistan

Jalil Ghassemi Nejad, Bae-Hun Lee, Byong-Wan Kim, Sang-Jip Ohh and Kyung Il Sung
College of Animal Life Sciences, Kangwon National University, Chuncheon 200-701, Korea

De-Jia Huang and Shyi-Kuen Yang
Department of Animal Science and Biotechnology, Tunghai University, Taichung 40704, Taiwan

Mohammad Khairul Alam, Yasumichi Ogata, Yukari Sato and Hiroaki Sano
Department of Animal Sciences, Faculty of Agriculture, Iwate University, Morioka 020-8550, Japan

M. R. Abdollahi and V. Ravindran
Institute of Veterinary, Animal and Biomedical Sciences, Massey University, Palmerston North 4442, New Zealand

B. J. Hosking and D. Ning
AsiaPac Biotechnology Co. Ltd., Dongguan City, Guangdong 523808, China

Q. H. Shang, Z. B. Yang, W. R. Yang, Z. Li, G. G. Zhang and S. Z. Jiang
Department of Animal Sciences and Technology, Shandong Agricultural University, Tai'an, Shandong 271018, China

Q. Y. Tian, Z. K. Zeng, Y. X. Zhang, S. F. Long and X. S. Piao
State Key Laboratory of Animal Nutrition, Ministry of Agriculture Feed Industry Centre, China Agricultural University, Beijing 100193, China

Lizhi Wang, Zhisheng Wang, Huawei Zou and Quanhui Peng
Animal Nutrition Institute, Sichuan Agricultural University, Ya'an 625014, China
Key Laboratory of Low Carbon Culture and Safety Production in Cattle in Sichuan, Ya'an 625014, China

Y. Konca, S. Buyukkilic Beyzi and M. Kaliber
Agricultural Faculty, Animal Science Department, Erciyes University, Kayseri 38039, Turkey

T. Ayaşan
East Mediterranean Agricultural Research Institute, Adana 01321, Turkey

A. Bozkurt Kiraz
Agricultural Faculty, Animal Science Department, Harran University, Şanlıurfa 38039, Turkey

In H. Kim
Department of Animal Resources and Science, Dankook University, Cheonan 330-714, Korea

Glenmer B. Tactacan
JEFO Innovation and Development, JEFO Nutrition Inc., SaintHyacinthe, Quebec, J2S 7B6, Canada

Seung-Yeol Cho
Eugenebio, Suwon 443-380, Korea

Jin H. Cho
Department of Animal Science, Chungbuk National University, Cheongju 361-763, Korea

P. Gunun and S. Sirilaophaisan
Department of Animal Science, Faculty of Natural Resources, Rajamangala University of Technology-Isan, Sakon Nakhon Campus, Phangkhon, Sakon Nakhon 47160, Thailand

M. Wanapat and A. Cherdthong
Tropical Feed Resources Research and Development Center (TROFREC), Department of Animal Science, Faculty of Agriculture, Khon Kaen University, Khon Kaen 40002, Thailand

N. Gunun and W. Kaewwongsa
Program in Animal Production Technology, Faculty of Technology, Udon Thani Rajabhat University, Udon Thani 41000, Thailand

K. Phesatcha and M. Wanapat
Tropical Feed Resources Research and Development Center (TROFREC), Department of Animal Science, Faculty of Agriculture, Khon Kaen University, Khon Kaen 40002, Thailand

J. C. Han, H. X. Qu, C. M. Zhang and Y. F. Yan
Department of Animal Science, College of Life Science, Shangqiu Normal University, Shangqiu, Henan 476000, China

G. H. Chen, J. G. Wang and J. L. Zhang
Department of Animal Science, College of Life Science, Shangqiu Normal University, Shangqiu, Henan 476000, China
College of Animal Husbandry and Veterinary Science, Henan Agricultural University, Zhengzhou, Henan 450002, China

Y. H. Cheng
Department of Biotechnology and Animal Science, National Ilan University, I-Lan 26047, Taiwan

C. Kaewtapee, N. Krutthai and C. Bunchasak
Department of Animal Science, Faculty of Agriculture, Kasetsart University, Bangkok 10900, Thailand

M. Manafi, M. Hedayati and S. Khalaji
Department of Animal Science, Faculty of Agricultural Sciences, Malayer University, Malayer 65719-95863, Iran

Fangfang Li, Yujing Zhu, Lan Ding and Yong Zhang
Department of Animal Nutrition, Faculty of Animal Sciences and Veterinary, Shenyang Agricultural University, Shenyang 110866, China

Muhammad Sarwar
Institute of Animal Sciences, University of Agriculture, Faisalabad 38000, Pakistan

Zafar Ullah
Institute of Animal Sciences, University of Agriculture, Faisalabad 38000, Pakistan
Department of Animal Sciences, University of Sargodha, Sargodha 40100, Pakistan

Gulraiz Ahmed
Sadiq Feeds (Pvt.) Ltd., Rawalpindi 46300, Pakistan.

Mehr un Nisa
Department of Food Science, Nutrition and Home Economics, Government College University, Faisalabad 38000, Pakistan

Y. F. Liu, H. B. Zhao, X. M. Liu, W. You, H. J. Cheng, F. C. Wan, G. F. Liu, X. W. Tan, E. L. Song and X. L. Zhang
Institute of Animal Science and Veterinary Medicine, Shandong Academy of Agricultural Sciences, Ji'nan, Shandong 250100, China
Shandong Key Lab of Animal Disease Control and Breeding, Ji'nan, Shandong 250100, China

Amit Kini and Custan Fernandes
Vinayak Ingredients (India) Private Limited, Lower Parel (West), Mumbai - 400 013, India

Dayaram Suryawanshi
Omega Laboratories, Satara, Maharashtra, 415521, India

H. K. Kang, S.-B. Park and C. H. Kim
Poultry Science Division, National Institute of Animal Science, RDA, Cheonan 331-801, Korea

B. K. An
Department of Animal Science and Technology, Konkuk University, Seoul 05029, Korea

S. H. An
Team of An Educational Program for Specialists in Global Animal Science, Brain Korea 21 Plus Project, Konkuk University, Seoul 05029, Korea

D. W. Kim
Animal Nutrition & Physiology Team, National Institute of Animal Science, RDA, Wanjugun 55365, Korea

Richard A. Zinn
Department of Animal Science, University of California, Davis, CA 95616, USA

José Ortiz, Martín Montaño, Alejandro Plascencia and Noemí Torrentera
Instituto de Investigaciones en Ciencias Veterinarias, UABC, Mexicali, Baja California 21380, México

Jaime Salinas
Facultad de Medicina Veterinaria y Zootecnia UAT, Cd. Victoria, Tamaulipas 87000, México

Ashraf A. Biswas, Lovelia L. Mamuad, Seon-Ho Kim, Yeon-Jae Choi and Sang-Suk Lee
Ruminant Nutrition and Anaerobe Laboratory, Department of Animal Science and Technology, Sunchon National University, Suncheon 540-742, Korea

Sung Sill Lee
Division of Applied Life Science (BK21+, IALS), Gyeongsang National University, Jinju 52828, Korea

Gui-Seck Bae
Department of Animal Science and Technology, Chung-Ang University, Anseong 456-756, Korea

Kichoon Lee
Department of Animal Sciences, The Ohio State University, Columbus, OH 43210, USA

Ha-Guyn Sung
Department of Animal Science & Technology, Sangji University, Wonju 26339, Korea
Ruminant Nutrition and Anaerobe Laboratory, Department of Animal Science and Technology, Sunchon National University, Suncheon 540-742, Korea

A. Estrada-Angulo, B. I. Castro, F. G. Ríos and H. Dávila-Ramos
Veterinary and Animal Science School, University Autonomous of Sinaloa, Culiacán 1084, Sinaloa, México

M. A. López-Soto, C. R. Rivera-Méndez, A. Barreras, J. D. Urías-Estrada and A. Plascencia
Research Institute of Veterinary Sciences, University Autonomous of Baja California, Tijuana 21100, México
Veterinary and Animal Science School, University Autonomous of Sinaloa, Culiacán 1084, Sinaloa, México

R. A. Zinn
Department of Animal Science, University of California, Davis, CA 95616, USA

J. Tang, M. Xie and S. S. Hou
Institute of Animal Sciences, Chinese Academy of Agricultural Sciences, Beijing 100193, China

Z. G. Wen and P. L. Yang
Feed research institute, Chinese Academy of Agricultural Sciences, Beijing 100081, China

S. A. Lee, C. Kong and B. G. Kim
Department of Animal Science and Technology, Konkuk University, Seoul 05029, Korea

O. Adeola
Department of Animal Sciences, Purdue University, West Lafayette, IN 47907, USA

Index

www.ingramcontent.com/pod-product-compliance
Lightning Source LLC
LaVergne TN
LVHW060603210926
848485LV00005B/1363
* 9 7 8 1 6 4 7 4 0 2 5 9 4 *